Influenza

Influenza

Edwin D. Kilbourne, M.D.

Distinguished Service Professor
Mount Sinai School of Medicine of the
City University of New York
New York, New York

Plenum Medical Book Company • New York and London

Library of Congress Cataloging in Publication Data

Kilbourne, Edwin D. (Edwin Dennis), 1920–
Influenza.

Includes bibliographies and index.
1. Influenza. 2. Influenza viruses. I. Title. [DNLM: 1. Influenza. 2. Orthomyxoviridae. WC 515 K483i]
RC150.K6 1987 616.2′03 87-2510
ISBN 978-1-4684-5241-9 ISBN 978-1-4684-5239-6 (eBook)
DOI 10.1007/ 978-1-4684-5239-6

Softcover reprint of the hardcover 1st edition 1987

Plenum Medical Book Company is an imprint of
Plenum Publishing Corporation
233 Spring Street, New York, N.Y. 10013

This book is dedicated to the memory of Frank L. Horsfall, Jr., mentor and friend, in whose laboratory at the Rockefeller Institute for Medical Research I first met the virus of influenza and learned the methods of quantitative biology.

Preface

My lifetime encompasses the postwar subsidence in the early 1920s of the greatest influenza pandemic in history, direct encounters with FM1 virus at Fort Monmouth in 1947, the care of influenza patients in the 1950s, the pursuit of the influenza virus through the modern pandemics of 1957 and 1968, and a present in which the genes of the virus have dissembled in the DNA of vaccinia virus and *Escherichia coli* through the wand of "high tech."

If my corpus could be fossilized for archival and archaeological purposes, it would be found to contain immune cells branded with the imprint of the "swine" influenza virus of post-1918 and brain cells no less imprinted with memories of the abortive return of its descendant during America's bicentennial. But before that unlikely event, I wanted to try to make some sense out of this baffling disease and its viruses—expecting no definitive revelations but hoping for a sharper definition of problems. Hence this book.

It is an audacious act in these days of specialization to essay a book such as this singlehandedly, but I have done so for selfish reasons. I wanted to reexamine old questions about the nature of influenza and its epidemics in the light of the dazzling advances in molecular biology of the past few years. No virus has been better studied, but few diseases are less well understood. The influenza virus glycoproteins have become models for biologists interested in membrance assembly and function; knowledge of their tertiary (and even quaternary) structure surpasses that available for most other proteins. The rapidly mutating RNAs of influenza virus have captured the interest and fired the imagination of evolutionary geneticists. Cellular immunologists have discovered this enveloped virus and, in collaboration with molecular geneticists, have studied the recognition of single viral proteins introduced by cloning vectors.

I marvel, properly, at all this activity and applaud it, but the physician in me asks, what about *influenza*? What about the disease? At this time, after a professional lifetime of laboratory investigations and 2 years of intensive scholarship, I cannot tell you *why* people get sick with influenza or *how* they do. How this elegantly characterized virus causes aches and pains and fever is simply not understood. Nor do we fully comprehend viral virulence or the genesis of pandemics. But we shall. The only prophecy that I dare to make is that we are on the threshold of deep insights into pathogenesis that will come from the convergence of presently parallel studies of the molecules of both virus and host. I hope that this book can expedite this convergence by illuminating gray areas and drawing attention to studies outside the mainstream of virology.

One day we shall "see influenza plain."

Edwin D. Kilbourne

New York

Acknowledgments

I take pleasure in acknowledging the unfailing generosity of colleagues throughout the world who have shared their research findings with me in advance of publication. They include the following, none of whom should be held responsible for my interpretation of their work:

Gordon L. Ada, Gillian M. Air, Robert B. Belshe, Robert F. Betts, George C. Brownlee, Doris J. Bucher, Robert M. Chanock, Purnell W. Choppin, Mary Lou Clements, Richard W. Compans, Robert B. Couch, Nigel J. Dimmock, Walter R. Dowdle, R. Gordon Douglas, Jr., Bernard C. Easterday, Francis A. Ennis, John P. Fox, Walter Gerhard, Mary Jane Gething, Paul W. Glezen, R. Edgar Hope-Simpson, Alan P. Kendal, Hans-Dieter Klenk, Robert M. Krug, Robert A. Lamb, William Graeme Laver, Hunein F. Maassab, Koichiro Maeno, Gordon Meiklejohn, Brian R. Murphy, Debi P. Nayak, Michael B. A. Oldstone, Peter Palese, James C. Paulson, Marcel Pons, Carol S. Reiss, Rudolf Rott, Christoph Scholtissek, Jerome L. Schulman, Irene T. Schulze, John J. Skehel, Masahiro Ueda, Colin Ward, Robert G. Webster, D. C. Wiley, Peter F. Wright, Nicholas G. Wrigley, and V. M. Zhdanov.

I am deeply indebted to Marilyn Tartaglia for her dedicated and tireless efforts in the preparation of the major portion of this manuscript and to Valerie M. Josephson, whose editorial skills facilitated its completion.

E.D.K.

Contents

I. INTRODUCTION

1. History of Influenza

II. THE INFLUENZA VIRUSES

2. Taxonomy and Comparative Virology of the Influenza Viruses

III. INFLUENZA: INFECTION AND DISEASE

7. Influenza in Man

IV. EPIDEMIOLOGY, SURVEILLANCE, AND CONTROL

10. The Epidemiology of Influenza

I

Introduction

1

History of Influenza

Influenza. . . possesses a special interest. . . being of all epidemics the most extensively diffused and apparently the least liable to essential modification either by appreciable atmospheric changes or by hygienic conditions under the control of man. . . . Unlike cholera, it outstrips in its course the speed of human intercourse. . . . There is a grandeur in its constancy and immutability superior to the influences of national habits.
(Thompson, 1852)

INTRODUCTION

Because influenza is defined as much by its epidemiology as by its symptomatology, and because its causative virus rapidly and progressively changes, historical considerations are uniquely important in defining and evaluating this disease. The continual antigenic evolution of the virus is soon followed by specific and reciprocal changes in the immune status of infected human populations so that the pattern of influenza each year or decade reflects the experience of the community during the preceding one.

THE CREDIBILITY OF HISTORICAL EVIDENCE

As we probe, still, at the terminus of the 20th century, for the origin of pandemics, we seek answers to many of the same questions raised in earlier centuries about meteorological conditions, disease outbreaks in animals, and other associated natural phenomenon. But the credibility of historical records is, in general, inversely proportional to their antiquity. With influenza, remote observations are clouded not only by temporal distance but by a background of poxes, murrains, plagues, agues, pestilence, and famine that often obscured the perception of influenza as a distinct and specific entity. To make matters worse, reliable chroniclers of disease seem to have been few, so that one is often left at the mercy of inept or imaginative observers. Then, too, influenza lacks the pathognomonic stigmata of smallpox, cholera, or poliomyelitis to permit its uniquivocal recognition by the casual observer.

On the other hand, investigators of the history of influenza during the present century have been aided greatly by modern techniques of virus isolation (since 1933) and identification of specific antibodies in the sera of elderly people (serological archaeology). Thus, in recognition of these variable levels of evidence in the documentation of outbreaks, the history of influenza is considered here in three periods.

1. Pre-1889, based on historical records describing the epidemic occurrence of influenzalike disease and its symptomatology.
2. 1889–1933, defined not only by written observations but by serological studies of antibodies in the human population, the footprints of influenza virus invasion.
3. Post-1933, the modern period of influenza research following the first isolation of influenza virus from man and the introduction of the technology for precise and specific study of the disease and identification of epidemics.

PREHISTORY: THE ORIGIN OF INFLUENZA

Influenza as we now know it is a contagious short-lived infection with a short incubation period. The survival and persistence of the causative virus, therefore, depends on a continuing chain of transmission from person to person. Although introduction of pandemic viruses may occur from animal sources (see Chapters 9 and 10), the viruses that cause human influenza appear to be obligate parasites of man. Therefore, it seems probable that the infection evolved in man no earlier than the development of large or socially linked communities of sufficient size to provide a renewing population of susceptibles. The historical records reviewed in the following sections have been interpreted by some authors as indicating an increased frequency of major outbreaks of influenza beginning in the latter part of the 19th century. If, in fact, this reading of the record is correct, it is consistent with the global increase in population density, urbanization, and crowding that has followed the industrial revolution.

INFLUENZA PRIOR TO 1889

The retrospective identification of epidemics from fragmentary historical records and anecdotal evidence is fraught with difficulty. In the case of influenza, the task is made more difficult by the ambiguous symptomatology of the illness and, paradoxically, by its confusion with respiratory infections of lesser severity on the one hand or with serious febrile diseases on the other. However, if one assumes that no major change in the disease has occurred in the last 500 years, then it is clear that the acute, rapidly spreading "catarrhal fevers" of the past were influenza. It is interesting to examine the descriptions of the older epidemics for any clue that they may provide to the evolution of this pervasive infection. The summary of major epidemics of "catarrhal fevers" and "influenza" shown in Table 1-1 is based principally on studies of British epidemics compiled by Theophilus Thompson (1852) and Charles Creighton (1891). All the epidemics listed have been noted by both authors save for the epidemic of 1758 listed only by Thompson. Thompson has been more selective in his listings, however, and I have agreed with his judgment on the basis of the published evidence. To his list I have added only an epidemic in 1693 credibly described by Creighton.

Studies in the 20th century suggest that influenza is an "unvarying disease caused by a varying virus" (Kilbourne, 1975). Even in the uniquely virulent epidemic of 1918, a typical case of illness resembled the 3-day fever characteristic of more recent epidemics. With this bias, case descriptions of influenza as a brief, acute, prostrating, self-limited respiratory illness are readily recognized in earlier writings, especially if such illness occurs in widespread outbreaks for short periods of time and is attended by an increase in mortality.

Past observers varied in clinical acumen and discrimination so that one must question the mention of "parotid swelling" (mentioned twice), "gastrodynia," and "double tertian fever" as being truly characteristic symptoms of the epidemics described, particularly when the incidence of such complications is not given. However, erysipelas and scarlatina as manifestations of streptococcal infections that not uncommonly complicate present-day influenza are reasonably linked with past epidemics. This negative bias to accepting unusual symptoms may, of course, lead to the discarding of evidence of interesting variability in viral virulence and tropism, but it seems the wisest course despite the fact that parotitis, for example, has been reported in connection with influenza A virus infections within the last decade (Brill and Gilfillan, 1977; Krilov and Swenson, 1985).

Notations on outbreaks of disease in animals in relation to influenza epidemics are of obvious interest because of present hypotheses on the animal source of modern pandemics (see Chapters 9 and 10). It is notable that concurrent respiratory disease in horses was frequently remarked in the 18th- and 19th-century epidemics. The epidemic of 1762 was preceded by "horse colds" in 1760. Although today we recognize the independence of outbreaks of influenza in horses and man, certainly in earlier times more of the human population was in close proximity to horses than at present. Hirsch (1883) points out no fewer than nine recorded epidemics of influenza throughout the world between 1693 and 1873 that were associated with epidemics of what appears to have been equine influenza.

It must be stressed that minor epidemics of less certain identity that were probably influenza, some with an impact on mortality, are not listed in Table 1-1. It is probable that, as in modern times, the virus continued to circulate and to cause some morbidity between major epidemics. Indeed, Hirsch's detailed and scholarly review of the world's experience with influenza identifies outbreaks in 132 different years between 1173 and 1875 (Hirsch, 1883).

INFLUENZA 1889–1933: SEROARCHEOLOGY AND THE RECYCLING OF ANTIGENS

Among epidemics of the latter half of the 19th century, the epidemic of 1889–1890 stands out not only in severity but in its wavelike extension in annual recurrences into the immediately succeeding years. In London, the epidemic in-

Table 1-1. Influenza: A Summary of Major Epidemics Affecting Great Britain, 1510–1889[a]

Epidemic year(s)	Season (England)	Prevalence[b]	Temporally associated animal disease	Anomalous symptoms
1510	—	Pandemic	Murrain in cattle	"Gastrodynia"
1557	Fall	Pandemic	—	Double tertian fever
1580	Fall	European	Murrain in beasts (Kent)	Parotid swelling
1658	Spring	Britain, Europe	—	"Cephalic affection"
1675	Fall	Britain, Europe	—	—
1688[c]	—	Dublin	"Nasal defluxin" in horses	—
1693[d]	Fall	Britain, Europe	Antecedent nasal discharge in horses	—
1710	Spring	Britain, Europe	—	"Quick pulse"
1729	Winter	England, Europe	—	—
1732–3	Spring–winter	Pandemic	Cough in horses	Parotid swelling
1737–8	Fall	England, Europe, North America	"Disease among horses"	—
1743	Spring	Britain, Europe	"Cough among horses"	—

1758	Fall	North America, Europe, Britain	—	—
1762	Fall	Britain, Europe	Antecedent "horse colds" (1760)	Variable mortality
1767	Summer	Britain, Europe, North America	Dogs and horses, horse cold	
1775	Fall	Pandemic	Disease in dogs and horses	Prurigo, erysipelas, pustules
1782	Spring	Pandemic	—	
1802–3	Spring	Britain, Europe	Cattle and domestic animals	"Interchanging with scarlatina"
1833	Spring	Pandemic	Concurrent disease in horses	—
1836–7	Winter	Pandemic	—	—
1847–8	Fall	Pandemic	—	—
1889–90	Winter	Pandemic	—	Increased mortality age 20–60 (Jan. 1890)

[a]From compilations by Thompson (1852) and Creighton (1883) Minor epidemics, some with impact on mortality, are not listed It is probable that as in modern times some influenza occurred each year
[b]Prevalence from data of Thompson (1852), Creighton, (1883), and Hirsch (1883) Pandemic implies historical evidence of involvement of more than two continents
[c]Evidence for widespread prevalence lacking, included because of associated animal respiratory disease
[d]Described only by Creighton (1883)

creased in severity with each successive wave so that mortality was highest in 1892 (Burnet and Clark, 1942). Another notable feature of this epidemic was the occurrence in 1890 of an increase in the proportion of deaths in the 20- to 40-year-old age group, ordinarily considered to be a unique characteristic of the 1918 pandemic. Total death rates attributed to influenza in England and Wales increased abruptly with the advent of the 1889 epidemic and persisted over the previous rate of fewer than ten per 100,000 through the 1930s (League of Nations' epidemiological report cited by Burnet and Clark, 1942).

Testing of the sera of aged people with modern influenza virus antigens has provided indirect evidence (1) of the prior circulation of contemporary strains of virus and their sequential appearance and (2) of the etiology of earlier epidemics. Such retrospective investigations are obviously limited by the life-span of man and the date on which blood samples have been taken. Within these constraints, our backward reach has extended to 1857. Ninety percent of subjects born between 1857 and 1877 were found to have antibodies to Hong Kong (H3) virus prior to its epidemic reappearance in 1968; 26% had preepidemic antibody to virus of the H2 ("Asian") subtype that caused the pandemic of 1957 (Masurel and Marine, 1973). This and other studies summarized by Davenport (1977) provide evidence for the sequential appearance of H2 and H3 viruses in the last decade or so of the last century.

Recently, serological evidence for the circulation of H1N1 viruses between 1908 and 1918 has been presented (Masurel and Heijtink, 1983; Rekart et al., 1982), which, taken together with evidence for H3 prevalence at that time and earlier, implies that the cocirculation of H1 and H3 subtypes since 1977 is not without precedent.

It must be emphasized, however, that serological data of this kind are of limited value in reconstructing the past. Their chief value lies in showing that past epidemics were almost certainly caused by influenza viruses closely related antigenically to present-day strains. Therefore, the specificity of past diagnoses assumed from historical records is confirmed. But human antibodies measured by conventional tests can delineate only shared determinants in viral surface antigens. Indeed, even this evidence indicates that the H3 virus of early in this century possessed a different (equine virus) neuraminidase antigen than the post-1968 strains (Kendal et al., 1973).

Serological archaeology assumes particular importance in its contribution to the understanding of the great 1918 pandemic. There is little doubt that a virus antigenically similar to the swine influenza virus first isolated by Shope in 1931 was a cause of that disease. The brief but epidemic reappearance of a swine influenza virus in American military recruits in 1976 confirms the pathogenic potential of swine influenza virus for man. Indeed, epidemiologic evidence suggests that swine influenza appeared as a new epizootic in swine at the time of the early 1918 epidemic (reviewed by Shope, 1958). Hence, the virus is most likely human in origin (see Chapters 9 and 10 for further discussion).

Human antibody profiles and the occurrence of a brisk epidemic in 1929 in the United States point to that year as the time of probable introduction of an antigenically significant H1 variant of swine influenza virus that was probably the prototype of the virus first isolated from man in 1933 by Smith, Andrewes, and Laidlaw (1933). Judging by excess death rates from influenza and pneumonia (Fig. 1-1) (Collins, 1957), the United States was affected only by minor epidemics through the 1930s and 1940s save for a notable outbreak in 1943–44. Of the pervasive but mild pandemic of 1946–47, more is said in the next section.

From a combination of serological and epidemiologic evidence, a highly provisional reconstruction of the viruses prevalent in man prior to the first isolation of virus in 1933 can be proposed (Table 1-2).

INFLUENZA POST-1933

Modern knowledge of the disease began with the demonstration that a specific cultivable virus was the causative agent. Since that time, prototype strains from most epidemics have been preserved, so that the history and evolution of the virus, as well as the disease, is a subject of growing interest for investigation by the techniques of molecular biology. The subsequent isolation of influenza B virus in 1940 and influenza C virus in 1960 permitted further discrimination of the virology of influenza epidemics. Apparently homogeneous outbreaks were sometimes found to have dual etiology (Kilbourne et al., 1951). No doubt, some of the puzzling characteristics of ancient epidemics, particularly with respect to immediately recurrent infections in the same population, are explicable as the result of sequential infection with these antigenically unrelated viruses (see below).

In 1957 the occurrence of an undisputed pandemic in the virologic era was a landmark in the history and study of influenza. From cases of Asian influenza, virus similar in structural and biological properties to all post-1933 isolates were recovered. This global epidemic in immunologically virgin populations caused enormous morbidity, and deaths from virus pneumonia uncomplicated by secondary bacterial infections were reminiscent of 1918. The etiologic link of 1918 and earlier pandemics to modern influenza viruses was thus strongly supported.

Another pandemic, born also in the Far East, was soon to follow. In 1968 Hong Kong influenza (H3N2) virus attacked a population partially immunized by

Table 1-2. Proposed[a] and Actual Prevalence of Influenza A Viruses 1889—1985[b]

Years of prevalence	Virus subtype	Like virus first isolated in	Vernacular name
1889–1901	H2N2[a]	1957	Asianlike
1901–1918	H3N8[a]	1968	Hong-Kong-like
1908–1918	H1N1[a]	1947	A-prime-like
1918–1929	H1N1[a]	1930	Swine-virus-like
1929–1946	H1N1[a]	1933	WS- and PR8-like
1946–1957	H1N1	—	A prime
1957–1968	H2N2	—	Asian
1968–	H3N2	—	Hong Kong
1976	H1N1	1930	Swine virus
1977–	H1N1	1950	Russian

[a]Proposed on basis of serological evidence
[b]Modified from data and analysis of Davenport (1977)

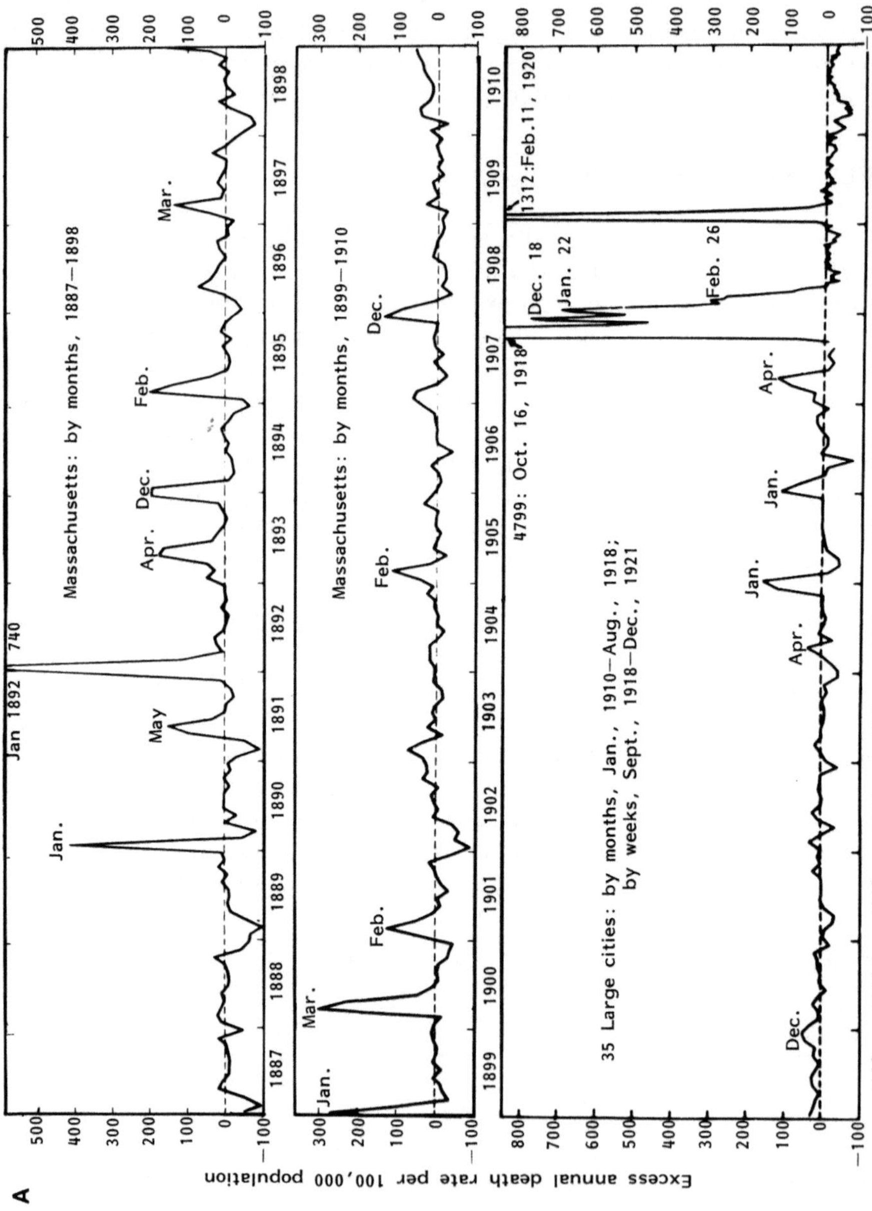
A
Excess annual death rate per 100,000 population
Massachusetts: by months, 1887–1898
Jan 1892 740
Jan.
May
Apr.
Dec.
Feb.
Mar.
Massachusetts: by months, 1899–1910
Jan.
Mar.
Feb.
Feb.
Dec.
35 Large cities: by months, Jan., 1910–Aug., 1918; by weeks, Sept., 1918–Dec., 1921
4799: Oct. 16, 1918
1312: Feb. 11, 1920
Dec. 18
Jan. 22
Feb. 26
Dec.
Apr.
Jan.
Jan.
Apr.

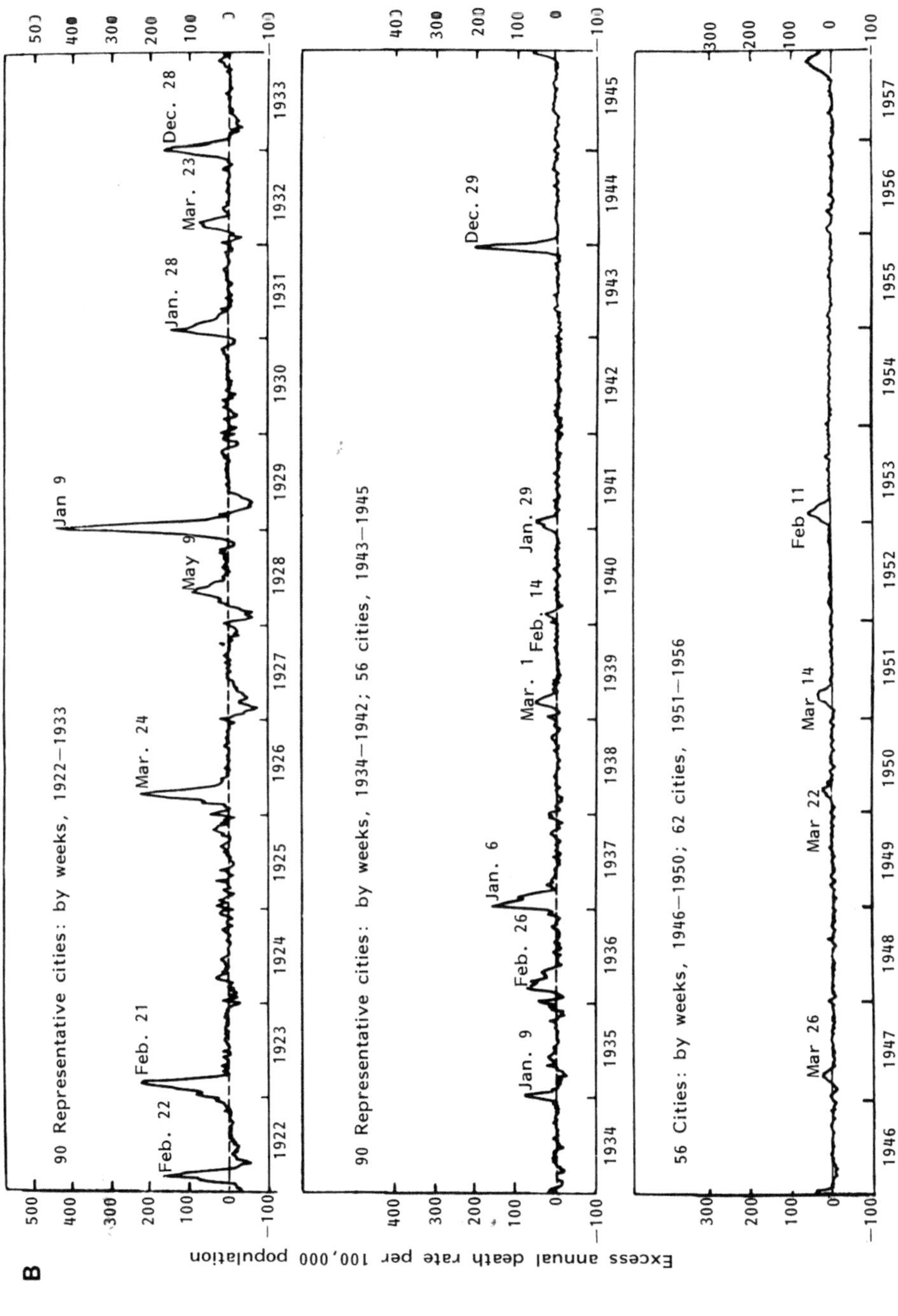

Figure 1-1. Excess annual death rates for Massachusetts (A) and for representative cities in the United States (B) from 1887 to 1957. (From Collins, 1957.)

the related N2 neuraminidase antigen of the antecedent Asian H2N2 virus and has persisted to this day following a rapid march across the world.

In 1976 an abortive outbreak of influenza at Fort Dix was caused by a strain of swine influenza virus indistinguishable from viruses then circulating in pigs. This event, which caused great concern and resulted in an unprecedented national immunization program, was historically significant on several accounts. (1) The Fort Dix epidemic was caused by the simultaneous circulation of two antigenically different subtypes of influenza A virus [the Hong Kong (H3N2) and swine (H1N1) viruses]. All previous evidence had pointed to the sequential replacement of one major antigenic variant by another. (2) The potential of the swine influenza virus (thought to be the descendant of the agent of the 1918 pandemic) to cause epidemic, transmissible human disease was clearly demonstrated. (3) The virus, although antigenically novel to all those less than 50 years of age, failed to establish itself in the human population, and (4) this event provided further evidence for the recycling of influenza A virus antigens in man.

The following year yet another major antigenic variant appeared—also of the H1N1 subtype—with the advent of so-called "Russian" influenza. The origin of this virus, however, was subsequently traced to mainland China. Of special interest was the fact that this virus was truly a recurrent or recycled strain, virtually identical by genetic analyses with virus recovered in 1951 in the early years of prevalence of the significant H1 antigenic mutation that had appeared in 1946–47 (see The Pandemics, below).

As might have been predicted, this virus produced epidemics primarily in the young as it quickly circled the globe. But once again, this new major variant did not displace the virus of H3N2 subtype that had persisted since 1968. Remarkably, but not surprisingly, genetic analyses of viruses isolated since 1977 have shown some to be recombinants (reassortants) of H3N2 and H1N1 viruses. (This phenomenon and its significance are discussed in Chapter 10.)

EARLY HISTORY REVISITED

From our present perspective, gained from analysis of these recent epidemics with the techniques of modern virology, we can look back at the epidemics of past centuries with greater insight. We are still obliged to speculate, of course, but the speculation will be better informed.

Man and horse may have traded or reassorted viruses in the 17th, 18th, and 19th centuries; cocirculating viruses of different subtypes may have been responsible for epidemics characterized by variable virulence; pandemics probably represented times when new subtype viruses were introduced. The variable severity of such pandemics may have reflected partial immunity of populations to either hemagglutinin or neuraminidase antigen, variation in intrinsic viral virulence, or seasonal differences in occurrence.

In summary, our newer understanding of the potential of the virus makes us less inclined to reject as "anomalous" the patterns and variation of old epidemics called "influenza."

THE PANDEMICS

If influenza smolders between outbreaks and bursts into flame in the winter, then its pandemics are global conflagrations that can burn with searing intensity in any season to envelop those of all ages.

Influenza is not a tidy or predictable disease, and those who attempt to label it and put it in a box do so at risk. After the pandemic of 1968 it appeared that a neat 11-year pattern of pandemic recurrences was evolving, if one called the 1946–47 outbreak (attended by low mortality) a pandemic. Thus, influenza "A prime" (1946) was followed by Asian influenza (1957) and Hong Kong influenza in 1968. All were caused by influenza A viruses markedly changed in antigenicity, which appeared to supplant completely preexisting subtypes and rapidly to infect all susceptible members of the population before burning themselves out to leave way for the next variant (see Table 1-3). The swine influenza virus tried to poke its nose into the tent in 1976, and the Russian H1N1 influenza invaded the young in 1977, but the approximate decennial timetable fell apart in 1979 when no pandemic appeared. Even those intrigued by the 11-year pattern recognized that three pandemics represented a short series and that it was a long stretch between either 1920 or 1929 and 1946.

As we continue to search for some glimmer of order, we can note the reappearance of vanished viruses (or at least their major antigens). Again, a periodicity of 68-year cycles equal to the human life-span can be found for the H2 and H3

Table 1-3. Pandemics in the Era of Modern Virology

	1946–47	1957	1968
Virus	H1N1	H2N2	H3N2
Vernacular name	A prime	Asian	Hong Kong
First virus isolation	Australia	Singapore	Hong Kong
Probable origin	?S E Asia	China	China
Antecedent virus	H1N1[a]	H1N1	H2N2
Difference of virus from immediately antecedent viruses	+	+ + +	+ +
Replaced all antecedent strains	+	+	+
Global circulation in 2 years	+	+	+
Significant effect on mortality	+	+	+
Initial spring-summer wave	?	+	+
Population immunity	Partial[b]	Low	Partial[c]
Prior circulation of virus (recycling)	+	+	+
Virus tenure	1946–1957	1957–1968	1968–

[a]Formerly H0N1
[b]Based on antecedent exposure to antigenically related H1N1 viruses
[c]Based on previous exposure to N2 neuraminidase

antigens: 1889–1957 and 1900–1968, respectively (Kilbourne, 1977). This would suggest that virtually all immune individuals must be gone before an old antigen can reenter were it not for the successful reappearance of H1N1 Russian virus after an interval of only 26 years. But, in fact, the return of H1N1 in 1977, although attended by worldwide distribution of disease, involved principally the young and therefore was not truly pandemic. Soon, we may learn whether or not swine influenza virus, only transiently found in man since 1918, will reappear.

Pandemics Defined

Taken literally, the term "pandemics" defines an epidemic that involves all people. In ordinary context the term also implies that the epidemic is pervasive within a relatively brief period (1 or 2 years) and that the disease is severe. On this latter point, disagreement can occur. Influenza is a profoundly incapacitating disease from which complete recovery is the rule. The case fatality rate rarely exceeds 0.01%, yet, because of the high morbidity, large numbers of people are killed, recorded as "excess mortality" over usual or preexisting rates. Indeed, excess mortality is a hallmark of influenza. The relatively low mortality associated with the 1947 epidemic [less than that of the 1943–44 epidemic (Sartwell and Long, 1948)] has caused some to question its categorization as a pandemic—a point to be discussed further in the following sections.

Other characteristics disputably typical of influenza pandemics include cyclical occurrence, a tendency to invade the population in successive waves, and a greater independence of season (Francis, 1958). In fact, early summer outbreaks of influenza were documented in the nonpandemic years of 1943 and 1950–51 (Andrewes, 1951). Thus, these criteria are neither absolute nor restrictive in defining a pandemic.

Within the time of modern virology, pandemics have been associated with (1) a virtual absence of antibody in the population to the hemagglutinin antigen of the emergent virus and (2) the immediate replacement of viruses of the antecedent influenza A subtype by the pandemic virus.

The Modern Pandemics

Influenza 1918

Although the pandemic of 1918 antedated modern virology and immunology, it was the first pandemic subjected to careful bacteriological and pathological scrutiny. In 1889 the germ theory of disease was not yet generally accepted, and tissue staining techniques were still under development for postmortem studies of disease pathology.

The impact of this deadliest pandemic in history was awesome. The disease killed 20 million people, 500,000 in the United States alone, and overshadowed the final days of a world war.

> Nothing else—no infection, no war, no famine—has ever killed so many in so short a period. . . . The destruction caused by Spanish influenza is memorialized in reams of

> published statistics in every technologically advanced nation that was not in a state of chaos in 1918 Singlehandedly, flu thrust the year of 1918 back into the previous century. . . Not since the 1890s had the death rate in New Orleans, Chicago, and San Francisco been as high, and the 1918 death rate in Philadelphia was higher than at any time since the typhoid and smallpox epidemics of 1876 (Crosby, 1976).

But it is only from the personal accounts of those who lived through and suffered in this revisitation by the Middle Ages that its true nature can be conveyed.

• A medical artist, sent to make a watercolor sketch of the peculiar violet cyanosis seen in some patients, accomplished his mission but followed the patient to the grave a few days later (Hewer, 1979).

• Describing the epidemic at Fort Devens, Massachusetts, a young physician wrote,

> these men start with what appears to be an ordinary attack of . influenza, . and when brought to the hospital they very rapidly develop the most vicious type of pneumonia that has ever been seen. Two hours after admission they have the mahogany spots over the cheek bones, and a few hours later you can begin to seen the cyanosis extended from the ears and spreading all over the face, until it is hard to distinguish the colored man from the white. It is only a matter of a few hours then until death comes, and it is simply a struggle for air until they suffocate. It is horrible . We have an averaging about 100 deaths per day (Grist, 1979).

• In Capetown "because coffins were in short supply, many were buried in blankets in mass graves" (Phillips, 1978).

• In October, 1918, the *Ottawa Journal* reported that "street cars rattled down Bank Street with windows wide open and plenty of room inside. Schools, vaudeville theaters, movie palaces are dark; pool halls and bowling alleys, deserted" (Morton, 1973).

• A young nurse stationed at Reading, England, recalled,

> It happened so suddenly. In the morning we received an order to open up a new unit for flu and by night we'd moved into a converted convent. Almost before the desks were out the stretchers were in—60-80 to a classroom. We could hardly squeeze between the cots and oh, they were so sick! They came from a nearby airforce base. . some had been lying unattended for days. They all had pneumonia. We knew those whose feet were black wouldn't live" (Morton, 1973).

• In Philadelphia,

> A widow of great wealth living with her servants in a large home in the suburbs was taken ill and died without being able to secure the services of either doctor or nurse.
>
> The medical students not assigned to the emergency hospital, using cars bearing medical insignia would motor into the city's slums and stop, to be immediately surrounded by a crowd imploring them to see friends or relatives sick nearby. There was a frightful mortality among pregnant women (Starr, 1976).

Philadelphia was among the most severely devastated of American cities.

> Visiting nurses often walked into scenes resembling the plague years of the 14th century. They drew crowds of supplicants—or people would shun them for fear of the white gowns and gauze masks they often wore. One nurse found a husband dead in the

> same room where his wife lay with newly born twins. It had been 24 hours since the death and the births, and the wife had no food but an apple which happened to lie within reach (Crosby, 1976).

This small sampling of accounts of the pandemic should suffice to paint a picture of the confusion, disruption of public life, and personal tragedy of which this pandemic was made—a picture that is difficult to connote with the stark terms "morbidity" and "mortality."

So dreadful was the 1918 epidemic, especially in its impact in impoverished parts of the world (Patterson and Pyle, 1983; Pool, 1973), that it is sometimes referred to as "unique" or "unprecedented" and possibly caused by a highly virulent agent unlike contemporary strains of influenza virus. Even had the virus then been isolated and were now available for study, the question of intrinsic viral virulence still might be unanswerable; 1918 was a time of world conflict and civilian disruption and a time when bacterial pneumonia occurred in epidemic form in military camps (Kilbourne, 1960). Yet, probably the virus was unusually virulent despite the fact that in most of those affected the disease was a typical short-lived illness (for further discussion see Chapters 6 and 7).

Influenza 1946–47

The inclusion here of the 1946–47 epidemic as a pandemic is based on laboratory and vaccine trial evidence of the sort not available before first isolation of the virus from man in 1933. The saturation of the world's population by this H1N1 virus that occurred in epidemics in 1946–47 and 1948–49 was attended by lower mortality even than occurred in 1943—a year of antigenic change in the virus (Sartwell and Long, 1948). The change in 1943, however, was of lesser magnitude than that which occurred in 1946–47 (Chu et al., 1950), which may explain the greater diffusion of the 1947 variant and its crowding out of all earlier H1N1 viruses in a population with only low levels of antibody to the new strain (Meiklejohn and Bruyn, 1949; Rasmussen et al., 1948).

The first documented epidemic of what later was christened influenza A prime occurred in Melbourne adolescents in May, 1946 (Anderson and Burnet, 1947). Notably, there was no impact on the adult community. Antigenic analysis of influenza strains of previous years disclosed that the virus had been present in Melbourne in October, 1945 (Anderson and Burnet, 1947). Serological evidence points also to the possible presence of the virus in the Caribbean, also in 1945 (Rasmussen et al., 1948).

Description of the epidemic as "mild" is based on the low overall mortality. Certainly, morbidity in the American military was high (Sartwell and Long, 1948), and the disease was as prostrating as typical influenza (Kilbourne and Loge, 1950). The disease was widespread in Europe in 1948–49 in countries that had escaped the brunt of the 1947 outbreak (Chu et al., 1950). A later mutant of the A prime virus caused high mortality in Liverpool in 1951 (Semple, 1951).

Historically, the important lesson of the 1947 epidemic, pandemic or not, was a clear demonstration that influenza A virus could undergo antigenic change of sufficient degree to escape and circumvent community immunity—whether naturally acquired or vaccine engendered (Francis et al., 1947)—induced by earlier strains.

Influenza 1957

Emanating from central China, the first indisputable pandemic of the virologic era girdled the globe between February, 1957 and January, 1958. Following coincident epidemics in Hong Kong and Singapore in April, the disease spread rapidly throughout the Orient. In the meantime, shipboard outbreaks led to the seeding of populations in the Middle East, Australia, Africa, and subsequently Europe and North and South America (Fig. 1-2). Notably, significant although focal epidemics occurred during the summer months north of the Equator that were followed by extensive morbidity and mortality in the fall in Europe and North America.

Within a month of the Singapore and Hong Kong outbreaks, studies in the United States, England, and Australia identified the new (H2N2) virus as a major antigenic variant, and worldwide spread of extensive disease was predicted (Langmuir, 1961). Thus, the change in virus was recognized early by laboratory surveillance, the pandemic was predicted, and the prophecy was fulfilled. Influenza epidemiology had come of age.

The special significance of the 1957 epidemic was the clear demonstration that a contemporary virus identifiable as an influenza virus by conventional techniques could indeed cause a pandemic, which, if less lethal than 1918, was attended by extensive morbidity and sharp increase in mortality. Careful and intensive clinical and pathological studies demonstrated that the virus, unaided by bacterial coinvaders, could produce fatal pneumonia (reviewed by Louria et al., 1959; Kilbourne, 1959). One indirect measure of the impact of this epidemic was the reduction of elective surgery necessitated by the decline in blood donations during the epidemic. In a manner reminiscent of the pandemics of 1889 and 1918, Asian influenza produced significant mortality in its later waves, especially in the United States in 1960. But total mortality remained but a fraction of that seen in the earlier pandemics (Fig. 1-1).

Influenza 1968

Only 11 years after the invasion of Asian influenza, its virus (H2N2) disappeared simultaneously with the advent of yet another virus of Far Eastern origin (H3N2) that rapidly acquired the sobriquet of "Hong Kong influenza" from the site of initial virus isolation. The epidemic apparently originated in early summer in southeastern China. The new virus caused a large epidemic in Hong Kong, infecting 15% of the population, then spread rapidly throughout the East in a pattern similar to that in 1957. Beyond the Far East, subsequent epidemics were more circumscribed, as worldwide dissemination of the virus occurred within a year, and high excess mortality in the first year was observed only in the United States (Cockburn et al., 1969).

The causative virus of the epidemic was initially identified as an "A2" (H2) variant by serological tests that did not discriminate between hemagglutinin and

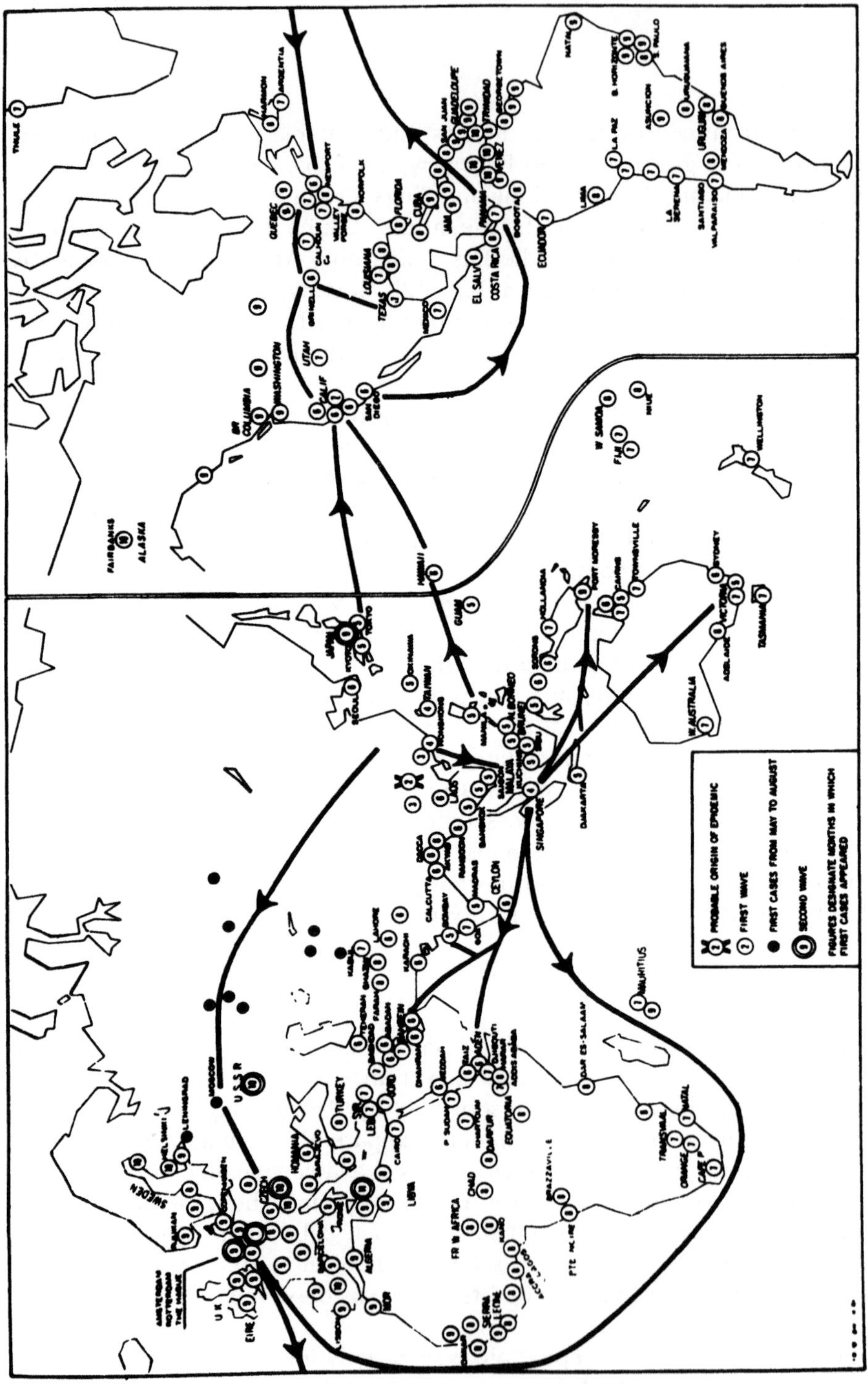

Figure 1-2. Spread of influenza in the 1957 "Asian" pandemic.

neuraminidase viral antigens. The use of hybrid reassortant viruses as test reagents promptly demonstrated, however, that the hemagglutinin of the Hong Kong agent was new (H3) but that it retained the neuraminidase (N2) of the Asian subtype viruses (Schulman and Kilbourne, 1969). Thus, the virus was not totally novel, a circumstance that might have blunted its impact (for further discussion see Chapter 7).

A remarkable characteristic of the virus was its infectiousness for laboratory workers and its capacity to infect lower animals. Indeed, the virus became established in swine in several parts of the world (Harkness et al., 1972; Nakajima et al., 1982; Ottis et al., 1982).

As the second major pandemic in 11 years, the Hong Kong epidemic was a sobering reminder of the omnipresent pandemic potential of "the last great plague" (Kilbourne, 1968, 1975) and prompted concern about the adequacy of current control measures for the handling of the disease. The surveillance network of WHO had proved its value, but true control of the disease remained a distant objective (see Chapter 8).

Influenza Post-1968: Major Antigenic Changes in the Virus Are Not Necessarily Followed by Pandemics

Since 1968, regional epidemics caused by virus of H3N2 subtype have continued to occur even after the abortive circulation of H1N1 swine influenza virus at Fort Dix, New Jersey in 1976 and the return of A-prime-like Russian H1N1 viruses in 1977.

On the face of it, the experience of the last 16 years has been quite unremarkable—resembling epidemiologically the wake of 1918-20—with a series of epidemics occurring at irregular intervals with variable impact on excess mortality rates (Fig. 1-1). But the insights of modern virology have revealed not only the cocirculation of viruses of both H3N2 and H1N1 subtypes but of reassortant viruses, which probably arose in man, that contain genes of both subtypes. These findings lend some credibility to the supposition from serological evidence of prior cocirculation of viruses of those two serotypes before World War I and remind us of the impossibility of definitively analyzing past epidemics without specific etiologic information.

Remarkably, H1N1 viruses did not immediately supplant or even predominate over H3N2 strains after their emergence, perhaps because of the relative insusceptibility of older members of the community. Indeed, H3N2 virus predominated in the United States in 1977-78, 1980-81, and 1982-83. However, in the 1983-84 season, isolations of H1N1 virus exceeded those of H3N2 by 9:1.

Influenza 1976: The First Epidemic of Swine Influenza Virus Infection in Man

Details of the appearance of swine influenza virus in military recruits in Fort Dix in early 1976 are discussed in Chapters 7 and 8. Suffice it to note here that this epidemic was of historic importance in demonstrating (1) the potential of this animal virus for replicating and inducing transmissible disease in man, (2) that cocirculation of two subtype variants of influenza A virus could occur, and (3) that introduction of a new viral subtype into a susceptible population need not be followed by pandemic spread of virus.

Influenza 1977

In the late spring of 1977, viruses were isolated from influenza outbreaks in Liaoning province of northern China that proved to be of H1N1 serotype indistinguishable from strains isolated in the early 1950s. These outbreaks were concentrated in primary and middle schools (Kung et al., 1978). Thus began a pandemic of the young as virus spread rapidly into south China and across its borders. By November the virus had reached Hong Kong, Singapore, and the USSR (Schild, 1978). Epidemics in Russia were the first to receive worldwide attention and led to the description of the new influenza as "Russian flu." Within 1 year the virus had become globally disseminated, causing disease principally in those less than 20 years of age. No country reported increased mortality rates associated with the virus in the first year of its appearance (Schild, 1978).

Viruses of H1N1 subtype continue to cocirculate with H3N2 strains at this writing.

REFERENCES

Anderson, S. G., and Burnet, F. M., 1947. Sporadic and minor epidemic incidence of influenza A in Victoria 1945–1946. I. Phase behaviour of influenza A strains in relation to epidemic characteristics, *Aust. J. Exp. Biol. Med.* **25**:235–242.

Andrewes, C. H., 1951, Epidemiology of influenza in the light of the 1951 outbreak, *Proc. R. Soc. Med.* **44**:803–804.

Brill, S. J., and Gilfillan, R. F., 1977, Acute parotitis associated with influenza type A, *N. Engl. J. Med.* **296**:1391–1392.

Burnet, F. M., and Clark, E., 1942, *Influenza. A Survey of the Last 50 Years in the Light of Modern Work on the Virus of Epidemic Influenza. Monograph from the Walter and Eliza Hall Institute of Research in Pathology and Medicine*. Macmillan, Melbourne, pp. 69–99.

Chu, C. M., Andrewes, C. H., and Gledhill, A. W., 1950, Influenza in 1948–1949, *Bull. WHO* **3**:187–214.

Cockburn, W. C., Delon, P. J., and Ferreira, W., 1969, Origin and progress of the 1968–69 Hong Kong influenza epidemic, *Bull. WHO* **41**:345–348.

Collins, S. D., 1957, *Influenza in the United States, 1887–1956. Public Health Monograph No. 48*, U.S. Government Printing Office, Washington.

Creighton, C., 1891, *A History of Epidemics in Britain, AD 664–1666*, Cambridge University Press, Cambridge.

Crosby, A. W., 1976, *Epidemic and Peace*, Greenwood Press, Westport, CT.

Davenport, F. M., 1977, Reflections on the epidemiology of myxovirus infections, *Med. Microbiol. Immunol.* **164**:69–76.

Francis, T., Jr., 1958, Influenza. Epidemiology. Influenza as a military problem, *Preventive Medicine in World War II*, Vol. IV, *Communicable Diseases* (J. B. Coates, Jr., E. C. Hoff, and P. Hoff, eds.), The Surgeon General, Dept. of the Army, Washington, pp. 85–128.

Francis, T., Jr., Salk, J. E., and Quilligan, J. J., 1947, Experience with vaccination against influenza in the Spring of 1947, *Am. J. Public Health* **37**:1013–1016.

Grist, N. R., 1979, Pandemic influenza 1918, *Br. Med. J.* **2**:1632–1633.

Harkness, J. W., Schild, G. C., Lamont, P. H., et al., 1972, Studies on relationships between human and porcine influenza. 1. Serological evidence of infection in swine in Great Britain with an influenza A virus antigenically like human Hong Kong/68 virus, *Bull. WHO* **46**:709–719.

Hewer, C. L., 1979, 1918 influenza epidemic, *Br. Med. J.* **1**:199.

Hirsch, A, 1883, *Acute Infective Diseases. Handbook of Geographical and Historical Pathology*, Vol. I (C. Creighton, trans.), New Sydenham Society, London.

Kendal, A. P , Minuse, E., Maassab, H. F., et al., 1973, Influenza neuraminidase antibody patterns of man, *Am. J. Epidemiol.* **98**:96–103.

Kilbourne, E. D., 1959, Studies on influenza in the pandemic of 1957–58. III. Isolation of influenza A (Asian strain) viruses from influenza patients with pulmonary complications. Details of virus isolation and characterization of isolates, with quantitative comparison of isolation methods, *J. Clin Invest* **38**:266–274

Kilbourne, E. D.,, 1960, The severity of influenza as a reciprocal of host susceptibility, in: *Ciba Foundation Study Group No. 4. Virus Virulence and Pathogenicity* (G. E W. Wolstenholme and C. M. O'Connor eds.), Little, Brown, Boston, pp. 58–77.

Kilbourne, E D., 1963, in: Infections primarily of the respiratory tract, *Preventive Medicine and Public Health* (W. G. Smillie and E. D. Kilbourne, eds.), Macmillan, New York, p. 192.

Kilbourne, E D , 1975, *The Influenza Viruses and Influenza*, Academic Press, New York.

Kilbourne, E D , 1977, Influenza pandemics in perspective, *J A.M.A* **237**:1225–1228.

Kilbourne, E. D , and Loge, J P., 1950, Influenza A prime A clinical study of an epidemic caused by a new strain of virus, *Ann. Intern. Med* **33**:371–379.

Kilbourne, E D , Anderson, H. C , and Horsfall, F. L., Jr., 1951, Concurrent infection with influenza A and B viruses in a single epidemic, *J. Immunol.* **67**:547–558.

Krilov, L R., and Swenson, P., 1985, Acute parotitis associated with influenza A infection, *J. Infect. Dis* **152**:853.

Kung, H. C , Jen, K. F., Yuan, W. C , et al., 1978, Influenza in 1977: recurrence of influenzavirus A subtype H1N1, *Bull. WHO* **56**:913–918.

Langmuir, A D., 1961, Epidemiology of asian influenza, *Am Rev Respir. Dis* **83**:2–14.

Louria, D. B , Blumenfeld, H. L., Ellis, J. T., et al., 1959, Studies on influenza in the pandemic of 1957–58. II. Pulmonary complications of influenza, *J. Clin. Invest.* **38**:213–265.

Masurel, N., and Heijtink, R. A., 1983, Recycling of H1N1 influenza A virus in man—a haemagglutinin antibody study, *J. Hyg.* **90**:307–402.

Masurel, N., and Marine, W. M., 1973, Recycling of Asian and Hong Kong influenza A virus hemagglutinins in man, *Am. J Epidemiol.* **97**:44–49.

Meiklejohn, G., and Bruyn, H. B., 1949, Influenza in California during 1947 and 1948, *Am. J. Public Health* **39**:44–49.

Morton, G, 1973, The pandemic influenza of 1918, *Can. Nurse* **69**:25–27.

Nakajima, K., Nakajima, S., Shortridge, K. F., et al., 1982, Further genetic evidence of maintenance of early Hong Kong-like influenza A(H3N2) strains in swine until 1976, *Virology* **116**:562–572

Ottis, K., Sidoli, L., Bachmann, P. A., et al , 1982, Human influenza A viruses in Pigs: Isolation of a H3N2 strain antigenically related to A/England/42/72 and evidence for continuous circulation of human viruses in the pig population, *Arch. Virol.* **73**:103–108.

Patterson, K. D., and Pyle, G. F., 1983, The diffusion of influenza in sub-Saharan Africa during the 1918–1919 pandemic, *Soc. Sci. Med.* **17**:1299–1307.

Phillips, H., 1978, Black October: Cape Town and the Spanish influenza epidemic of 1918, in: *Conference on the Economics of Health Care in Southern Africa*, University of Capetown, Capetown, S.A., paper #36.

Pool, D. I , 1973, The effects of the 1918 pandemics of influenza on the Maori population of New Zealand, *Bull. Hist. Med.* **47**:273–281.

Rasmussen, A F , Jr., Stoker, J. C., and Sinadel, J. E., 1948, The army experience with influenza 1946–1947. II. Laboratory aspects. *Am. J. Hyg.* **47**:142–149.

Rekart, M , Rupnik, K., Cesario, T. C., et al., 1982, Prevalence of hemagglutination inhibition antibody to current strains of the H3N2 and H1N1 subtypes of influenza A virus in sera collected from the elderly in 1976, *Am J. Epidemiol.* **115**:587–597.

Sartwell, P E., and Long, A. P., 1948, The Army experience with influenza, 1946–1947. I. Epidemiological aspects, *Am. J. Hyg.* **47**:135–141.

Schild, G., 1978, An 'old' flu virus returns, *New Scientist* **79**:949–950.

Schulman, J. L., and Kilbourne, E. D., 1969, Independent variation in nature of the hemagglutinin and neuraminidase antigens of influenza virus: Distinctiveness of the hemagglutinin antigen of Hong Kong/68 virus, *Proc. Natl. Acad. Sci.* **63**:326–333.

Semple, A. B., 1951, Epidemiology of the influenza epidemic in Liverpool in 1950–51, *Proc. R. Soc. Med.* **44**:794–796.

Shope, R. E., 1958, Influenza. History, epidemiology, and speculation. (The R. E. Dyer Lecture), *Public Health Rep.* **73**:165–178.

Smith, W., Andrewes, C. H., and Laidlaw, P. P., 1933, A virus obtained from influenza patients, *Lancet* **2**:66–68.

Starr, I., 1976, Influenza in 1918, *Ann. Intern. Med.* **85**:516–518.

Thompson, T., 1852, *Annuals of Influenza or Epidemic Catarrhal Fever in Great Britain from 1510 to 1837* Sydenham Society, London.

II

The Influenza Viruses

2

Taxonomy and Comparative Virology of the Influenza Viruses

INTRODUCTION

Viral taxonomy has evolved slowly (and often contentiously) from a time when viruses were identified at the whim of the investigator by place names, names of persons (investigator or patient), sigla, Greco-Latin hybrid names, host of origin, or name of associated disease. Originally studied by pathologists and physicians, viruses were first named for the diseases they caused or the lesions they induced. Yellow fever virus turned its victims yellow with jaundice, and the virus now known as poliovirus destroyed the anterior horn cells or gray (polio) matter of the spinal cord. But the close kinship of polioviruses with coxsackievirus B1, the cause of the epidemic pleurodynia, is not apparent from names derived variously from site of pathogenic lesion and place of original virus isolation (Coxsackie, New York).

In practice, many of the older names of viruses remain in common use, but the importance of a formal, more regularized approach to viral classification has been increasingly recognized by students and practitioners of virology as more basic information has become available about the nature of viruses. Accordingly, the International Committee on Taxonomy of Viruses has devised a generally accepted system for the classification and nomenclature of viruses that serves the needs of comparative virology in formulating regular guidelines for viral nomenclature. Although this nomenclature remains highly diversified and at times inconsistent, retaining many older names, it affirms "an effort. . . toward a latinized nomenclature" (Matthews, 1981) and places primary emphasis on virus structure and replication in viral classification.

TAXONOMY OF INFLUENZA VIRUSES

"Influenza" is the name conferred not by taxonomists or scientists but apparently by its victims in the middle of the 18th century (Creighton, 1891). Whether this Italian word referred to "influenza di freddo" [influence of the cold (Francis, 1959)] or astrological influence (International Committee on Taxonomy of Viruses, 1982) is unclear, but surely the word is appropriate for its potentially devastating effects if not its causation. The viruses causing influenza have been alphabetically named in the chronological order of their isolation and definition as influenza A, B, and C viruses. (Along the way, deficiencies in basic knowledge led to brief consideration of Sendai virus, now classified as parainfluenza virus 1, as influenza D virus.)

As the first viruses recognized to possess in vitro enzymatic activity and to attach to an identifiable substrate, influenza viruses were christened myxo (mucin-reacting) viruses, later changed to orthomyxovirus as the paramyxoviruses were distinguished from and related to them on the basis of evolving knowledge of viral structure and function (see Table 2-1).

Taken literally, the term orthomyxovirus combines the Greek word *orthos* for "straight or correct" and another Greek word, *myxa*, for "mucous," with an Italian form of the Latin *influentia*. Pity the archeologist of the future who has to decipher this designation!

As with no other virus group, the epidemiologic orientation of influenza virus research has led to the acquisition of a vast inventory of virus strains and the need for a variety of prototype strains with which to conduct research. It has long been traditional, therefore, to identify new influenza strains in terms of place of origin, isolate number, and year of isolation, e.g., A/Puerto Rico/8/34, which has become, in laboratory vernacular, simply "PR8." Improved methods of antigenic analysis and determination of viral protein structure, although they have aided enormously in the sorting out of viruses causing human and animal disease, have also blurred the perception of host of origin—e.g., is "swine" influenza virus a human or animal virus?—as a cardinal taxonomic criterion.

Although some antigenic cross reactivity is sporadically demonstrable among them, antigenically distinct subtypes have been defined for 13 hemagglutinins and nine neuraminidases of human, swine, equine, and avian influenza A viruses (Table 2-2). Of these, only H1N1, H2N2, and H3N2 viruses have been isolated from epidemic human infections.

Among avian species various combinations of hemagglutinins and neuraminidases are found, suggesting, along with other evidence, genetic reassortment of the genes for these proteins. In recent years transmission of H3N2 virus, presumably from man to swine, has resulted in combination of this virus with traditional swine (H1N1) virus in Japan to produce hybrid H1N2 virus (Yasuhara et al., 1983).

Because conventional antigenic analyses of influenza viruses identify only their surface antigens, routine methods are obviously incapable of identifying genes for internal or nonstructural proteins and, therefore, can detect only hemagglutinin/neuraminidase antigenic hybrids or reassortants. The application of RNA

Table 2-1. Classification of the Influenza Viruses[a]

Taxonomic status	International name	English vernacular name	Viral envelope antigens[b]
Family	Orthomyxoviridae	Influenza virus group	—
Genus	*Influenza virus*	Influenza virus	—
Type	A	Influenza A virus	—
Subtypes (human viruses)			H1N1 H2N2 H3N2
Strain designation (example)		A/England/1/51 (H1N1)[c] A/Swine/Iowa/15/30 (H1N1)[d]	
Type	B	Influenza B virus	
Subtypes	None		
Strain designation (example)		B/Great Lakes/1/54	—[e]
Type	C[f]	Influenza C virus	
Subtypes	None		
Strain designation (example)		C/Paris/1/67	

[a]Data from Fourth Report of the International Committee on Taxonomy of Viruses (1982) and WHO Memorandum (1980)
[b]H, hemagglutinin, N, neuraminidase
[c]Type/place of isolation/strain number/date of isolation (subtype)
[d]Host of origin is given when original isolate is from a nonhuman host
[e]H and N antigens but no subtypes have been designated
[f]Classification of influenza C virus as a member of the genus *Influenza virus* is provisional

Table 2-2. Subtypes of Hemagglutinin and Neuraminidase Antigens of Influenza A Viruses[a]

Hemagglutinin subtype	Former designation(s)	Neuraminidase subtype	Former designation(s)
H1	H0,H1,Hswl	N1	N1
H2	H2	N2	N2
H3	H3,Heq2,Hav7	N3	Nav2,Nav3
H4	Hav4	N4	Nav4
H5	Hav5	N5	Nav5
H6	Hav6	N6	Nav1
H7	Heq1,Hav1	N7	Neq1
H8	Hav8	N8	Neq2
H9	Hav9	N9	Nav6
H10	Hav2		
H11	Hav3		
H12	Hav10		
H13	—[b]		

[a]WHO Memorandum (1980)
[b]Proposed as new H subtype, avian source (Hinshaw et al., 1982, 1983)

gel electrophoresis, oligonucleotide mapping, and RNA–RNA hybridization techniques have identified such reassortants as A/California/10/78 (H1N1) that clearly have inherited genes from both H1N1 and H3N2 parental human viruses (Young and Palese, 1979; Bean et al., 1980). No provision currently exists for classification of natural reassortants.

Laboratory-produced reassortant viruses (widely used as vaccine strains or reference reagents), if not antigenically hybrid and thus obviously reassortants, bear a postscript "R" after the strain donating the surface antigens. For example, X-31 [laboratory designation (Kilbourne, 1969; Kilbourne et al., 1971)], an early high-yielding vaccine strain widely employed in chemical and immunologic studies of the virus, is properly designated A/Aichi/2/68 (H3N2) (R). This designation, however, without reference to the original papers describing its origin, gives no hint that six of the eight RNAs are derived from PR8 [A/PR/8/34 (H1N1)] (Baez et al., 1980). If antigenically hybrid, reassortants should be identified with respect to the strain of origin of the hemagglutinin and neuraminidase antigens, e.g., A/Bel/42 (H1)–Singapore/1/57 (N2) (R) (WHO Memorandum, 1980).

In short, present taxonomic labels, although useful, do not substitute for precise description as an identification of influenza virus strains in formal scientific communications. No less important is citation of the laboratory of origin. PR8 virus strains from different laboratories may be antigenically different (Paucker, 1960), and Cambridge and Mount Sinai strains of PR8 differ cytopathogenically in human conjunctival cell cultures (J. L. Schulman and E. D. Kilbourne, unpublished data).

The 50-year genealogy of another classical prototype strain, WSN, illustrates not only the hundreds of laboratory passages that it has undergone but the variety of host tissues in which the virus has replicated since leaving the respiratory tract in man in 1933 (see Fig. 6-1). This virus, long used in genetic studies (reviewed by Kilbourne, 1963), defies conventional classification, and, indeed, few working with the virus today know its exact history.

RELATION OF INFLUENZA VIRUSES TO OTHER ENVELOPED VIRUSES WITH RNA GENOMES

A wide variety of viruses have evolved using linear single-stranded RNA as their genetic moiety and host-derived envelopes as their limiting membranes. Such viruses contain virus-coded spikelike projections on their surfaces and share replication characteristics to the extent that pseudotypes—virus particles with mixed envelope antigens—may be formed during mixed infection. However, data in Table 2-3 emphasize critical differences as well as superficial similarities among them. RNA recovered from toga- and coronaviruses is infective, being of positive or message sense (Baltimore, 1971), whereas the strategy of influenza virus replication requires transcription with an endogenous, virus-coded transcriptase of a negative-sense RNA genome to form mRNA. Replication of the Retroviridae is fundamentally different from that of all other enveloped viruses, emphasizing again the superficiality of surface structure as a taxonomic criterion, both literally and figuratively.

Table 2-3. Taxonomic Relationship of Influenza Viruses (Orthomyxoviruses) to Other Enveloped, Single-Stranded RNA Animal Viruses[a]

Genome strategy	Family or group	Genus [example(s)]	Species example (English vernacular name)	Host(s)
Positive sense	Togaviridae	Arbovirus group A	Sindbis virus	Vertebrates, arthropods
		Arbovirus group B	Yellow fever virus	Vertebrates; arthropods
		Rubella virus	Rubella virus	Man
		Mucosal disease virus group	Hog cholera virus	Vertebrates
	Coronaviridae	Coronavirus	Human coronavirus	Man
Negative sense	Paramyxoviridae	Paramyxovirus group	Parainfluenza virus 1 (Sendai)	Man, mouse
		Measles–rinderpest distemper group	Measles virus	Man
		Respiratory syncytial virus group	Respiratory synctial virus	Man
	Orthomyxoviridae	Influenza virus group	Influenza A virus	Vertebrates
	Rhabdoviridae	Rabies virus group	Rabies virus	Vertebrates, arthropods
	Bunyaviridae	Bunyamwera supergroup	Bunyamwera virus	Vertebrates; arthropods
	Arenaviridae	LCM virus group	Lymphocytic choriomeningitis virus	Vertebrates
DNA replication step	Retroviridae	(RNA tumor viruses)	—	Vertebrates

[a]Data derived from Fourth Report of the International Committee on Taxonomy of Viruses (1982)

More closely related structurally are the paramyxoviruses, some of which share with influenza viruses an external virus-coded neuraminidase. Comparison of the influenza and parainfluenza viruses (Table 2-4) reminds one of the original disease-related basis for viral taxonomy in that both of these viruses with hemagglutinating and neuraminidase activity primarily attack respiratory tissue with the production of acute respiratory tract disease. Indeed, homologies of regions of the HA and NA proteins with the HN protein of a parainfluenza virus have been noted (Blumberg et al., 1985). But paramyxoviruses, especially measles virus, can establish persistent infection, a characteristic ascribed to swine influenza virus by Shope (1935–1936) but not yet verified as a mechanism of influenza pathogenesis. The characteristic differences in influenza and parainfluenza viruses with respect to epidemiologically significant antigenic variation is not totally explicable on the basis of differences in their capacity to participate in genetic reassortment.

INFLUENZA VIRUSES AS SEGMENTED GENOME VIRUSES

No system of nomenclature relates the diverse viruses that carry their genetic information in discrete segments. However, viruses possessing this type of genome have the capacity for genetic reassortment with related strains and, therefore, the potential for rapid evolution of essentially new viruses derived from biparental contribution analogous to sexual reproduction. Potential advantages to viruses capable of genetic reassortment include the facilitation of interspecific infection and adaptation to extend viral host range (Kilbourne, 1981). Other advantages are cited in Chapter 6. The taxonomic implications of reassortment

Table 2-4. Comparison of Influenza and Parainfluenza Viruses

Viral property	Influenza viruses	Parainfluenza viruses
Genome	Segmented: eight molecules of linear ss RNA	One molecule of linear ss RNA
Genome strategy	Negative sense	Negative sense
Genetic Reassortment	+	0
Require host cell nuclear functions	+	0
Virion size (diameter)	80–124 nm	150 nm or more
Enveloped, budding virus	+	+
Hemagglutinin and neuraminidase	Two separate glycoproteins[a]	Both functions in single glycoprotein
Epidemiologically significant antigenic variation	+	Unproved
Persistent infection	Unproved	+

[a]One glycoprotein in influenza C virus

Table 2-5. Prototype Segmented Genome Viruses of Vertebrates and Their Host Range[a]

Genus	Subgroup	Prototype	RNA	Segments	Gene products	Host range
Reovirus	Reovirus	Reovirus type I	ds	10	11	Mammals, birds
Reovirus	Orbivirus	Bluetongue virus	ds	10	9	Mammals, insects
Reovirus	Rotavirus	Human rotavirus	ds	11	10-12	Mammals
Birnaviruses[b]	—	Infectious pancreatic necrosis virus	ds	2	6	Fish
Birnaviruses[b]	—	Infectious bursal disease virus	ds	2	6	Chickens
Bunyavirus	—	Bunyamwera virus	ss	3	5	Mammals, insects
Orthomyxovirus	—	Influenza A virus	ss	8	10	Mammals, birds
Arenavirus	—	LCM virus	ss	2	3-4	Mammals

[a]Revised from Kilbourne (1981)

[b]Bisegmented ds RNA animal virus group

are in its potential for the instant creation of new viruses as genes from two or more viruses are mixed in varying proportions, an event now found to have occurred with cocirculating influenza viruses in nature (Young and Palese, 1979; Bean et al., 1980).

Examples of the varied viruses of vertebrates with segmented RNA genomes are listed in Table 2-5, which includes both single- and double-stranded viruses containing two to 11 RNA segments that code for three to 12 gene products. These viruses are widely distributed in nature and infect a variety of hosts, including insects, fish, birds, and mammals. The role of segmented genomes in viral evolution, adaptation, and survival has not yet been demonstrated, but by facilitating genetic interchange, they appear to enhance the evolutionary and adaptive advantage of viruses possessing them (Kilbourne, 1981). Natural reassortment of viruses other than influenza viruses has not been reported.

REFERENCES

Baez, M., Palese, P., and Kilbourne, E. D, 1980, Gene composition of high-yielding influenza vaccine strains obtained by recombination, *J. Infect. Dis.* **141**:362–365.

Baltimore, D., 1971, Expression of animal virus genomes, *Bacteriol. Rev.* **35**:235–241.

Bean, W. J., Jr., Cox, N. J., and Kendal, A. P., 1980, Recombination of human influenza A viruses in nature, *Nature* **284**:638–640.

Blumberg, B., Giorgi, C., Roux, L., et al., 1985, Sequence determination of the Sendai virus HN gene and its comparison to the influenza virus glycoproteins, *Cell* **41**:269–278

Creighton, C., 1891, *A History of Epidemics in Britain, AD 664–1666,* Cambridge University Press, Cambridge.

Francis, T., Jr., 1959, Influenza, in: *Viral and Rickettsial Infections of Man,* (T. M. Rivers and F. Horsfall, Jr., eds.), 3rd ed. J. B. Lippincott, Philadelphia, p. 633.

Hinshaw, V. S., Air, G. M., Gibbs, A. J., et al, 1982, Antigenic and genetic characterization of a novel hemagglutinin subtype of influenza A viruses from gulls, *J. Virol.* **42**:865–872

Hinshaw, V. S., Air, G. M., Schild, G. C., et al., 1983, Characterization of a novel hemagglutinin subtype (H13) of influenza A viruses from gulls, *Bull. WHO* **61**:677–679.

International Committee on Taxonomy of Viruses, 1982, Fourth report, *Intervirology* **17**:106–107.

Kilbourne, E. D., 1963, Influenza virus genetics, *Prog. Med. Virol.* **5**:79–126.

Kilbourne, E. D., 1969, Future influenza vaccines and the use of genetic recombinants, *Bull. WHO* **41**:643–645.

Kilbourne, E. D., 1981, Segmented genome viruses and the evolutionary potential of asymmetrical sex, *Perspect. Biol. Med.* **25**:66–77.

Kilbourne, E. D., Schulman, J. L., Schild, G. C., et al., 1971, Correlated studies of a recombinant influenza-virus vaccine. I. Derivation and characterization of virus and vaccine, *J. Infect. Dis.* **124**:449–462.

Matthews, R. E. F., 1981, The classification and nomenclature of viruses: Summary of results of meetings of the International Committee on Taxonomy of viruses in Strasbourg, August 1981, *Intervirology* **16**:53–60.

Paucker, K., 1960, Apparent antigenic change in a laboratory line of influenza A virus during continued propagation in chick embryos, *J. Immunol.* **85**:148–162.

Shope, R. E., 1935–1936, The influenzas of swine and man, *Harvey Lect.* **31**:183–213.

WHO Memorandum, 1980, A revision of the system of nomenclature for influenza viruses, *Bull WHO* **58**:585–591.

Yasuhara, H., Hirahara, T., Nakai, M , et al., 1983, Further isolation of a recombinant virus (H1N2, formerly Hsw1N2) from a pig in Japan in 1980, *Microbiol. Immunol* **27**:43–50.

Young, J. F., and Palese, P., 1979, Evolution of human influenza A viruses in nature: Recombination contributes to genetic evolution of H1N1 viruses, *Proc. Natl. Acad. Sci. U.S.A.* **76**:6547–6551.

3

Viral Structure and Composition

INTRODUCTION

The influenza viruses comprise three types of viruses that are closely related in structure, chemical composition, and biological activity. These viruses, types A, B, and C, are pleomorphic enveloped viruses bearing surface projections (peplomers or spikes). Influenza A and B viruses contain four major structural proteins and three proteins with RNA replicase and transcriptase activity (Table 3-1). Eight segments of linear, single-stranded RNA are contained in separate nucleocapsids—rodlike structures of nucleoprotein (NP) (Heggeness et al., 1982)—with which the replicase and transcriptase proteins PB_1, PB_2, and PA are in close association in an RNP complex. The interior of the virus particle is delimited by repeating units of the matrix or membrane (M) protein in all three viral subtypes. The external spikes of influenza A and B viruses number 400 to 500 (Ruigrok et al., 1984) and comprise two different glycoproteins, the hemagglutinin (HA) and neuraminidase (NA). In the case of influenza C virus, only seven RNA segments and a single external glycoprotein have been demonstrated.

SIZE AND MORPHOLOGY OF INFLUENZA VIRUS PARTICLES

As observed by electron microscopy, influenza virus particles are 80–120 nm in diameter (Fig. 3-1); the mass of an influenza A or B virus particle is estimated by scanning transmission electron microscopy to be 174×10^6 daltons (Ruigrok et al., 1984). These measurements are confirmed by small-angle neutron-scattering curves of influenza B virus that indicate as well a lipid bilayer centered at a radius of 425 Å that is 40 Å to 50 Å thick and constitutes 25–28% of the virion mass. The surface glycoproteins comprise 40–46% of the total mass of the virus particle. Surprisingly little protein is found by this technique in the interior of

Table 3-1. The Structural Proteins of Influenza A Viruses

Proteins	Coded by RNA segment	Average molecular weight	Polypeptide length (no. amino acids)	Molecules per virion	Structural features	Function
PB1[a]	2[b]	86,300–86,500	757	20–60	Basic protein (see text)	Transcriptase activity; initiation
PB2[a]	1	85,800	759	20–60	Basic protein	Host cell RNA cap binding; transcriptase activity
PA[a]	3	82,400–83,000	716	20–60	Acidic protein	Transcriptase activity; chain elongation
HA	4	~77,000	~548	500	Glycoprotein; 3D structure known	Binds virion to cell receptor; fusion and entry
HA1		~50,000	~326			
HA2		~27,000	~222			
NP[a]	5	56,101	482–498	1000	Coiled filaments	Encapsidates RNA in RNPs; possible role in transcription
NA	6	~56,000 (tetramer ~220,000)	466	100	Glycoprotein; 3D structure known	Virion release and desialization of virion surface proteins
M	7	27,861	252	3000		Bounds viral core; major virion protein

[a]Comprise the RNA transcriptase–replicase complex
[b]Numbered in inverse order of migration of glyoxal-treated RNA on polyacrylamide gels (i.e., decreasing order of size)

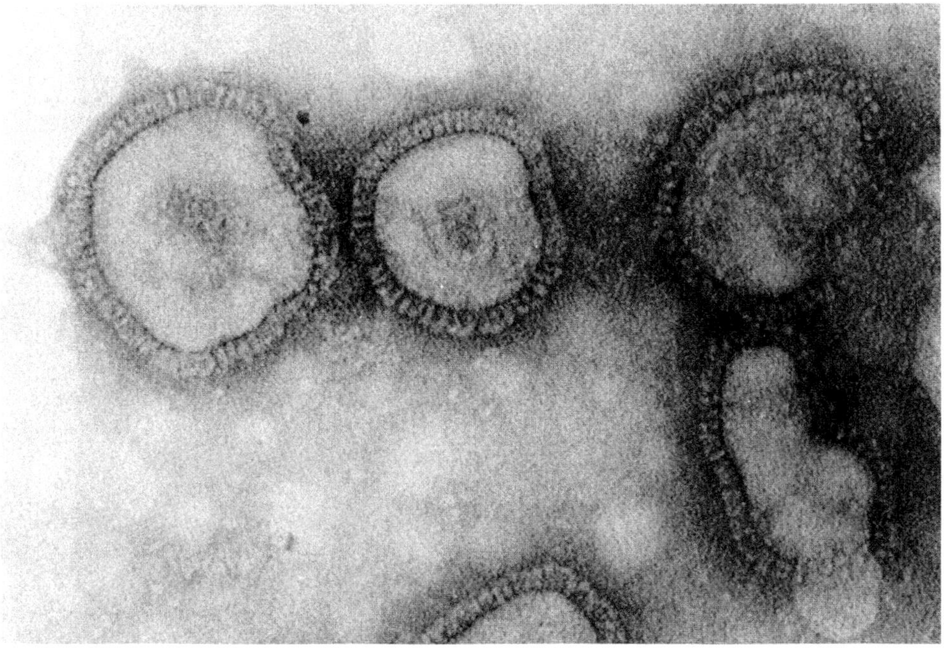

Figure 3-1. Influenza virus particles as observed by electron microscopy. Magnification × 160,000; negative staining with 4% sodium phosphotungstate at pH 7.0.(Courtesy of Erickson, 1977.)

the virus particle, suggesting that many particles do not contain the full complement of ribonucleoprotein complexes (Cusack et al., 1985).

Particles of standard laboratory viruses are mainly spherical; newly isolated strains contain many elongated or filamentous forms (Fig. 3-2). The pleomorphism of influenza viruses is in part an artifact of preparation and negative staining. Freeze-dried and shadowed preparations show less variability in particle size and shape, but filamentous forms are still observed (Nermut and Frank, 1971). Although filament formation can be induced by surface-active agents (Blough, 1963), it is clear that filaments represent morphological variation characteristic of a given strain of virus. Filaments bear surface spikes like those of spherical particles (Fig. 3-2), and filamentousness is genetically determined, as has been shown by virus adaptation (Choppin et al., 1960) and reassortment (Kilbourne and Murphy, 1960; Kilbourne, 1963) experiments. Highly filamentous strains replicate to lower titer in the allantoic sac of the chick embryo (the standard host for influenza virus production), and, indeed, the two characteristics appear to be linked (Kilbourne and Murphy, 1960) and probably determined by the gene(s) coding for the M protein(s) (E. D. Kilbourne, unpublished data).

The linkage of high yield and spherical morphology in laboratory strains well adapted to chicken embryos suggests that this may be the morphological form of virus predominant in naturally infected hosts. In this view, the filamentous nature of viruses newly derived from man may represent inefficient replication during early passage in an alien host, i.e., the chicken embryo. However, in one

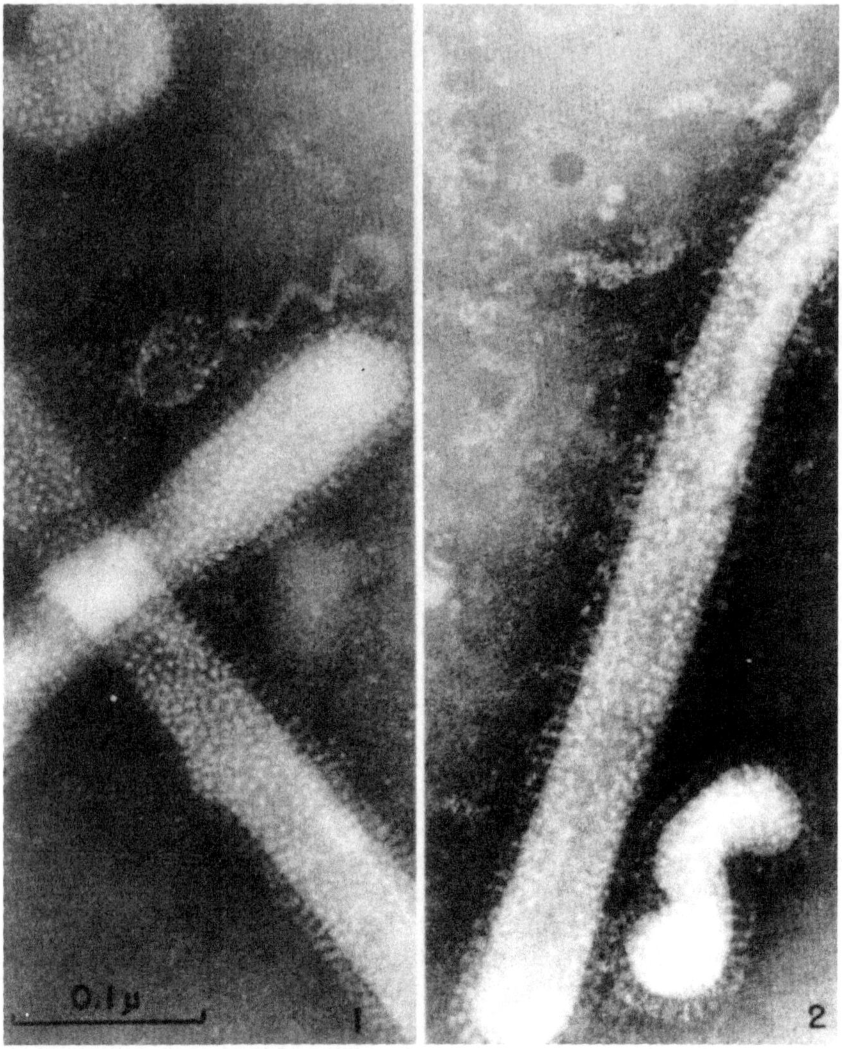

Figure 3-2. Filamentous influenza virus particles showing surface projections or spikes. Negative staining with 2% phosphotungstic acid. (From Choppin et al., 1961. Reproduced with permission, from *Virology* **13**:549.)

study of equine and swine influenza viruses, material derived directly from infected animals was described as "filamentous," although the ratio of spherical to filamentous forms was not given (Wilson et al., 1977). Further studies are needed to resolve this question. The possibility remains that either form may predominate in natural infection depending on the genetic characteristics of the given strain.

The hemagglutinin and neuraminidase spikes projecting from the surface of influenza A and B virus particles differ in morphology. The HA spikes are rod shaped, about 16 nm long and 5 nm wide, and are triangular when viewed end on (Wrigley, 1979). Spikes of NA are mushroom shaped, consisting of a head attached to the particle by a hydrophobic stalk. Because the mean ratio of HA/NA spikes is about 5 : 1, NA spikes are difficult to resolve in electron microscopy of untreated virus particles. However, it is clear from correlated biochemical and morphological analyses of particles from which HA has been selectively removed that (1) viral strains differ in their proportion of HA to NA per virion and (2) NA spikes are, indeed, mushroom shaped in situ and appear to be evenly distributed over the surface of the virus particle (Erickson and Kilbourne, 1980) (Fig. 3-3). Earlier observations have suggested that the spikes were clustered in discrete areas (Compans et al., 1969; Kendal and Madeley, 1970; Nermut, 1972). Clumping, observed with HA antibodies (Compans et al., 1969; Kendal and Madeley, 1970), might have been an artifact reflecting fluidity of membranes (Singer and Nicolson, 1972) and the tendency of antibodies reacting with membrane proteins to form patches and caps (Taylor et al., 1971). However, a recent study with immunogold staining under conditions said to exclude "capping" reports a uniform distribution of HA and a patchy distribution of NA spikes (Murti and Webster, 1986).

In contrast to influenza A and B viruses, influenza C virus has only a single surface glycoprotein, which appears to possess both HA and receptor-hydrolyzing activity. This structural difference may be a requirement for the formation of a surface lattice of glycoprotein molecules in hexagonal arrays, observed most frequently with influenza C virus (reviewed by Air and Compans, 1983). A single morphological unit has been noted at each of the six vertices of the hexagonal arrays (Herrler et al., 1981). Influenza C virus spikes are comparable in dimension to hemagglutinin spikes of influenza A and B viruses (i.e., 8–10 × 4–5 nm).

Wrigley (1979) questions the analogy with the icosahedral structure of the nonenveloped viruses, pointing out that particle assembly by budding "precludes a rigid quasi-crystalline structure." He sees the influenza virus particle as "a squashy particle in a minimum energy configuration in suspension (i.e., near spherical) but capable of considerable distortion under external stresses" (see Fig. 3-4). However, purified particles from certain zonal sucrose gradient fractions are homogeneous and spherical (Fig. 3-5) and have the highest infectivity/viral protein ratio (Ruigrok et al., 1985).

The Viral Envelope

The influenza virus envelope consists of a host-derived lipid bilayer lined by the viral M protein into which are inserted the hydrophobic regions of the HA and NA external glycoproteins (Choppin and Compans, 1975; Wrigley, 1979; Skehel et al., 1980; Karadaghi et al., 1984).

The lipid bilayer serves to stabilize the virus particle (reviewed by Wrigley, 1979), a stabilization that is probably enhanced by binding of surface glycoproteins and M protein on opposite sides of the membrane (Karadaghi et al., 1984). Removal of spikes leaves a naked but intact particle (Fig. 3-6).

The shell of M protein beneath and in apposition to the lipid membrane is 6 nm thick (Apostolov and Flewett, 1969) and is variously estimated to consist of 800 to 4000 molecules per virus particle (Fig. 3-7).

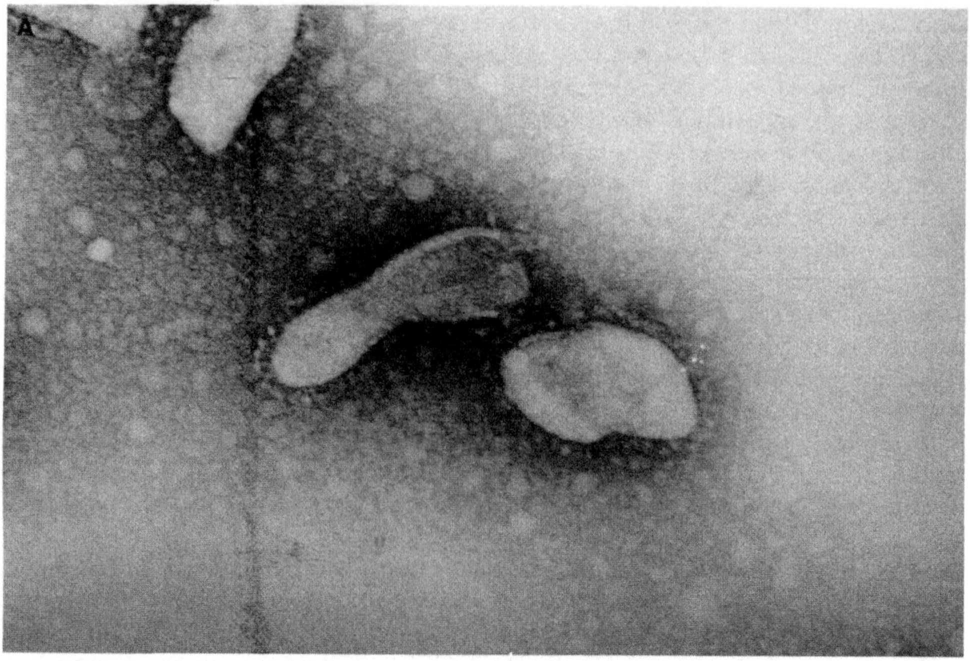

Figure 3-3. Ultrastructure of influenza virus after removal of HA spikes with trypsin. The square heads of the remaining NA spikes are attached to the virus particles by short filaments. The NA spikes are evenly distributed over the viral surface. (From Erickson and Kilbourne, 1980. Reproduced with permission, from *Virology* **107**:325, 326.)

Internal Structure

The exact arrangement of viral proteins within the interior or core of the virus particle is unknown. The principal protein of the interior is the viral nucleoprotein (NP), 1000 molecules of which are contained in each particle (Wrigley, 1979). The eight RNA segments of the virus genome are encapsidated, probably individually, by NP, identifiable in disrupted virus preparations as flexible rods of varying lengths (50–150 nm) (Compans et al., 1972; Pons et al., 1969). Nucleocapsid segments are right-handed helices that are subject to significant structural modification in response to environmental changes, e.g., changes in salt concentration (Heggeness et al., 1982). There is some evidence that in the native virus particle larger helical structures (Fig. 3-8) are present, perhaps formed by the orderly aggregation of smaller segments (Murti et al., 1980). Whether or not the RNP segments are ordered is relevant to questions of genome assembly and problems with genetic reassortment (Murti et al., 1980; Heggeness et al., 1982).

RNA associated with NP in the RNP complex is accessible to ribonuclease and therefore partially exposed but is an integral part of the RNP structure. Each RNP has the appearance of a disk 5 nm thick and 7 nm in diameter. RNA is probably coiled, with each loop representing about 25 nucleotides of RNA. The coil

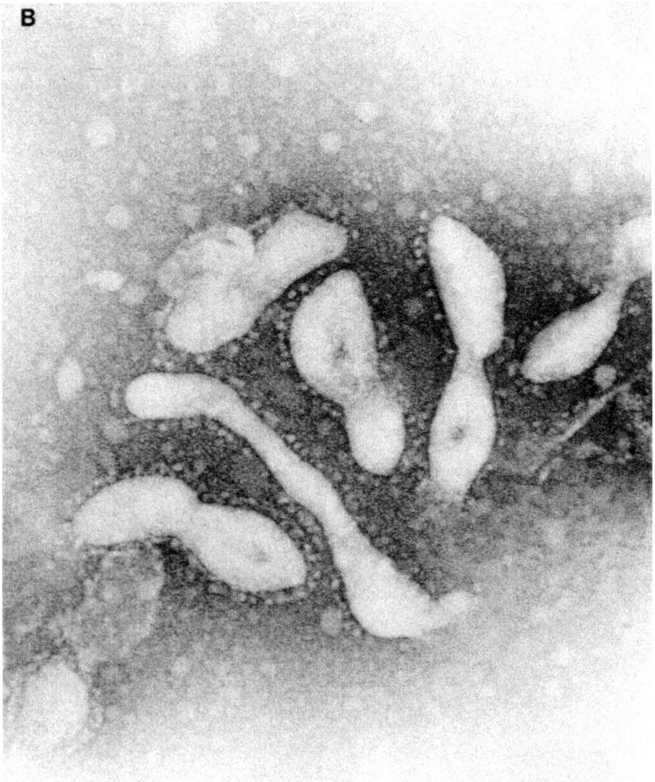

Figure 3-3. *(Continued)*

is arranged as a double-helical ribbon with about 5 1/2 loops per turn (Jennings et al., 1983) (Fig. 3-9).

The structural relationship of the three largest viral proteins (PB_1, PB_2, and PA) to the RNP is as yet undefined except to say that all are intimately associated with the viral RNA (Wrigley, 1979). Together, the three "P proteins" constitute only 1.1–2.2% of structural proteins of the virus, and only a few molecules of each polypeptide are present in each virion (Kawakami et al., 1981) (Table 3-1). The few complete nucleotide sequences of these proteins available have demonstrated close homology even between human and avian strains of PB_2 and PA (Roditi and Robertson, 1984; Robertson et al., 1984). On the other hand, genes of PB_1 and PB_2, basic proteins of nearly identical size and involved in similar functions, show a complete lack of sequence homology. Thus, they appear not to have evolved from a single ancestral gene (Winter and Fields, 1982).

Little information on secondary structure is available on the P proteins. As basic proteins, PB_1 and PB_2 are rich in arginine and other basic amino acids. A prediction of secondary structure for PB_1 cites the presence of a "four-helical-supersecondary structure" and "four polypeptide double helices" or antiparallel pleated sheets (Sivasubramanian and Nayak, 1982).

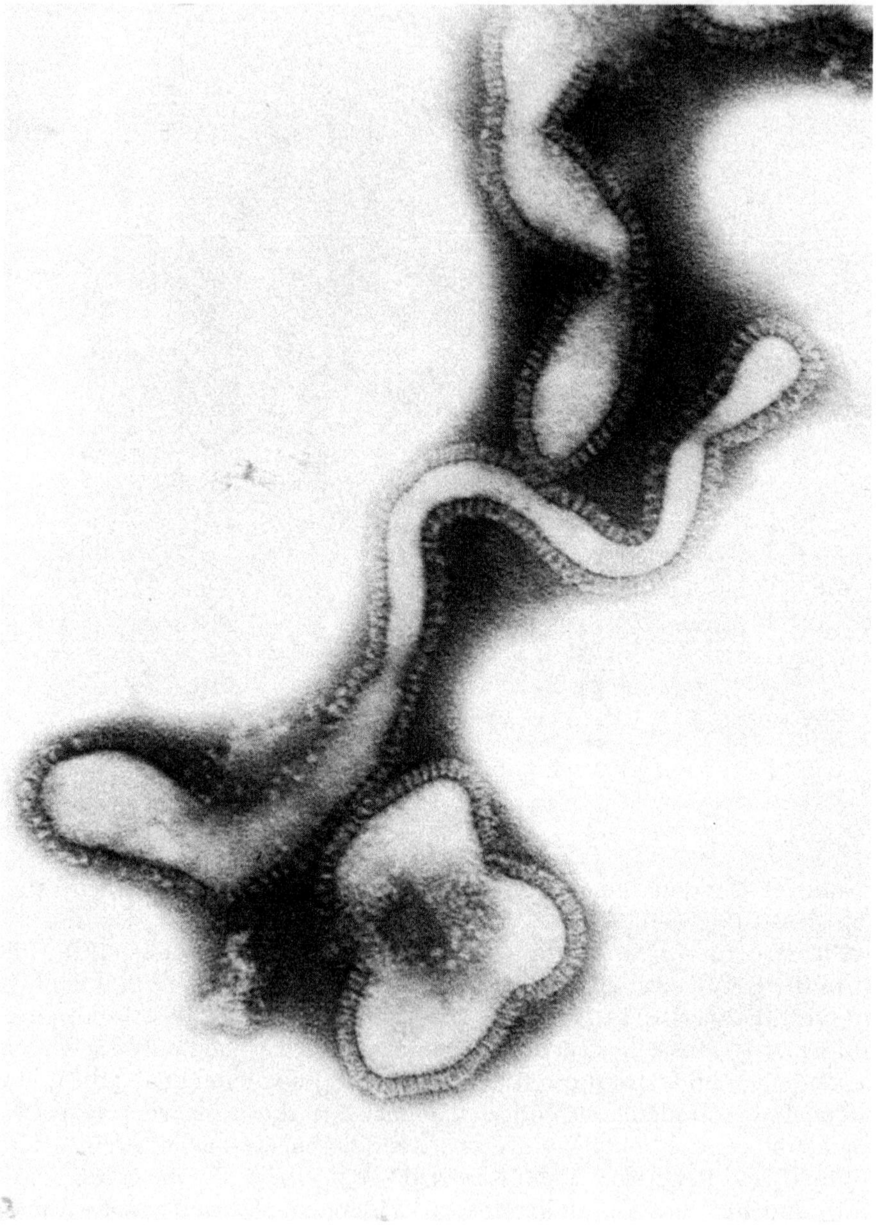

Figure 3-4. Pleomorphism of influenza virus particles prepared by conventional methods for electron microscopy. Negative staining with 4% sodium phosphotungstate. (Courtesy of Erickson, 1977.)

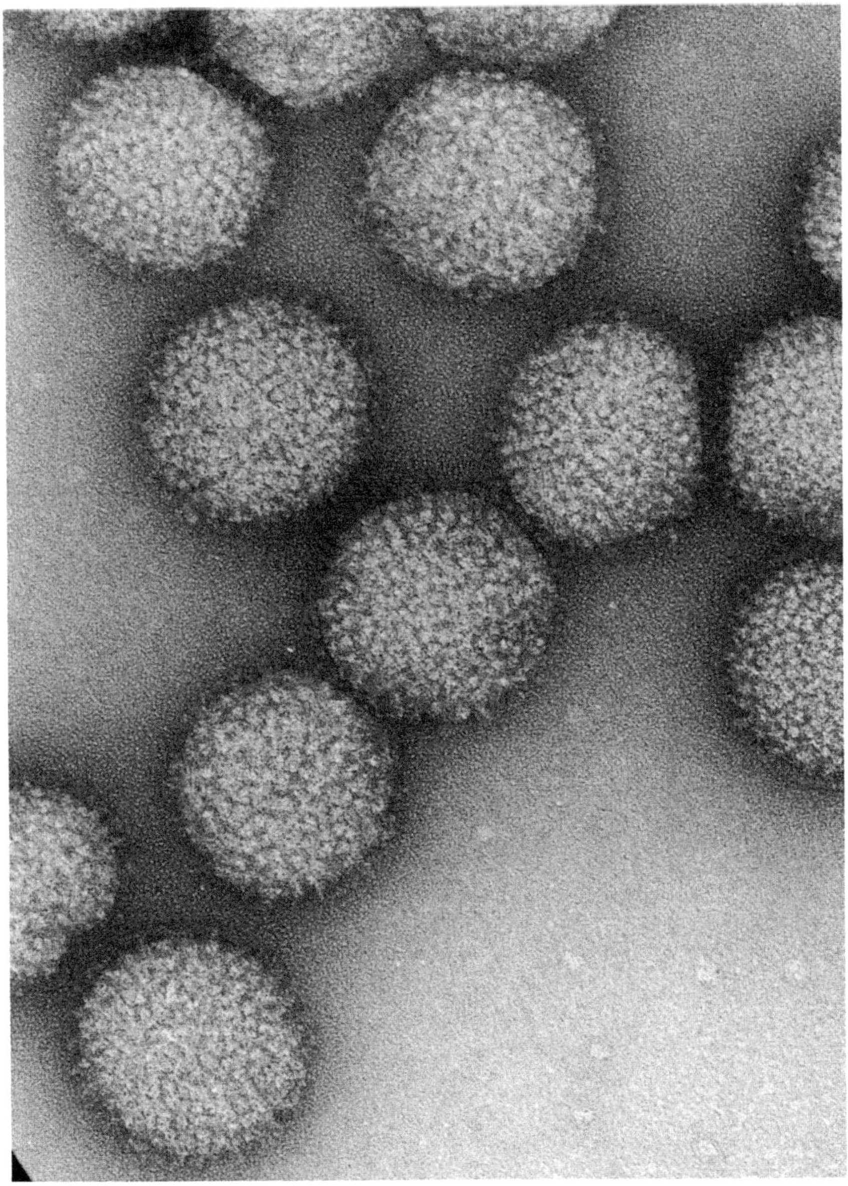

Figure 3-5. Well-preserved, uniformly spherical influenza B virus particles stained with uranyl acetate at pH 4.2. (Courtesy of R. W. H. Ruigrok.)

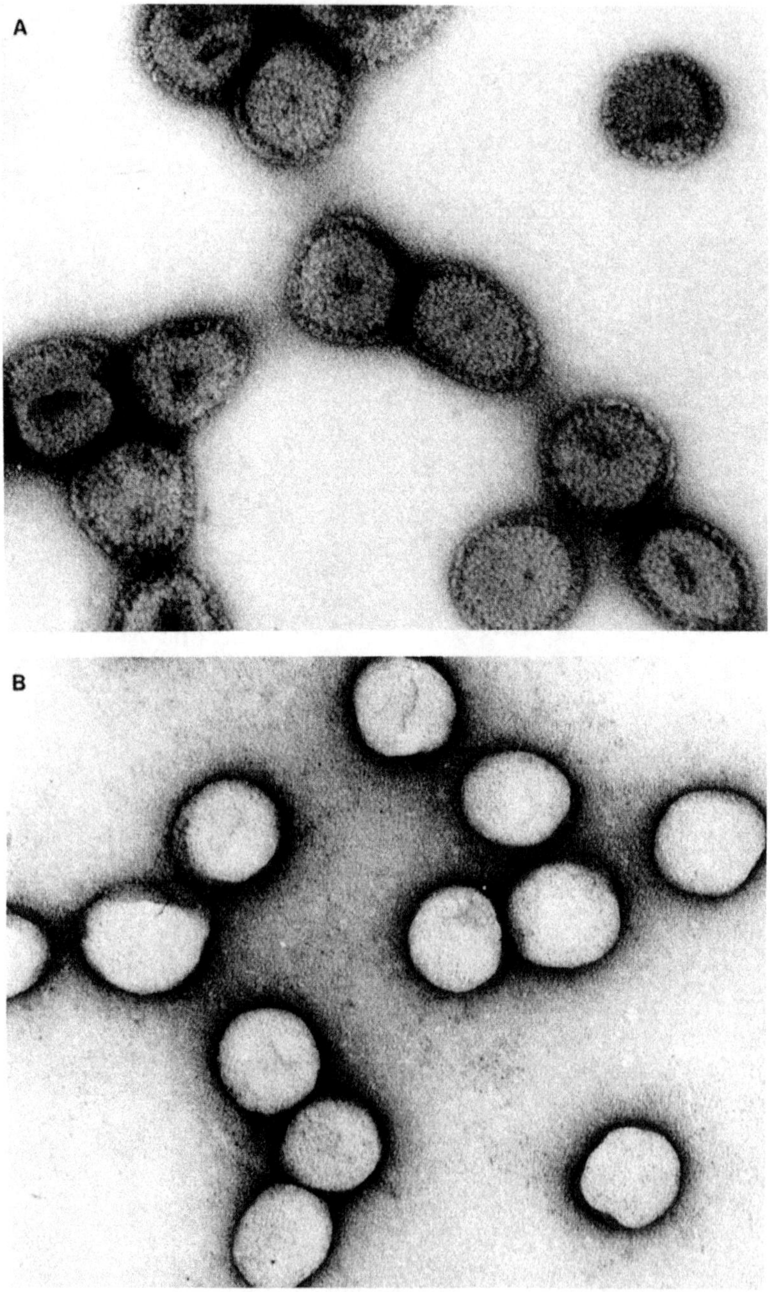

Figure 3-6. Influenza virus particles before (A) and after (B) removal of surface spikes with bromelain. ×150,000; 1% uranyl acetate stain. (From Kharitonenkov et al., 1978. Reproduced with permission, from *FEBS Lett.* **96:**120.)

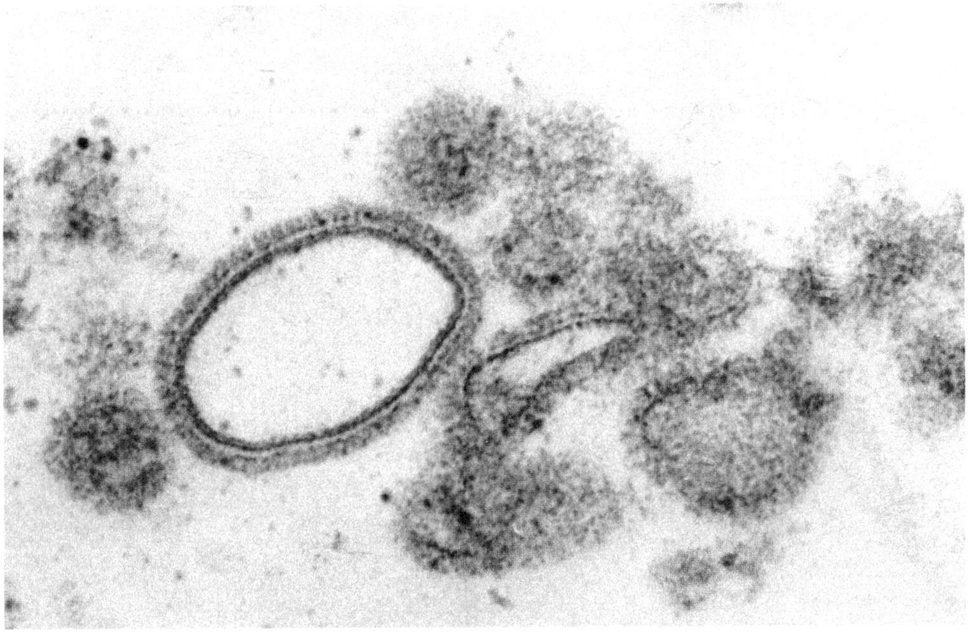

Figure 3-7. Fixed and sectioned particles of influenza A virus showing electron-dense inner layer of M protein. (From Apostolov and Flewett, 1969. Reproduced with permission, from *J. Gen. Virol.* **4**:366, Plate I.)

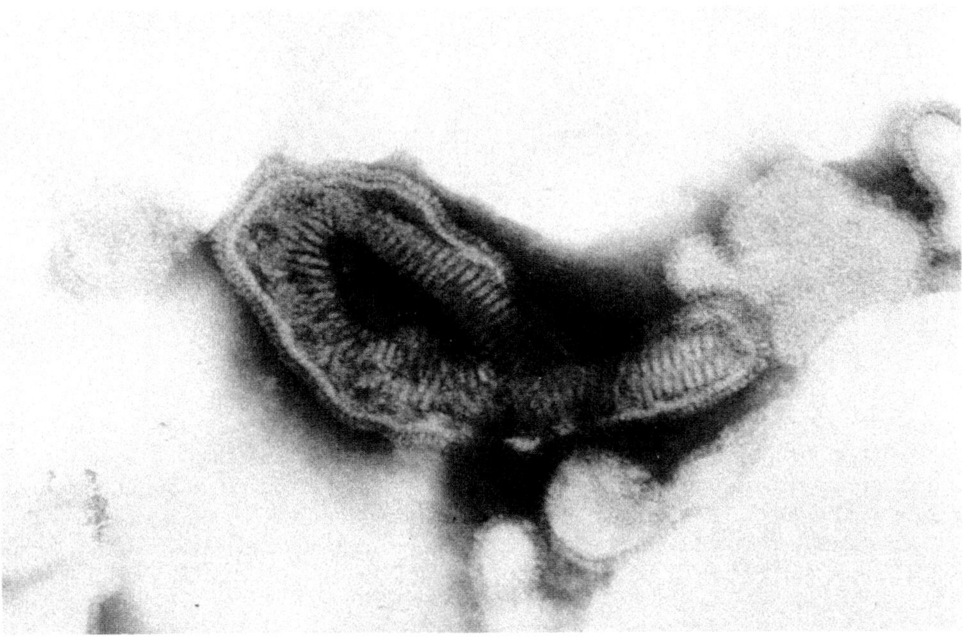

Figure 3-8. Tightly coiled RNP in disrupted virus particles. (Courtesy of Erickson, 1977.)

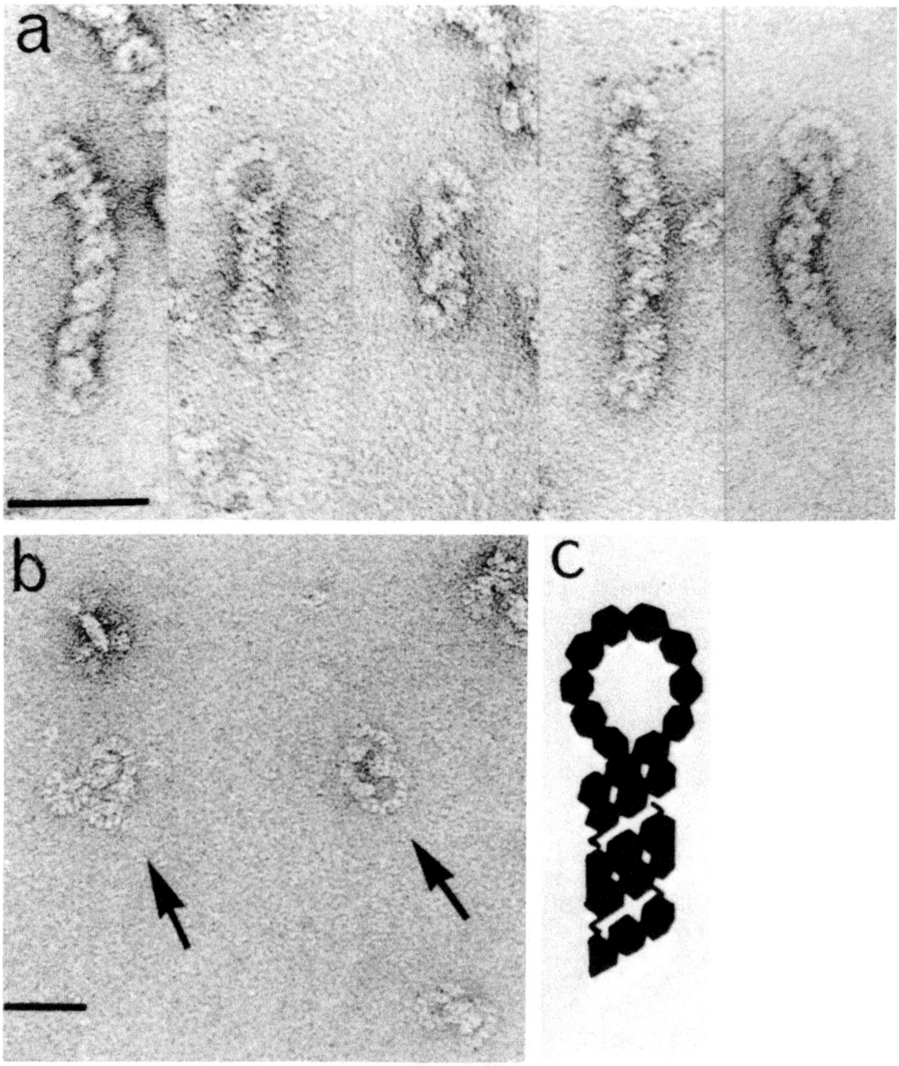

Figure 3-9. Electron micrograph of viral RNPs (see text). a: Variable loop size and double helical ribbon structure. b: Segment 8 RNPs opened by increased salt concentration. c: Diagrammed structure of segment 8 RNP as a polymer of 35–40 NP subunits forming a ribbon wound into a right-handed helix. About 25 nucleotides of RNA are associated with each NP subunit. (From Jennings et al., 1983. Reproduced with permission, from *Cell* **34**:624. Copyright © M.I.T.)

STRUCTURE OF THE ENVELOPE GLYCOPROTEINS

Evidence from genetic reassortment yielding antigenic hybrids of HA and NA has demonstrated clearly that in both influenza A (Laver and Kilbourne, 1966) and B viruses (Tobita and Kilbourne, 1974), HA and NA functions are carried by different viral proteins. From H1N2 antigenic hybrids, Laver and Valentine (1969) isolated morphologically different viral spikes, one type associated with HA and the other with NA activity. These proteins have been intensively studied in the past decade, and through crystallography and x-ray diffraction studies their three-dimensional structure has been resolved recently in a series of classical studies. The structure of no other viral proteins is better understood, and certainly no other viral proteins are more important in the epidemiology of infection. A full review of the complex structure of these proteins is beyond the scope of this book and would be duplicative of recent excellent reviews by those directly involved in the structural research (Wiley et al., 1984; Ward, 1981). The sections that follow summarize essentials of the studies as well as other equally important studies locating the principal antigenic regions of the HA and NA molecules. The contribution of the structural studies to the understanding of viral antigenic variation, virulence, and evolution becomes apparent in Chapters 6 and 10.

Hemagglutinin Structure

Definitive structural studies have been carried out only on the HA of a single H3 strain, the HA of a 1968 prototype, A/Aichi/2/68 (H3N2), carried in the reassortant virus X-31 (Kilbourne, 1969). However, complete and partial amino acid sequences are available for viruses of other HA subtypes, these sequences demonstrate conservation of structurally important residues, including six disulfide bridges, demonstrating that the H3 model has general applicability (Wilson et al., 1981). Furthermore, comparison of two influenza B virus sequences with the H3 three-dimensional model shows that these viruses of different type share most of the important structural features of type A viruses (Krystal et al., 1983).

The influenza virus HA is a trimer of 224,640 molecular weight, triangular in cross section, measuring 13.5 nm from membrane to tip, and varying in radius from 1.5 to 4.0 nm. Posttranslational cleavage either during or after viral replication removes an arginine to divide the chain into two disulfide-linked chains, HA_1 (328 amino acids) and HA_2 (221 amino acids) (Wilson et al., 1981; Verhoryen et al., 1980; Ward and Dopheide, 1979). This cleavage is required for viral infectivity and HA-mediated fusion activity (Lazarowitz and Choppin, 1975; Klenk et al., 1975; Huang et al., 1980).

As illustrated in Fig. 3-10, the trimeric glycoprotein comprises two structurally distinct regions: a triple-stranded coiled coil of α helices that extends from the viral membrane and a globular region of eight-stranded β-sheet structure. This latter region is distal to the cylindrical triple-stranded α helices (coiled coils) of the fibrous stalklike region that stabilizes the trimer and functions as a stem that unites the globular region with the viral membrane. At the membrane end of the molecules is a five-stranded β sheet containing the N terminus of HA_1 and the C terminus of HA_2.

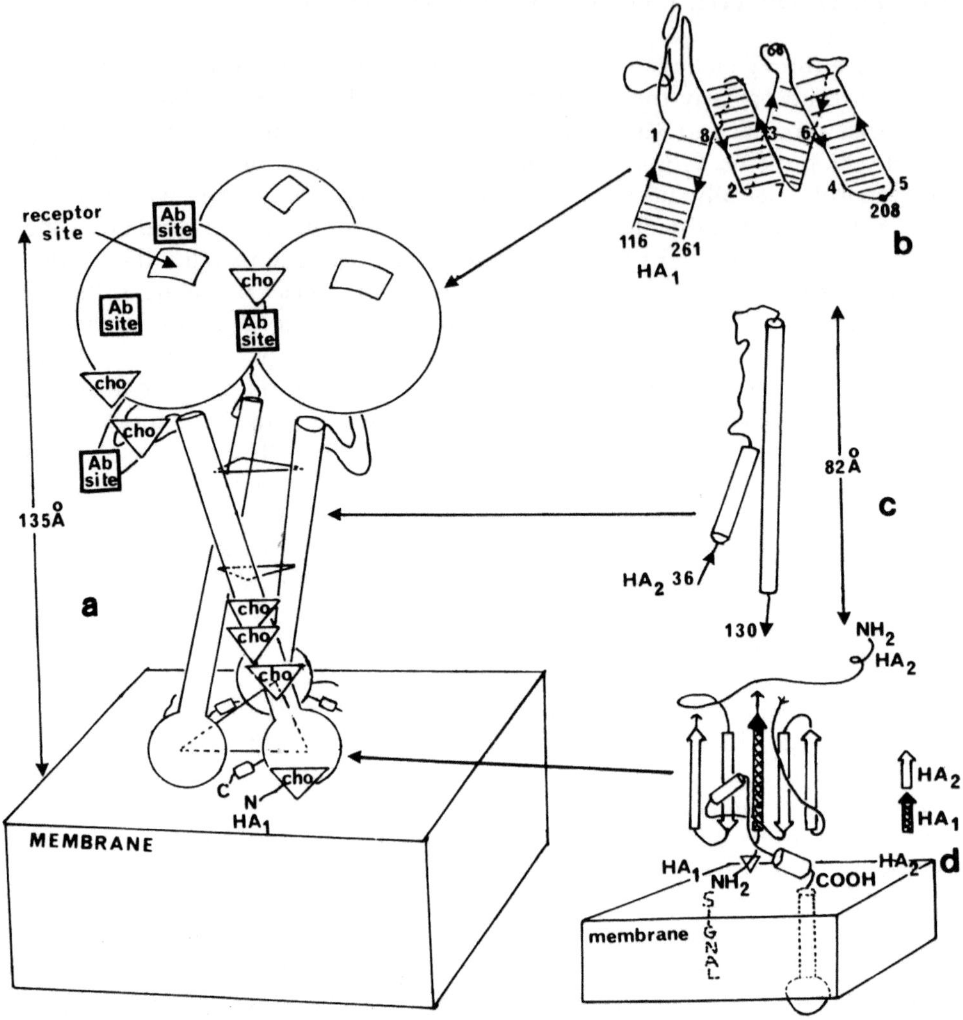

Figure 3-10. Diagrammatic representation of HA trimer inserted into the viral membrane (see text). (From Wilson et al., 1981. Reprinted by permission from *Nature* **289**:368. Copyright © 1981 by Macmillan Journals Limited.)

With this knowledge of three-dimensional structure it has been possible to correlate changes in amino acid sequence associated with viral antigenic variation with topographical regions of the hemagglutinin. On this basis, four antigenic domains have been identified on the H3 model (Wiley et al., 1981), three verified by electron microscopy of HA-monoclonal antibody complexes (Wrigley et al., 1983).

Only minor local structural distortion of HA structure associated with a single amino acid substitution is required for an antigenic change sufficient to allow the virus to escape neutralization by a monoclonal antibody (Knossow et al., 1984). However, studies of the antigenic structure of the H1 prototype virus strain A/PR/8/34 H1N1 have defined five immunodominant antigenic areas, one of which is dependent on conformational changes of the HA monomer as it is presented in trimeric form (Caton et al., 1982). Another important finding of the latter study was the potential influence of carbohydrates in modulating antigenic activity on regions of the HA molecule. Monomeric and trimeric forms of the HA are antigenically distinguishable (Nestorowicz et al., 1985).

Neuraminidase Structure

The morphological difference between HA and NA spikes was first clearly shown by Laver and Valentine (1969), who demonstrated that detergent-released HA spikes were rodlike and that NA spikes consisted of mushroomlike oblong structures with centrally attached fibers or stalks. The unusual three-dimensional structure of this viral enzyme (E.C. 3.2.1.18 acylneuraminyl hydrolase) is now known (Varghese et al., 1983), and its catalytic and antigenic sites have been provisionally defined (Colman et al., 1983). As with the HA, structure has been determined principally by x-ray diffraction of crystalline protein.

Influenza virus (N2) neuraminidase proves to be a tetrameric protein with circular fourfold symmetry. In the three-dimensional structure of the box-shaped ($100 \times 100 \times 60$ Å) head, each monomer consists of six β sheets of identical topology arranged in propeller formation (Varghese et al., 1983) (Fig. 3-11).

Like the HA, the NA is embedded in the viral membrane by a hydrophobic region. Unlike the HA and most other membrane glycoproteins, the membrane-embedded stalk of the NA represents the N-terminal end of the molecule (Fields et al., 1981; Blok et al., 1982).

Studies of the amino acid sequence of the tetrameric head released from the viral membrane by protease demonstrates conservation of sequence of N2, N1, and influenza B virus NAs at crucial sites, strongly suggesting a common structure (reviewed by Varghese et al., 1983).

The four catalytic sites of the enzyme have been identified on the distal surface of the tetrameric head (Colman et al., 1983). These sites comprise a large pocket on the molecular surface. This morphology is characteristic of the catalytic sites of enzymes that catalyze the removal of end groups (Colman and Ward, 1985). Earlier studies have shown that the tetramer is required for enzymatic activity (Bucher and Kilbourne, 1972; Kendal and Eckert, 1972; Lazdins et al., 1972; Wrigley et al., 1973). Influenza neuraminidase in intact virus or as isolated tetramer shows multiphasic (non-Michaelis–Menten) kinetics. The four active centers of the NA are 40 Å apart, and thus, "like other proteins showing interdependent ligand binding to identical subunits, the (catalytic) effect would have to be mediated in some way through the subunit interface" (Colman and Ward, 1985).

The dependence of NA inhibition by antibody on substrate size suggested to Fazekas (1963) some years ago that the catalytic site of the enzyme was not antigenic and that inhibition of enzyme activity by antibody was mediated by steric

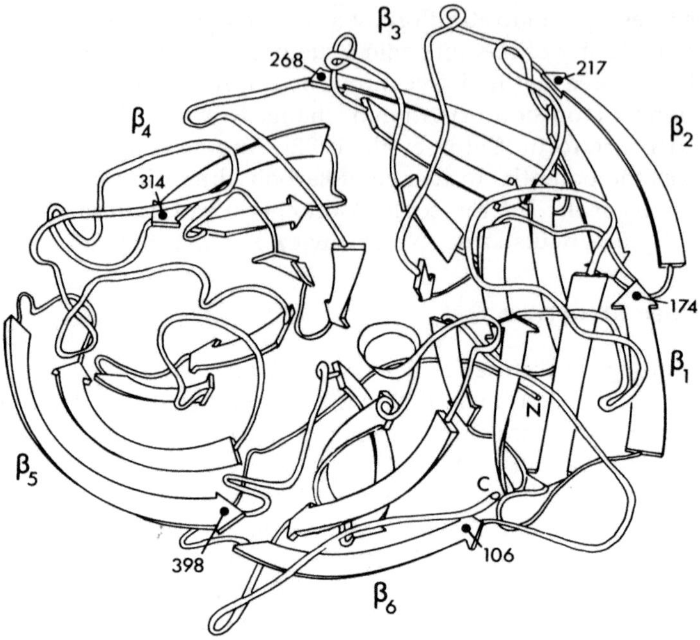

Figure 3-11. Diagram of subunit folding of N2 neuraminidase. The molecular fourfold symmetry axis is indicated, and the N-terminal segment emanating from that axis is on the underneath side of the subunit. The view is down the symmetry axis from above. The C terminus is on the outside surface near a subunit interface. The six β sheets of the propeller fold are near 4 o'clock, 2 o'clock, 12 o'clock, 10 o'clock, 8 o'clock, and 6 o'clock, respectively. The N and C termini are indicated, as are residue numbers at the C-terminal end of each sheet. (From Varghese et al., 1983. Reprinted by permission from *Nature* **303:**36. Copyright © 1983 by Macmillan Journals Limited.)

hindrance (see Fig. 3-12). Structural studies demonstrate that all of seven segments of the polypeptide chain identified as potential antigenic sites are sufficiently close to inhibit activity with macromolecular substrates. Antigenic determinants "form a nearly continuous surface across the top of the NA monomer encircling the catalytic site" (Colman et al., 1983).

Carbohydrate

Carbohydrate, exclusive of the ribose of virion RNA, accounts for 5–8% of the mass of the particle (Choppin and Compans, 1975) and is contained in the two glycosylated viral proteins hemagglutinin and neuraminidase as well as in mucopolysaccharides and glycolipids (Ward et al., 1983). Carbohydrate represents 16–24% of the hemagglutinin by weight (Waterfield et al., 1979) and 46% of the neuraminidase (Groome et al., 1977).

Oligosaccharide side chains are attached to the hemagglutinin by *N*-glycosidic linkages to asparagine (Asn) residues at glycosylation sites of Asn-Xaa-Thr or Asn-

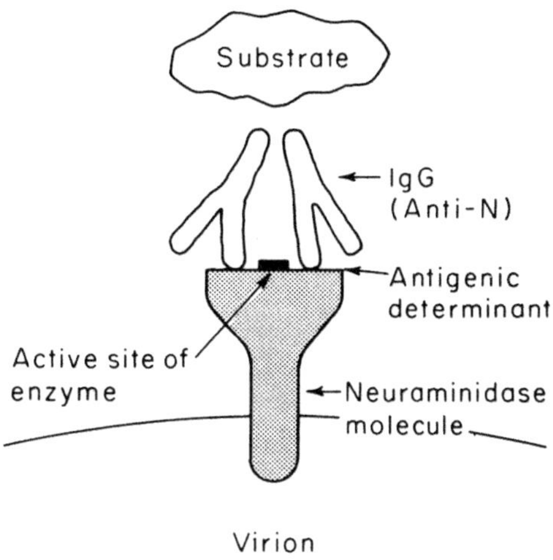

Figure 3-12. Steric inhibition of neuraminidase function by antibody. A single hypothetical neuraminidase spike is shown with two bivalent antibody molecules attached. Experimental evidence indicates that attachment of IgG will sterically block access of high-molecular-weight substrates to the enzymically active site, which is discrete from the antigenic determinants. Anti-N, antineuraminidase. (From Kilbourne et al., 1975. Reproduced with permission, from *Perspectives in Virology,* Vol.IX, M. Pollard (ed.), p. 105, Academic Press, New York.)

Xaa-Ser sequences (Keil et al., 1979; Waterfield et al., 1979; Ward and Dopheide, 1980; Wilson et al., 1983). The number and location of glycosylation sites vary from one viral subtype to another and even among strains within a subtype. Study of 14 type A strains (of four subtypes) and one type B virus revealed 26 probable glycosylation sites among these strains. In individual strains, the number of potential glycosylation sites varied from five to nine (Wilson et al., 1983). Actual side-chain units average four to six in number (reviewed by Jackson et al., 1981).

Oligosaccharide side chains are of two types: type I (complex) contain glucosamine, mannose, galactose, and fucose; type II (simple) contain only glucosamine and mannose. Each glycosylation site has a typical set of heterogeneous oligosaccharides (Keil et al., 1984a,b). Certain of the side chains are antigenic and constitute the cross-reactive host antigen of the virus (Brown et al., 1982) recognized in earlier studies of the virus (summarized by Laver and Webster, 1966). Differences among viruses in oligosaccharide location and composition are both host-cell and virus-strain dependent (Nakamura and Compans, 1979; Klenk, 1980).

Oligosaccharide patterns differ substantially among hemagglutinins, particularly those of different subtypes. Because oligosaccharide side chains of HA1 and HA2 differ in number and composition in different strains, influenza A viruses can be characterized and compared by analysis of these differences (Klenk et al., 1983). On the other hand, it should be emphasized that glycosylation sites at the base of the HA spike are highly conserved among HA subtypes (Keil et al., 1984a,b) and may contribute to the stability of the hemagglutinin structure. Potential protease cleavage sites are partially shielded by carbohydrate, which also may contribute to molecular stability (Wilson et al., 1983).

Carbohydrates can play a role in antigenic variation by preventing antibody binding to antigenic sites (Skehel et al., 1984).

The viral neuraminidase contains the four oligosaccharides (glucosamine, mannose, galactose, and fucose) found in the hemagglutinin (Ward et al., 1983). As in the hemagglutinin, these carbohydrate units are attached to the polypeptide by *N*-glycosidic linkage to asparagine (Ward et al., 1983; Colman and Ward, 1985). In addition, *N*-acetylgalactosamine occurs uniquely in neuraminidase and is antigenically related to chick embryo host antigen. The virus-chorioallantoic membrane cross reaction probably is mediated through terminal galactose and fucose residues (Ward et al., 1983).

About half of neuraminidase-associated carbohydrate is in the membrane-attached "stalk" of the molecule (Allen et al., 1977), which bears two (N1) or three (N2) glycosylation sites (Colman and Ward, 1985). The N2 heads contain five potential glycosylation sites, four of which have been shown to be glycosylated; N1 heads have three potential glycosylation sites, all of which are probably glycosylated (Colman and Ward, 1985). The structural role of carbohydrate in the neuraminidase is not yet clear, but as with the hemagglutinin, it may contribute to resistance to proteolysis.

Lipid

Lipids comprise 20–24% of the virus particle and are located principally in the viral membrane. Phospholipids (10–13%) and cholesterol (6–8%) predominate, with glycolipids representing a smaller fraction (Choppin and Compans, 1975). Most lipids are acquired from the plasma membrane of the cell during the process of virus budding, and thus the composition of lipids in the viral membrane reflects closely that of the host cell (Kates et al., 1961; reviewed by Choppin and Compans, 1975). However, influenza viral membranes appear to be significantly richer in (glyco)sphingolipids than host-cell membranes in egg-grown viruses (Huang, 1976); earlier the selective incorporation of fatty acyl chains had been described (Blough and Lawson, 1968).

An interesting recent development has been the demonstration that certain viral proteins, including the influenza virus hemagglutinin, contain fatty acids bound as acyl chains. Specificity for the incorporation of fatty acids into viral membrane proteins occurs at the level of the polypeptide acyltransferases (Schmidt, 1984). Influenza virus neuraminidase is nonacylated. The contribution of acylation to protein structure and function is unknown. However, all acylated viral glycoproteins have fusion activity (Schmidt, 1982).

RNA

Influenza A and B viruses contain eight, and influenza C virus seven, discrete segments of RNA. The total molecular weight of the RNAs is 4.9×10^6 for influenza A and 5.3×10^6 for influenza B; they comprise approximately 1% of the viral mass. Within the virus each RNA segment is encapsidated in RNP, a complex of RNA, NP, and the three proteins involved in RNA replication—PB_1, PB_2, and PBA. RNA segments range in size as shown in Table 3-1 (Sleigh et al., 1979; reviewed by Lamb, 1983).

Most RNA segments are transcribed as monocistronic messages to code for single polypeptides. In influenza A viruses, the two smallest RNAs each code for two proteins. The coding assignments of the RNAs are presented in Chapter 4 on viral replication, in which nonstructural proteins coded by the virus and present only in the host cell are discussed.

Terminal nucleotide segments of the individual RNAs of influenza A, B, and C viruses are highly conserved at their 5′ and 3′ ends and show partial inverted complementarity (reviewed by Desselberger et al., 1980). These structural features are probably significant in relation to the common requirements of the RNAs for recognition and replication by transcriptive and replicative enzymes.

Defective virus particles (Chapter 4) contain extra RNA molecules less than 1000 nucleotides in length (reviewed by Nayak, 1980).

STRUCTURAL DIFFERENCES AMONG INFLUENZA A, B, AND C VIRUSES

The structural similarities of influenza A, B, and C viruses far exceed their differences and appear to justify their classification together in the same virus group (Chapter 2). However, even minor differences have been carefully scrutinized as a possible clue to their different epidemiologic behavior (Chapter 11).

Influenza A and B viruses are remarkably similar in structure and composition, although the RNA segments and most proteins of influenza B virus are larger (Racaniello and Palese, 1979) (Table 3-2). Nucleotide sequencing of cDNA copies of genes for HA has shown conservation of structural features in HA (Krystal et al., 1983), NP (Londo, 1983), and NA (Shaw et al., 1982) proteins in influenza A and B viruses. The seventh RNA of influenza A virus codes for a nonstructural protein, M_2, not yet found in influenza B virus-infected cells. RNA segment 6,

Table 3-2. Principal Structural Differences in Influenza A, B, and C Viruses

	Influenza viruses		
	A	B	C
RNA			
M.W.	4.9×10^6	5.3×10^6[a]	—
Segments	8	8	7
Gene products	10	10	8
Structural proteins	7	7	6
Glycoproteins	2 (HA,NA)	2 (HA,NA)	1 HA-RDE[b]
Nonstructural proteins	3 (M_2,NS_1,NS_2)	3 (NS_1,NS_2,NB)	1 (NS)
Morphology	—	—	Ordered hexagonal arrays of surface spikes, many particles without RNPs

[a]All RNAs and gene products (except for M_2 protein) are larger in the case of influenza B virus

[b]The single glycoprotein g p 88 carries both hemagglutinating and receptor-destroying (RD) activity, but the enzyme is a neuraminate—O-acetylesterase (Herrler et al 1985)

which codes for the NA of influenza B virus, codes also, through a bicistronic mRNA, for a small glycoprotein (NB) found only in the cell (Shaw et al., 1983) and not identified with influenza A and C virus infections.

Influenza C virus differs significantly from influenza A and B viruses in having one less RNA segment and only a single, presumably multifunctional, glycoprotein (Table 3-2) as found in paramyxoviruses. The N-terminal sequence of the light chain of this protein (analogous to HA_2) is like that of N termini of paramyxovirus F_1 subunits, which are also concerned with fusion activity. The influenza C virus HA lacks primary sequence homology with the HAs of influenza A and B viruses but shares structural similarities with them (Nakada et al., 1984a; Pfeifer and Compans, 1984), as does its NP (Nakada et al., 1984b). A nonstructural (NS) protein has recently been identified as the gene product of influenza C virus RNA 7 (Nakada et al., 1985).

The morphological feature of ordered hexagonal arrays of glycoprotein on the virus particle surface (Herrler et al., 1981) (Table 3-2) is not always observed (Martin et al., 1977), and its significance is not understood.

TICK-BORNE VIRUSES STRUCTURALLY SIMILAR TO ORTHOMYXOVIRUSES

Certain viruses (prototype strains Dhori and Thogoto) isolated from ticks in India and Africa and originally considered as possible bunyaviruses have striking similarities to orthomyxoviruses (Clerx et al., 1983). They are enveloped viruses containing seven segments of single-stranded RNA of negative sense. They contain NP and a putative M protein and a single major surface glycoprotein, as well as three large proteins present in small amounts that are analogous to the P proteins of influenza viruses.

Unlike the influenza viruses, the tick-borne agents do not replicate in the chick embryo but do replicate to high titer in the brains of infant mice. They do not agglutinate chicken erythrocytes but do agglutinate goose red blood cells.

These viruses also have been isolated from vertebrates. Their role in disease and their definitive classification remain to be determined. Further studies of this group are urgently needed to clarify their evolutionary and possible ecological relationship to the influenza viruses.

REFERENCES

Air, G. M., and Compans, R. W., 1983, Influenza B and influenza C viruses, in: *Genetics of Influenza Viruses* (P. Palese and D. W. Kingsbury, eds.), Springer-Verlag, New York, pp. 280–304.

Allen, A. K., Skehel, J. J., and Yuferov, V., 1977, The amino acid and carbohydrate composition of the neuraminidase of B/Lee/40 influenza virus, *J. Gen. Virol.* **37**:625–628.

Apostolov, K., and Flewett, T. H., 1969, Further observations on the structure of influenza viruses A and C, *J. Gen. Virol.* **4**:365–370.

Blok, J., Air, G. M., Laver, W. G., et al., 1982, Studies on the size, chemical composition, and partial sequence of the neuraminidase (NA) from type A influenza viruses show that the N-terminal region of the NA is not processed and serves to anchor the NA in the viral membrane, *Virology* **119**:109–121.

Blough, H. A., 1963, The effect of vitamin A alcohol on the morphology of myxoviruses. I. The production and comparison of artificially produced filamentous virus, *Virology* **19**:349–358.

Blough, H. A., and Lawson, D. E. M., 1968, The lipids of paramyxoviruses: A comparative study of Sendai and Newcastle disease viruses, *Virology* **36**:286–292.

Brown, L. E., Ward, C. W., and Jackson, D. C., 1982, Antigenic determinants of influenza virus hemagglutinin—IX. The carbohydrate side chains from an Asian strain, *Mol. Immunol.* **19**:329–338

Bucher, D. L., and Kilbourne, E. D., 1972, A (N2) neuraminidase of the X-7 influenza virus recombinant: Determination of molecular size and subunit composition of the active unit, *J. Virol.* **10**:60–66

Caton, A. J., Brownlee, G. G., Yewdell, J. W., et al., 1982, The antigenic structure of the influenza virus A/PR/8/34 hemagglutinin (H1 subtype), *Cell* **31**:417–427

Choppin, P. W., and Compans, R. W., 1975, The structure of influenza virus, in *The Influenza Virus and Influenza* (E. D. Kilbourne, ed.), Academic Press, New York, pp. 15–47.

Choppin, P. W., Murphy, J. S., and Tamm, I., 1960, Studies of two kinds of virus particles which comprise influenza A2 virus strains. III. Morphological characteristics: Independence of morphological and functional traits, *J. Exp. Med.* **112**:945–952.

Choppin, P. W., Murphy, J. S., and Stoeckenius, W., 1961, Surface structure of influenza virus filaments, *Virology* **13**:548–550

Clerx, J. P. M., Fuller, F., and Bishop, D. H. L., 1983, Tick-borne viruses structurally similar to orthomyxoviruses, *Virology* **127**:205–219

Colman, P. M., and Ward, C. W., 1985, Structure and diversity of the influenza virus neuraminidase, *Curr. Top. Microbiol. Immunol.* **114**:177–225

Colman, P. M., Varghese, J. N., and Laver, W. G., 1983, Structure of the catalytic and antigenic sites in influenza virus neuraminidase, *Nature* **303**:41–44

Compans, R. W., Dimmock, N. J., and Meier-Ewert, H., 1969, Effect of antibody to neuraminidase on the maturation and hemagglutinating activity of an influenza A virus, *J. Virol.* **4**:528–534

Compans, R. W., Content, J., and Duesberg, P. H., 1972, Structure of the ribonucleoprotein of influenza virus, *J. Virol.* **10**:795–800

Cusack, S., Ruigrok, R. W. H., Krygsman, P. C. J., et al., 1985, Structure and composition of influenza virus: A small-angle neutron scattering study, *J. Mol. Biol.* **186**:565–582.

Desselberger, U., Racaniello, V. R., Zazra, J. J., et al., 1980, The 3′ and 5′-terminal sequences of influenza A, B and C virus RNA segments are highly conserved and show partial inverted complementarity, *Gene* **8**:315–328

Erickson, A. H., 1977, *Different trypsin sensitivity of the hemagglutinin from antigenically similar recombinant influenza viruses*, doctoral dissertation, Department of Microbiology, Mount Sinai School of Medicine, Graduate School of Biological Sciences, The City University of New York, New York

Erickson, A. H., and Kilbourne, E. D., 1980, Mutation in the hemagglutinin of A/N-WS/33 influenza virus recombinants influencing sensitivity to trypsin and antigenic reactivity, *Virology* **107**:320–330.

Fazekas de St. Groth, S., 1963, Steric inhibition: Neutralization of a virus-borne enzyme, *Ann. N.Y. Acad. Sci.* **103**:674–687.

Fields, S., Winter, G., and Brownlee, G. G., 1981, Structure of the neuraminidase gene from human influenza virus A/PR/8/34, *Nature* **290**:213–217.

Groome, N. P., Belyavin, G., Landsdell, A., et al., 1977, Structure and composition of the N2 neuraminidase of influenza virus. Effect of carbohydrate content on the validity of molecular weight estimations, *Biochim. Biophys. Acta* **495**:58–70.

Heggeness, M. H., Smith, P. R., Ulmanen, I., et al., 1982, Studies on the helical nucleocapsid of influenza virus, *Virology* **118**:466–470.

Herrler, G., Nagele, A., Meier-Ewert, H., et al., 1981, Isolation and structural analysis of influenza C virion glycoproteins, *Virology* **113**:439–451

Herrler, G., Rott, R., Klenk, H-D., et al., 1985, The receptor-destroying, enzyme of influenza C virus is neuraminate-O-acetylesterase. *EMBO Journal* **4**:1503–1506

Huang, R. T. C., 1976, Sphingolipids of influenza viruses, *Biochim. Biophys. Acta* **424**:90–97

Huang, R. T. C., Wahn, K., Klenk, H.-D., et al., 1980, Fusion between cell membranes and liposomes containing the glycoprotein of influenza virus, *Virology* **104**:294–302.

Jackson, D. C., Brown, L. E., and White, D. O., 1981, Antigenic determinants of influenza virus hemagglutinin. VI. Antigenic characterization of the oligosaccharide sidechains from HA of influenza

virus hemagglutinins, *J. Gen. Virol.* **52:**163–168.
Jennings, P. A., Finch, J. T., Winter, G., et al., 1983, Does the higher order structure of the influenza virus ribonucleoprotein guide sequence rearrangements in influenza viral RNA? *Cell* **34:**619–627.
Karadaghi, S. E., Zakomirdin, J. A., Shimane, C., et al., 1984, Interaction of influenza virus proteins with planar bilayer lipid membranes. I. Characterization of their adsorption and incorporation into lipid bilayers, *Biochim. Biophys. Acta* **778:**269–275.
Kates, M., Allison, A. C., Tyrrell, D. A. J., et al., 1961, Lipids of influenza virus and their relation to those of the host cell, *Biochim. Biophys. Acta* **52:**455–466.
Kawakami, K., Ishihama, A., and Hamaguchi, M. , 1981, RNA polymerase of influenza virus. I. Comparison of the virion-associated RNA polymerase activity of various strains of influenza virus, *J. Biochem.* **89:**1751–1757.
Keil, W., Klenk, H.-D., and Schwarz, R. T., 1979, Carbohydrates of influenza virus. III. Nature of oligosaccharide-protein linkage in viral glycoproteins, *J. Virol.* **31:**253–256.
Keil, W., Geyer, R., Niemann, H., et al., 1984a, The carbohydrates of the hemagglutinin of influenza virus, in: *Segmented Negative Strand Viruses* (R. Compans and D. Bishop, eds.), Academic Press, New York, pp. 289–298.
Keil, W., Niemann, H., Schwarz, R. T., et al., 1984b, Carbohydrates of influenza virus. V. Oligosaccharides attached to individual glycosylation sites of the hemagglutinin of fowl plague virus, *Virology* **133:**77–91.
Kendal, A. P., and Eckert, E., 1972, The preparation and properties of ^{14}C-carboxamidomethylated subunits from A/1957 influenza neuraminidase, *Biochim. Biophys. Acta* **258:**484–495.
Kendal, A. P., and Madeley, C. R., 1970, Flocculation of influenza virus by specific antineuraminidase antibody, *Arch. Ges. Virusforsch.* **31:**219–229.
Kharitonenkov, I. G., Siniakov, M. S., Grigoriev, V. B., et al., 1978, The length of the influenza virus spikes measured by photon correlation spectroscopy, *FEBS Lett.* **96:**120–124.
Kilbourne, E. D., 1963, Influenza virus genetics, *Prog. Med. Virol.* **5:**79–126.
Kilbourne, E. D., 1969, Future influenza vaccines and the use of genetic recombinants, *Bull. WHO* **41:**643–645.
Kilbourne, E. D., and Murphy, J. S., 1960, Genetic studies of influenza viruses. I. Viral morphology and growth capacity as exchangeable genetic traits. Rapid *in ovo* adaptation of early passage Asian strain isolates by combination with PR8, *J. Exp. Med.* **111:**387–406.
Kilbourne, E. D., Palese, P., and Schulman, J. L., 1975, Inhibition of viral neuraminidase as a new approach to the prevention of influenza, in: *Perspectives in Virology*, Vol. IX (M. Pollard, ed.), Academic Press, New York, pp. 99–113.
Klenk, H.-D., 1980, Processing of the haemagglutinin: Glycosylation and proteolytic cleavage, in: *Structure and Variation in Influenza Virus* (G. Laver and G. Air, eds.), Elsevier/North Holland, Amsterdam, pp. 213–222.
Klenk, H.-D., Rott, R., Orlich, M., et al., 1975, Activation of influenza A viruses by trypsin treatment, *Virology* **68:**426–439.
Klenk, H.-D., Keil, W., Niemann, H., et al., 1983, The characterization of influenza A viruses by carbohydrate analysis, *Curr. Top. Microbiol. Immunol.* **104:**247–257.
Knossow, M., Daniels, R. S., Douglas, A. R., et al., 1984, Three-dimensional structure of an antigenic mutant of the influenza virus hemagglutinin, *Nature* **311:**678–680.
Krystal, M., Young, J. F., Palese, P., et al., 1983, Sequential mutations in hemagglutinins of influenza B virus isolates: Definition of antigenic domains, *Proc. Natl. Acad. Sci. U.S.A.* **80:**4527–4531.
Lamb, R. A., 1983, The influenza virus RNA segments and their encoded proteins, in: *Genetics of Influenza Viruses* (P. Palese and D. W. Kingsbury, eds.), New York, pp. 21–69.
Laver, W. G., and Kilbourne, E. D., 1966, Identification in a recombinant influenza virus of structural proteins derived from both parents, *Virology* **30:**493–501.
Laver, W. G., and Valentine, R. C., 1969, Morphology of the isolated hemagglutinin and neuraminidase subunits of influenza virus, *Virology* **38:**105–119.
Laver, W. G., and Webster, R. G., 1966, The structure of influenza virus. IV. Chemical studies of the host antigen, *Virology* **30:**104–115.
Lazarowitz, S. G , and Choppin, P W., 1975, Enhancement of infectivity of influenza A and B viruses by proteolytic cleavage of the hemagglutinin polypeptide, *Virology* **68:**440–454.
Lazdins, I., Haslam, E. A., and White, D. O., 1972, The polypeptides of influenza virus. VI. Compo-

sition of the neuraminidase, *Virology* **49**:758–765.
Londo, D. R., Davis, A R., and Nayak, D. P., 1983, Complete nucleotide sequence of the nucleoprotein gene of influenza B virus, *J. Virol.* **47**:642–648.
Martin, M. L., Palmer, E. L., and Kendal, A P., 1977, Lack of characteristic hexagonal surface structure on a newly isolated influenza C virus, *J. Clin. Microbiol.* **6**:84–86.
Murti, K G., Bean, W. J., Jr , and Webster, R. G., 1980, Helical ribonucleoproteins of influenza virus: An electron microscopic analysis, *Virology* **104**:224–229.
Murti, K G., and Webster, R G , 1986, Distribution of hemagglutinin and neuraminidase on influenza virions as revealed by immunoelectron microscopy, *Virology* **149**:36–43
Nakada, S., Creager, R. S., Krystal, M., et al., 1984a, Influenza C virus hemagglutinin: Comparison with influenza A and B virus hemagglutinins, *J Virol* **50**:118–124
Nakada, S., Creager, R S , Krystal, M., et al., 1984b, Complete nucleotide sequence of the influenza C/California/78 virus nucleoprotein gene, *Virus Res.* **1**:433–441
Nakada, S , Graves, P N , Desselberger, U., et al., 1985, Influenza C virus RNA 7 codes for a nonstructural protein, *J Virol* **56**:221–226.
Nakamura, K., and Compans, R. W., 1979, Host cell- and virus strain-dependent differences in oligosaccharides of hemagglutinin glycoproteins of influenza A viruses, *Virology* **95**:8–23
Nayak, D. P , 1980, Defective interfering influenza viruses, *Annu Rev. Microbiol* **34**:619–644.
Nermut, M. V.,, 1972, Further investigation on the fine structure of influenza virus, *J Gen Virol* **17**:317–331
Nermut, M V., and Frank, H , 1971, Fine structure of influenza A2 (Singapore) as revealed by negative staining, freeze-drying and freeze-etching, *J. Gen Virol* **10**:37–51.
Nestorowicz, A , Laver, G., and Jackson, D. C., 1985, Antigenic determinants of influenza virus hemagglutinin. X A comparison of the physical and antigenic properties of monomeric and trimeric forms, *J Gen. Virol* **66**:1687–1695
Pfeifer, J. B , and Compans, R W , 1984, Structure of the influenza C glycoprotein gene as determined from cloned DNA, *Virus Res* **1**:281–296
Pons, M. W., Schulze, I. T , Hirst, G K., et al., 1969, Isolation and characterization of the ribonucleoprotein of influenza virus, *Virology* **39**:250–259.
Racaniello, V R , and Palese, P , 1979, Influenza B virus genome Assignment of viral polypeptides to RNA segments, *J Virol.* **29**:361–373.
Robertson, J. S., Robertson, M E St C., and Roditi, I. J , 1984, Nucleotide sequence of RNA segment 3 of the avian influenza A/FPV/Rostock/34 and its comparison with the corresponding segment of human strains A/PR/8/34 and A/NT/60/68, *Virus Res.* **1**:73–79
Roditi, I J , and Robertson, J S , 1984, Nucleotide sequence of the avian influenza virus A/fowl plague/Rostock/34 segment 1 encoding the PB2 polypeptide, *Virus Res.* **1**:65–71
Ruigrok, R. W H., Andree, P. J., Hooft Van Huysduynen, R A M., et al., 1984, Characterization of three highly purified influenza virus strains by electronmicroscopy, *J Gen. Virol* **65**:799–802.
Ruigrok, R. W. H., Krijgsman, P. C. J., deRonde-Verloop, F. M , et al., 1985, Natural heterogeneity of shape, infectivity and protein composition in an influenza A (H3N2) virus preparation, *Virus Res* **3**:69–76.
Schmidt, M. F. G., 1982, Acylation of viral spike glycoproteins: A feature of enveloped RNA viruses, *Virology* **116**:327–338.
Schmidt, M. F. G., 1984, The transfer of myristic and other fatty acids on lipid and viral protein acceptors in cultured cells infected with Semliki Forest and influenza virus, *EMBO J* **3**:2295–2300.
Shaw, M. W , Lamb, R A., Erickson, B W., et al., 1982, Complete nucleotide sequence of the neuraminidase gene of influenza B virus, *Proc. Natl. Acad. Sci. U.S.A.* **79**:6817–6821.
Shaw, M W., Choppin, P. W., and Lamb, R. A., 1983, A previously unrecognized influenza B virus glycoprotein from a bicistronic mRNA that also encodes the viral neuraminidase, *Proc Natl Acad Sci U S A.* **80**:4879–4883.
Singer, S. J , and Nicolson, G L., 1972, The fluid mosaic model of the structure of cell membranes, *Science* **175**:720–731
Sivasubramanian, N , and Nayak, D P., 1982, Sequence analysis of the polymerase 1 gene and the secondary structure prediction of polymerase 1 protein of human influenza virus A/WSN/33, *J Virol.* **44**:321–329.
Skehel, J J , Hay, A J , and Waterfield, M D , 1980, Influenza-virus, in: *Cell Membranes and Viral*

Envelopes, Vol. 2 (H. A. Blough and J. M. Tiffany, eds.), Academic Press, London, pp. 647–681.
Skehel, J. J., Stevens, D. J., Daniels, R. S., et al., 1984, A carbohydrate side chain on hemagglutinins of Hong Kong influenza viruses inhibits recognition by a monoclonal antibody, *Proc. Natl. Acad. Sci. U.S.A.* **81**:1779–1783.
Sleigh, M. J., Both, G. W., and Brownlee, G. G., 1979, A new method for the size estimation of the RNA genome segments of influenza virus, *Nucleic Acids Res.* **6**:1309–1321.
Taylor, R. B., Duffus, W. P. H., Raff, M. C., et al., 1971, Redistribution and pinocytosis of lymphocyte surface immunoglobulin molecules induced by anti-immunoglobulin antibody, *Nature (New Biol.)* **233**:225–229.
Tobita, K., and Kilbourne, E. D., 1974, Genetic recombination for antigenic markers of antigenically different strains of influenza B virus, *J. Virol* **13**:347–352.
Varghese, J. N., Laver, W. G., and Colman, P. M., 1983, Structure of the influenza virus glycoprotein antigen neuraminidase at 2.9 Å resolution, *Nature* **303**:35–40.
Verhoryen, M., Fang, R., Min Jou, W., et al., 1980, Antigenic drift between the hemagglutinin of the Hong Kong influenza strains A/Aichi/2/68 and A/Victoria/3/75, *Nature* **286**:771–775.
Ward, C. W., 1981, Structure of the influenza virus hemagglutinin, *Curr. Top. Microbiol. Immunol.* **94/95**:1–74.
Ward, C. W., and Dopheide, T. A. A., 1979, Primary structure of the Hong Kong (H3) hemagglutinin, *Br. Med. Bull.* **35**:51–56.
Ward, C. W., and Dopheide, T. A. A., 1980, The Hong Kong (H3) hemagglutinin. Complete amino acid sequence and oligosaccharide distribution for the heavy chain of A/Memphis/102/72, in: *Structure and Variation of Influenza Virus* (W. G. Laver and G. Air, eds.), Elsevier, Amsterdam, pp. 27–38.
Ward, C. W., Murray, J. M., Roxburgh, C. M., et al., 1983, Chemical and antigenic characterization of the carbohydrate side chains of an Asian (N2) influenza virus neuraminidase, *Virology* **126**:370–375.
Waterfield, M. D., Espelie, K., Elder, K., et al., 1979, Structure of the hemagglutinin of influenza virus, *Br. Med. Bull.* **35**:56–67.
Wiley, D. C., Wilson, I. A., and Skehel, J. J., 1981, Structural identification of the antibody-binding sites of Hong Kong influenza hemagglutinin and their involvement in antigenic variation, *Nature* **289**:373–378.
Wiley, D. C., Wilson, I. A., and Skehel, J. J., 1984, The hemagglutinin membrane glycoprotein of influenza virus, *Biol. Macromol. Assemblies* **1**:299–336.
Wilson, I. A., Skehel, J. J., and Wiley, D. C., 1981, Structure of the hemagglutinin membrane glycoprotein of influenza virus at 3 Å resolution, *Nature* **289**:366–373.
Wilson, I. A., Ladner, R. C., Skehel, J. J., et al., 1983, The structure and role of the carbohydrate moieties of influenza virus hemagglutinin, *Biochem. Soc. Trans.* **11**:145–147.
Wilson, J. C., Tudor, L. R., and Darlington, R. W., 1977, Morphology of influenza virions from the infected host, *Am. Soc. Microbiol. Abstr.*
Winter, G., and Fields, S., 1982, Nucleotide sequence of human influenza A/PR/8/34 segment 2, *Nucleic Acids Res.* **10**:2135–2143.
Wrigley, N. G., 1979, Electron microscopy of influenza virus, *Br. Med. Bull.* **35**:35–38.
Wrigley, N. G., Skehel, J. J., Charlwood, P. A., et al., 1973, The size and shape of influenza virus neuraminidase, *Virology* **51**:525–529.
Wrigley, N. G., Brown, E. B., Daniels, R. S., et al., 1983, Electron microscopy of influenza haemagglutinin-monoclonal antibody complexes, *Virology* **131**:308–314.

4

Replication of Influenza Viruses

REPLICATION SYSTEMS AND VIRAL QUANTITATION

Initial studies of influenza virus replication were carried out in small animals (ferrets and mice) in which the human influenza viruses were first isolated and propagated. Exploration of host range demonstrated that virus could infect the respiratory tract of a wide variety of mammalian hosts. Infection of immunologically responsive hosts simulated human infection and provided valuable information about viral pathogenesis (see Chapter 8), but the quantitative biology of influenza began with the introduction of the chick embryo as an experimental host system (Burnet, 1940). The cells of the entodermal surface of the allantoic sac of the developing embryo provide a homogeneous layer of highly susceptible cells maintained under controlled environmental conditions. From these cells, 900–3000 virus particles per cell (Horsfall, 1955) are released into the allantoic sac, from which they are readily harvested after chilling of eggs to minimize hemorrhage into the clear allantoic fluid. The HA of virus produced in the allantoic sac is largely cleaved by endogenous proteases (Lazarowitz et al., 1971), and, therefore, the ratio of infective to noninfective virus particles is relatively high.

The chick embryo remains the best host system for the production of large amounts of virus for biochemical studies or vaccine manufacture. For such studies, in which the primary concern is with extraction of viral macromolecules or the antigenic mass represented by the viral HA and NA, measurement of infectivity usually is of no concern. Therefore, the quantity of virus can be measured in HA units (considering HA activity to be a stable property) or as total viral protein of purified preparations.

As is the case with other viruses, quantitation of infective particles (virions) is most precisely accomplished by measurement of plaque-forming units (PFU) in monolayers of susceptible cells (see typical replication curves in Fig. 4-1). Details concerning such cell systems are presented in Chapters 5 and 7. At this point,

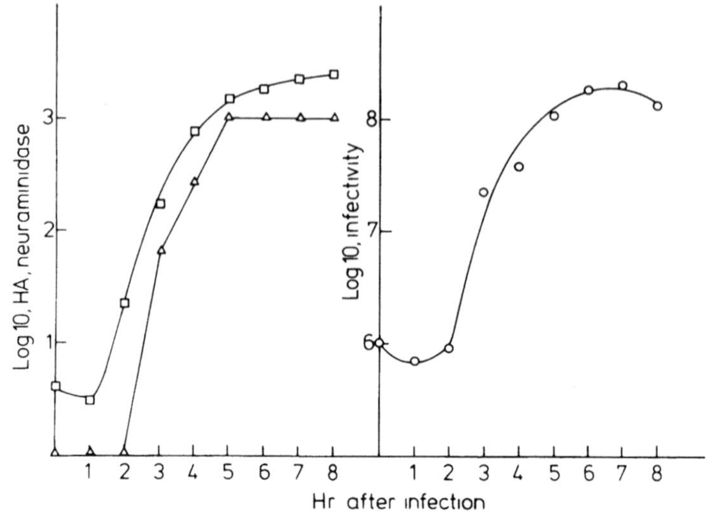

Figure 4-1. Representative replication curve of influenza virus in cell culture showing increase in cell hemagglutinin (HA), neuraminidase, and infective (plaque-forming) virus. (From Scholtissek and Klenk, 1975. Reproduced with permission, from *The Influenza Viruses and Influenza*, E. D. Kilbourne (ed.), p. 216, 1975, Academic Press, New York.)

it can be stated briefly that a limited number of virus–host cell systems are in current use for detailed quantitative studies of influenza viral replication. These have evolved as investigators have sought permissive systems in which replication is demonstrably efficient and the yield of infected virus is high. The systems include fowl plague virus (FPV) (H7N7) and WSN (H1N1) in chick embryo fibroblasts (CEF), WSN virus in the Madin–Darby bovine kidney (MDBK) cell line, and, more recently, the reassortant X-31 (and many other viruses) in MDCK cells in the presence of trypsin.

It is now clear that the presence of exogeneous proteases to insure cleavage of the HA can greatly enhance viral yields in cells that lack adequate amounts of endogeneous protease, thus rendering such cells more permissive for influenza viral replication. Therefore, it is likely that many other cell types than the ones now in use can be used for the study of influenza virus replication if the occasion demands. On the other hand, there remain cells in which influenza virus replication is truly abortive and in which the replication block is at an earlier stage than HA cleavage. Studies of abortive or defective replication are important, however, in providing insight into the process of productive viral replication and the kinetics of infection (see the following sections).

VIRAL GENES AND GENE PRODUCTS INVOLVED IN VIRUS REPLICATION: CODING ASSIGNMENTS OF INFLUENZA VIRUS GENES

The mapping of the influenza virus genome has been accomplished in a remarkably short period of time by a combination of correlative genetic, physicochemical, and biological techniques (reviewed by Kilbourne, 1983). Initial studies of this segmented genome virus were misleading in suggesting the existence of only eight gene products in virus-infected cells. These polypeptides were distributed in size to match the eight viral RNAs also identifiable on polyacrylamide gel electrophoresis. Therefore, it appeared that each RNA segment represented a single gene from which a single mRNA was transcribed as a monocistronic message for translation into a single viral protein. Subsequently, technical improvements in polyacrylamide gel electrophoresis have disclosed the presence of additional polypeptides in infected cells, and the complete nucleotide sequencing of all influenza virus genes has revealed additional (and overlapping) reading frames predictive of additional gene products. The present state of knowledge concerning influenza virus RNA segments and their coding assignments is summarized in Table 4-1 (from Mahy, 1983). Apparent differences in the coding assignments of influenza B and C virus RNAs are mentioned elsewhere in this book (Table 3-2) together with provisionally ascribed functions of the virion structural proteins (Table 3-1). Functions of the nonstructural proteins NS_1, NS_2, M_1, and NB are largely unknown and are discussed where appropriate in the following sections.

STAGES OF INFECTION AND REPLICATION

Viral Attachment to Host Cells: Adsorption

Infection is initiated by attachment of the HA spikes of influenza virus to neuraminic (sialic) acid-containing receptors on host cells. Neuraminic acid contained in glycoproteins and glycolipids is ubiquitously distributed in cells of many tissues, including those insusceptible to infection; however, the adsorption step, although necessary for infection to evolve, does not necessarily result in infection. Indeed, the binding of virus to the membranes of erythrocytes, in which repli-

Table 4-1. Influenza A Virus Genome RNA Segments and Their Coding Assignments[a]

vRNA segment	Length (nucleotides)	mRNA length (nucleotides)	Predicted nascent polypeptide length (amino-acid residues)	Encoded polypeptide
1	2341	2320	759	PB2
2	2341	2320	757	PB1
3	2233	2210	716	PA
4	1778	1756	566	HA
5	1565	1539	498	NP
6	1413	1391	454	NA
7	1027	1004	252	M_1
		314	97	M_2
		275	9	?
8	890	868	237	NS_1
		396	121	NS_2

[a]Based on genetic and sequence analysis of six influenza A viruses comprising representatives of H1, H3, and H7 HA subtypes Revised and abridged from Mahy (1983)

cation cannot ordinarily occur, is the basis of the hemagglutination reaction generally employed in the measurement of influenza viruses and antibody in vitro.

This is not to say that adsorption lacks specificity. Influenza viruses bind to oligosaccharides bearing *N*-acetylneuraminic acid in the terminal position. Influenza A viruses (including strains of the same HA subtype) differ with respect to their recognition of terminal *N*-acetylneuraminic acid $\alpha_{2,3}$-Gal and $\alpha_{2,6}$-Gal linkages (Carroll et al., 1981; Rogers and Paulson, 1983). Single amino acid substitutions in the influenza virus hemagglutinin can change receptor binding specificity and have aided in confirmation of the location of the receptor binding site in a pocket on the distal tip of the HA (Rogers et al., 1983) (see Fig. 3-5 in Chapter 3). The biological role of receptor binding specificity in influenza virus infections is almost certainly important but remains largely undefined. Recent investigations point to the importance of amino acid sequence in the viral receptor binding site for the capacity of the virus to replicate in an avian host (Naeve et al., 1984).

Neuraminidase, as the other surface glycoprotein in the virus, hydrolyzes the substrate to which the hemagglutinin binds and might reasonably be expected to participate in initial stages of infection. However, the bulk of evidence demonstrates the primary role of the hemagglutinin in viral attachment: (1) HA-specific, but not NA-specific, antibody neutralizes the virus prior to attachment to host cells and inhibits red blood cell agglutination; (2) selective removal of HA spikes, but not of NA spikes, prevents infection or hemagglutination. Even a virus containing a NA (N_9) that possesses anomalous hemagglutinating activity is neutralized only with HA-specific antibodies, providing further evidence that NA does not play a significant role in the initial adsorption step in infection (Laver et al., 1984).

Viral Entry

Following attachment, a virus particle may enter the host cell by any of several basic mechanisms: direct penetration, endocytosis, or fusion, or perhaps a combination of these processes (reviewed by Dimmock, 1982). Of these, direct penetration in the absence of virus–plasma membrane fusion or endocytosis has not been seen with influenza viruses. There exists strong evidence that influenza viruses can enter the cell either through fusion of their envelopes with the cell membrane or through their cellular uptake in endosomes. What remains uncertain is the relative importance of these routes of entrance in natural infection. Dimmock (1982) has emphasized the difficulty in studying early events in infection by morphological or biochemical methods because (in laboratory models) the high proportion of noninfective to infective virus particles makes it hard to relate observations made to the actual process of infection; i.e., are the virus particles observed in various intracellular sites potentially capable of participating in replication?

However, two recent discoveries have revealed previously unsuspected properties of influenza viruses (and consequent physiological requirements) that are obviously relevant to their capacity to enter and to infect cells. The first of these is the requirement that HA molecules on the virion be proteolytically cleaved in situ before infection (beyond the adsorption stage) can proceed. The second is the capability of influenza viruses to fuse with cell membranes—an activity long known for paramyxoviruses. With influenza viruses this property is most clearly manifest at low pH (Huang et al., 1981a). It now appears that proteolytic cleavage of the HA is a precondition for its fusion capacity (reviewed by Garten et al., 1981), thus linking fusion capacity and capacity of the virus to infect.

Proteolytic Activation of the HA and Virus–Cell Fusion

Synthesis of the HA is followed by its posttranslational cleavage at the plasma membrane by host-cell protease to yield two disulfide-bonded subunits, HA_1 and HA_2 (Lazarowitz et al., 1971). This cleavage can occur naturally during the course of infection during virus release or cell entry, or virus released from cells with HA in uncleaved form can be "activated" to infectivity by in vitro trypsin treatment (Lazarowitz and Choppin, 1975; Klenk et al., 1975; reviewed by Choppin and Scheid, 1980) (Fig. 4-2). Proteolytic cleavage is carried out in the host cell by trypsinlike proteases (Lazarowitz et al., 1973), and activation of infectivity involves cleavage of a peptide bond with arginine or lysine in carboxyl linkage (Klenk et al., 1977; Garten et al., 1981), creating a new N terminus on the HA_2 subunit, the C terminus of which is embedded in the viral membrane (Laver, 1971; Lazarowitz et al., 1971; Skehel and Waterfield, 1975). Structure of the N terminus is highly conserved and resembles that of the N terminus of the F_1 (fusion) polypeptide of paramyxoviruses, suggesting a similar role for the cleaved HA_2 N terminus. Indeed, structurally related synthetic oligopeptides block replication of influenza viruses in cell culture (Richardson et al., 1980). Thus, fusion activity of the HA depends on a specific amino acid sequence exclusively at the amino terminus of HA_2 (Garten and Klenk, 1983). A possible role for protein-bound fatty acids in fusion induction has been suggested (Schmidt and Lambrecht, 1985).

The importance of an acid environment in influenza-virus-mediated fusion is in inducing a conformational change in the molecule (Skehel et al., 1981; Sato et al., 1983) to allow interaction of the HA_2 terminus with the cell membrane.

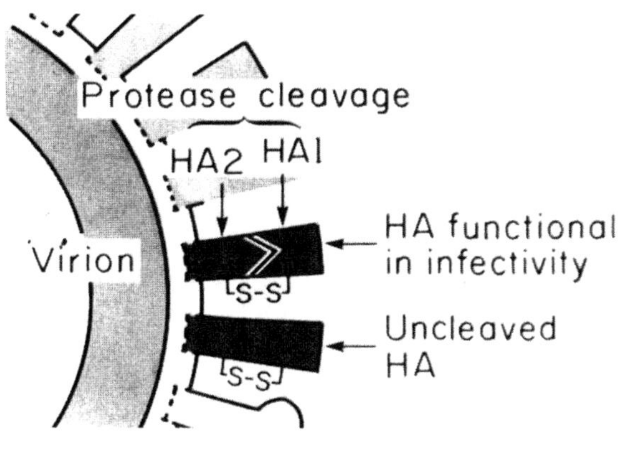

Figure 4-2. Protease cleavage of hemagglutinin (HA) leaves spike intact and joined by disulfide bonds. Cleavage is necessary for virus infectivity (see text)

This conformational change is detectable with monoclonal antibodies (Yewdell et al., 1983). Furthermore, monoclonal antibodies can block acid-induced conformational change and, as a consequence, inhibit influenza-virus-induced hemolysis (Kida et al., 1983).

The NA is probably not significantly involved in virus-mediated fusion, as shown by experiments utilizing selective destruction of NA on intact virus particles (Shibata et al., 1982) and by study of reconstituted vesicles or liposomes (Hosaka, 1983; Kawasaki et al., 1983) or reassortant viruses (Lenard et al., 1982). Furthermore, hemagglutinin expressed from a cloned HA gene promotes membrane fusion (White et al., 1982).

Virus–Cell Fusion

Whether or not viral entry into the cell occurs primarily by virus–cell plasma membrane fusion or by endocytosis, fusion appears to be an essential part of the entry process. Early studies by electron microscopy suggested that influenza viruses enter the cell by fusion of the virus to the external cell membrane (Fig. 4-3). Ingenious experiments in which either cells (Fidgen and Tisdale, 1981) or viruses (Huang et al., 1981b) were immobilized on solid substrates have shown that virus–cell fusion can occur under circumstances in which the possibility of endocytosis is precluded. Indeed, productive infection occurred in the latter case, even at neutral pH (Huang et al., 1981b). Therefore, external virus–cell fusion certainly is one mechanism by which viral entrance can occur.

Cell Entry and Uncoating through Endocytosis

That influenza viruses can enter cells by endocytosis has long been established (Fig. 4-4) (reviewed by Matlin et al., 1981). Recent studies have shown that this mode of entry is not inconsistent with, nor contradictory to, the apparent requirement for virus–cell fusion in infection. Rather, it has been postulated that fusion at acid pH occurs intracellularly after transport of viruses within endosomes to lysosomes (Matlin et., 1981; Yoshimura et al., 1982). Most recently, it has been shown (1) that prelysosomal endocytic vesicles (endosomes) have a low interior pH (Tycko and Maxfield, 1982; Maxfield, 1982), offering a reasonable site for virus–membrane interaction, and (2) in fact, uncoating of virus occurs in endosomes before reaching secondary lysosomes (Yoshimura and Ohnishi, 1984), as also is the case with Semliki forest virus (Marsh et al., 1983).

Details on viral uncoating and subsequent dissembly await further investigation. It is clear that RNP complexes must survive hydrolytic enzymes until primary transcription mediated by these complexes can occur.

Transcription and Replication of Influenza Virus RNA

As a virus containing negative (nonmessage) sense genomic RNA, the influenza virus also contains RNA-dependent enzymes required for transcription and replication of its genome. These RNA transcriptase and replicase functions are carried out by the largest virion proteins, PBl, PB2, and PA, as part of an RNA-

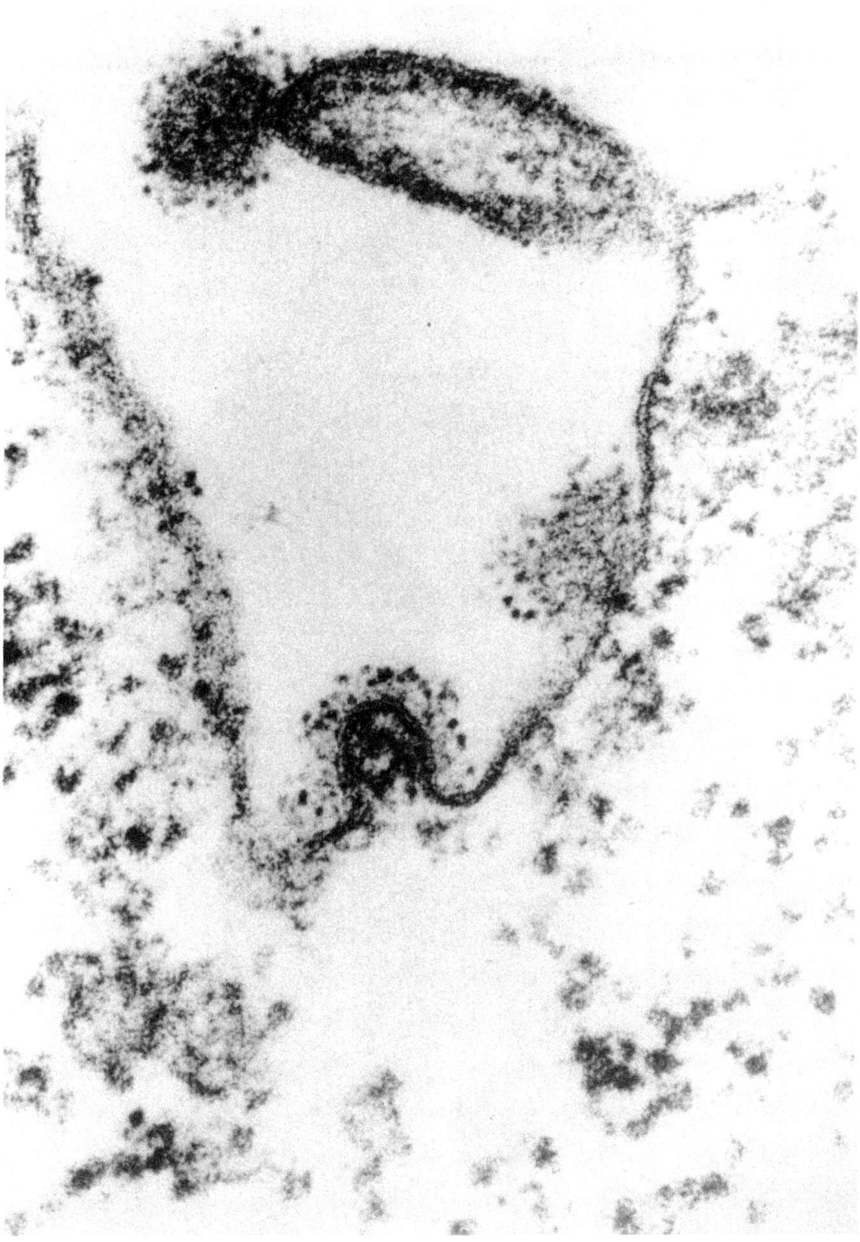

Figure 4-3. Fusion of influenza virus with the cell plasma membrane. Virus spikes are identified with ferritin-conjugated antibody. (From Matlin et al., 1981. Reproduced with permission, from *The Journal of Cell Biology* **91**:609. Copyright © 1981 by the Rockefeller University Press.)

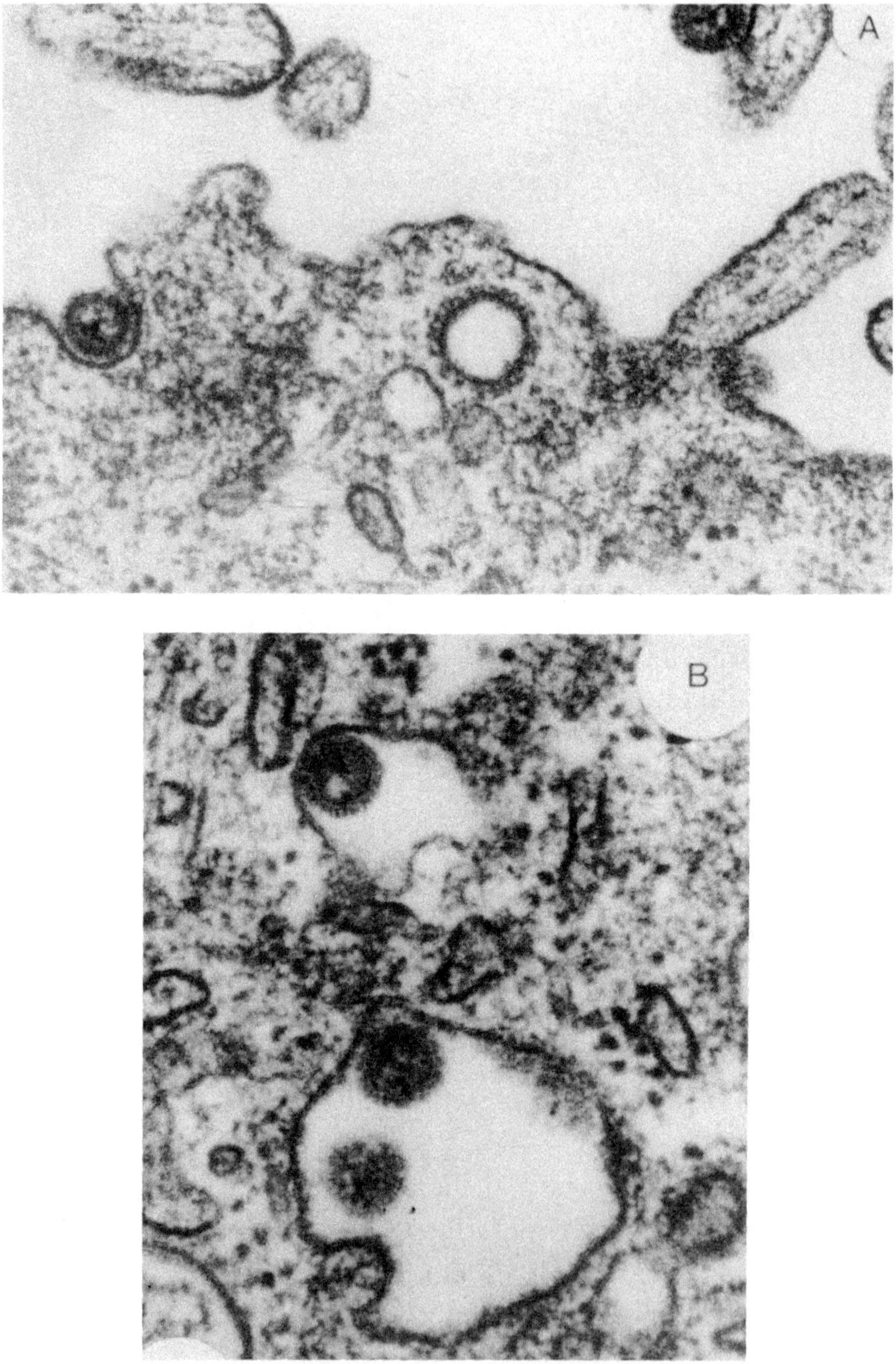
A
B

associated RNP polymerase complex that contains the NP protein as an important structural component. The exact role of each of these proteins in RNA and virus replication has not been determined definitively, but much has been learned recently of the process, especially regarding the initial step of primary transcription.

In primary transcription, mRNA is made from input virion RNA. The virus-coded mRNA molecules are partial transcripts that are capped at the 5′ end and polyadenylated at the 3′ end (Fig. 4-5). Secondary transcription is required for the production of full-length template nonpolyadenylated (A-) RNAs needed for synthesis of RNA for incorporation into newly replicated virus. Details of the process of primary transcription and background on the development of present knowledge can be found in comprehensive recent reviews (Krug, 1983; Mahy, 1983; Lamb and Choppin, 1983). The present discussion summarizes the principal conclusions drawn from studies in several laboratories.

Primary Transcription

A major contribution to the understanding of primary transcription has been the discovery by Krug and associates that influenza virus employs a unique mechanism for mRNA synthesis. This mechanism involves the capture and utilization of host-cell capped RNA fragments as primers for virus transcription (Plotch et al., 1981). The sequence of events is diagrammed in Fig. 4-6. Kawakami et al. (1985) have proposed recently that the Cap 1 structure stimulates the RNA polymerase by allosteric modulation rather than by priming transcription initiation.

Of the transcriptase proteins, PB2 has been identified as an endonuclease that recognizes and cleaves capped methylated host RNAs at a purine residue ten to 13 nucleotides from the cap. This event occurs in the nucleus, a unique site for mRNA synthesis of a nononcogenic RNA virus (Herz et al., 1981; Ulmanen et al., 1983). A requirement for host cell nuclear functions in influenza virus replication has been previously shown to be related to a need for an intact DNA-dependent RNA polymerase II system (Lamb and Choppin, 1977; Spooner and Barry, 1977). Apparently, specific subsets of host cell mRNAs are used in the priming of virus mRNA synthesis (Shaw and Lamb, 1984).

Initiation of transcription proceeds by incorporation of a G residue on the 3′ end of newly synthesized, capped primer fragments (Fig. 4-6, step 2). Stimulation of initiation requires specific interaction between the capped RNA and one or more of the proteins in the transcriptase complex (Krug, 1983).

Not all mRNAs of influenza virus are transcribed from colinear nucleotide sequences. Viral RNA (vRNA) segment 7 contains three reading frames. That for the longest (M_1) structural protein occupies 75% of the length of the segment, and the transcribed mRNA is colinear. Overlapping this is a second reading frame for a smaller (M_2) nonstructural protein of unknown function coded from an mRNA containing interrupted (presumably spliced) regions. A third mRNA is also interrupted and has the coding potential for only nine amino acids that are iden-

---

Figure 4-4. Entry of influenza virus particles by endocytosis. (A) Virus in smooth-surfaced pits and vesicles. (B) Virus in endosomes. (From Matlin et al., 1981. Reproduced with permission, from *J. Cell Biol.* **91**:607. Copyright © 1981 by the Rockefeller University Press.)

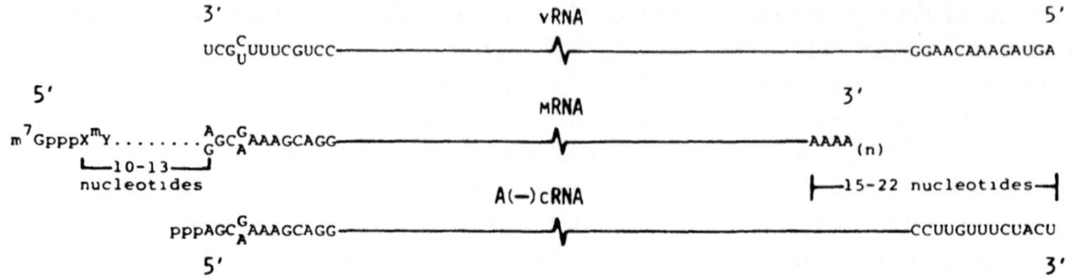

Figure 4-5. Schematic representation of influenza virus vRNA, mRNA, and A(−)cRNA. The conserved nucleotides at the ends of each RNA segment are shown. (From Lamb and Choppin, 1983. Reproduced with permission, from the *Ann. Rev. Biochem.* **52**:490. Copyright © 1983 by Annual Reviews Inc.)

tical to those at the carboxy terminus of the M_1 protein. No translation product of mRNA has yet been found. Similarly, RNA segment 7 contains overlapping reading frames that code for two nonstructural proteins, NS_1 and NS_2 (reviewed by Lamb and Choppin, 1983).

A second reading frame from RNA segment 6 of influenza B virus also is overlapping and codes for a small glycosylated protein not found in influenza A virus infection (Shaw et al., 1983). Lamb and Choppin (1983) have speculated that the use of overlapping reading frames represents an efficient use of genomes that may prove to be limited only to viruses with small genomes.

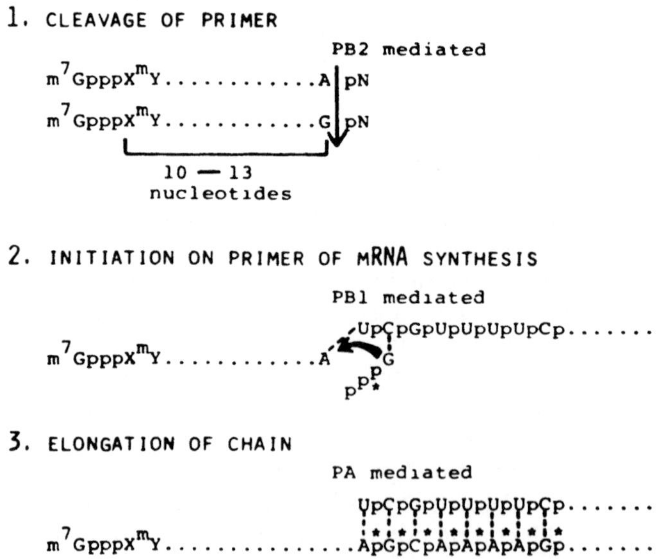

Figure 4-6. Mechanism for the priming of influenza viral RNA transcription by capped RNAs. (From Lamb and Choppin, 1983. Reproduced with permission, from *Annu. Rev. Biochem.* **52**:493.)

Less is known about elongation of the transcribed RNA. PBl appears to have the nucleotide binding site of the transcriptase and is responsible for both initiation and elongation (Romanos and Hay, 1984; Krug, 1983).

No specific function for the PA protein in primary transcription or later stages of virus replication has been established, although its presence in the cytoplasm suggests a role in vRNA synthesis.

Viral mRNA synthesis is under regulatory control: early in infection synthesis of mRNAs for NP and NS_1 proteins predominates; later mRNAs coding for HA, NA, and M increase. Synthesis of mRNAs of the P proteins is relatively low throughout all but the earliest stage of infection. The rate of synthesis of NS_1 mRNA declines, and that of M mRNA increases late in infection. Relative rates of mRNA synthesis correspond to the rates of synthesis of the respective proteins. Thus, viral gene expression is regulated principally at the transcriptional level.

Secondary Transcription and RNA Replication

Synthesis of template RNAs to be used in virion RNA synthesis is dependent on continued production of functional viral proteins. Template RNA complementary to vRNA represents a complete transcript of vRNA (Fig. 4-5) and lacks poly(A) (reviewed by Lamb and Choppin, 1983). Like viral mRNAs, template RNA appears to be synthesized in the nucleus (Herz et al., 1981; Beaton and Krug, 1984), occurring at fixed sites in the nuclear cage (Jackson et al., 1982). Synthesis is effected by the same transcription complex used for mRNA transcription without the requirement of a capped primer (Beaton and Krug, 1984).

In the infected cell, synthesis of template RNA is not regulated; equimolar amounts are synthesized throughout infection (Hay et al., 1977). These full-length transcripts represent only 5% of viral RNA transcripts in the cell. The dependence of their synthesis on coincident cellular or, more probably, viral protein synthesis suggests that a protein or proteins are required to enable the transcriptase complex (1) to initiate secondary transcription without a primer and (2) to continue transcription past the termination site used in mRNA synthesis (Krug, 1983).

Production of vRNA is much less well understood than synthesis of viral mRNA and template cRNA. Evidence is conflicting as to the temporal relationship of vRNA and mRNA synthesis. Little is known of the roles of the P proteins in vRNA synthesis. However, at least two are probably required (Mahy et al., 1981).

Synthesis of Viral Proteins

New viral proteins are detectable in infected cells within an hour of infection, associated with a decrease in host protein synthesis soon thereafter (Skehel, 1972). Shutoff of host protein synthesis is neither rapid nor complete, a fact that has complicated studies of viral protein synthesis. Not surprisingly, more is known about the synthesis and transport of the viral glycoproteins that have distinguishable biological activity and follow the synthetic patterns of other glycosylated and secretory proteins than is known concerning the P and smaller nonstructural proteins NS_2 and M_2 that are present in small amounts in the cell.

Like other glycoproteins, HA and NA are translated on membrane-fixed ribosomes. M and presumably other viral proteins are translated on free ribosomes.

A consensus of studies of viral protein synthesis recognizes an early phase in which synthesis of NP and NS_1 is predominant and a later phase characterized by a decline in the rate of synthesis of NS_1 and an increase in the rate of synthesis of M, HA, and NA (Skehel, 1973; Meier-Ewart and Compans, 1974; Lamb and Choppin, 1976; Hay et al., 1977; Inglis and Mahy, 1979). However, although they concur that NS_1 and NP are always made earlier than M and HA, Minor et al. (1979) emphasize that patterns of viral protein synthesis appear to be influenced by the particular virus–host cell system in which infection is studied.

The P Proteins

Synthesis of PB_1, PB_2, and PA continues at a low rate throughout the infection cycle. All three proteins are found in tight association with the nucleus of the infected cell, but PA is also found in the cytoplasm. The nuclear distribution of these proteins is consistent with their role in primary transcription, which is a nuclear event. The significance of the continued presence of PA in the cytoplasm suggests that it may be involved in another function, such as vRNA synthesis (Briedis et al., 1981). No specific role in transcription has yet been assigned to the PA protein.

NP

NP migrates promptly to the nucleus, in which it is demonstrable by immunofluorescent staining prior to the appearance of HA and NA in the cytoplasm (Breitenfeld and Shafer, 1957; reviewed by Maeno and Kilbourne, 1970). Subsequently, NP is apparent in the cytoplasm and finally at the periphery of cells (Maeno and Kilbourne, 1970) (Fig. 4-7). Along with the P proteins, NP is found in nucleocapsids (RNPs), which contain vRNA and remain in the nucleus. Other RNPs found in the cytoplasm are incorporated into virus particles (Krug, 1971, 1972). However, nuclear localization of NP is not dependent on complexing with RNA or P proteins; NP synthesized from cloned DNA in their absence is demonstrable in the nucleus (Lin and Lai, 1983).

Phosphorylation of a proportion of NP occurs rapidly after synthesis and is associated with transport to the nucleus. NP and RNP remaining in the cytoplasm are predominantly nonphosphorylated (Almond and Felsenreich, 1982). Phosphorylation of NP in vitro has been linked to in vitro virus transcription (Kamata and Watanabe, 1977), but this observation has not been confirmed (Almond and Felsenreich, 1982).

Rates of synthesis of individual RNPs parallel their rates of accumulation except for the RNP containing the RNA 7 (M protein genes), which, despite a tenfold faster rate of synthesis, accumulates at the same rate as other RNPs, probably related to its faster degradation (Rees and Dimmock, 1982).

Zhirnov and Bukrinskaya (1981) have reported two forms of NP of different molecular weights (NP 56 and NP 53), the smaller being a cleavage product of

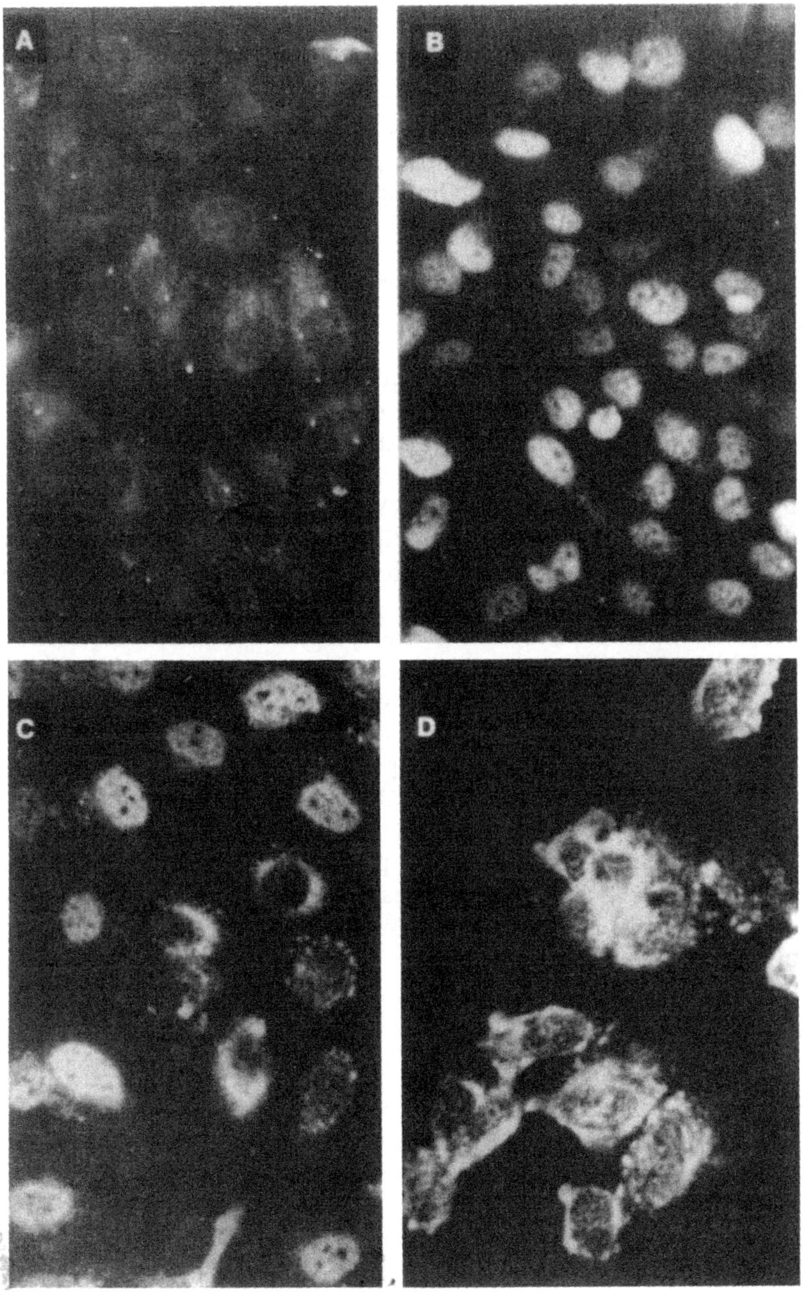

Figure 4-7. Immunofluorescent staining of NP protein at 2, 3, 7, and 17 hr after infection (panels a–d) Note initial nuclear localization followed by cytoplasmic appearance. (From Maeno and Kilbourne, 1970. Reproduced with permission, from *J. Virol.* **5**:158.)

the larger. The excised 3000-mol. wt. peptide is thought to represent the end terminus of the molecule destined to be involved in virus budding. Cleavage of NP has not been found in influenza viruses of nonhuman origin (Zhirnov and Bukrinskaya, 1984).

In summary, the NP is synthesized early and is associated with P proteins and viral RNA in transcriptase complexes (at which time its primary role may be structural). It is not required for transcriptase activity in vitro (Kawakami and Ishihama, 1983; Kato et al., 1985). NP also has an important structural role in integrating components of the virus particle. The importance of phosphorylation and cleavage remain to be determined.

The Nonstructural Proteins

NS_1 is synthesized early and in large quantities and is found in both the nucleus and cytoplasm. In the nucleus it has been demonstrated variably either in nuclei or nucleoplasm, depending on virus strain and cell type (Young et al., 1983; Krug and Soeiro, 1975). In the cytoplasm, NS_1 accumulates in paracrystalline inclusions in association with cellular RNA (Yoshida et al., 1981). This affinity for RNA may be related to its function, but NS_1 inclusions are not found with all influenza A viruses, and NS_1 proteins are heterogeneous with respect to charge (Petri et al., 1982). NS_1 is phosphorylated (Privalsky and Penhoet, 1978).

NS_2 has been reported to remain in the cytoplasm after synthesis (Breidis et al., 1981). However, earlier evidence for its presence in the nucleus (Krug and Etkind, 1973) has recently received strong support from immunofluorescence studies (Greenspan et al., 1985). Like NS_1, the role of NS_2 in viral replication has not been defined. The NB glycoprotein, apparently unique to influenza B virus and encoded by the RNA also coding for the NA, appears simultaneously with the NA and disappears from infected cells within 3 hr of synthesis (Shaw and Choppin, 1984). The influenza A virus M_2 protein appears to become associated with the cell membrane during viral synthesis, although it is not incorporated into the virion. It is synthesized by 2 hr post-infection and is rapidly transported to the cell surface (Zebedee et al., 1985).

Hemagglutinin

As a glycosylated protein inserted into the plasma membrane, the influenza viral hemagglutinin (HA) is synthesized according to the sequence found with cellular membrane glycoproteins. This involves cotranslational insertion into the membrane of the rough endoplasmic reticulum followed by transportation via the Golgi complex to the membrane for which the protein is destined—in the case of the HA, the plasma membrane of the cell (reviewed by Edwardson, 1984). In the process, the hydrophobic N-terminal signal peptide is removed by proteolysis, and the elongating peptide chain is glycosylated in a rough endoplasmic reticulum vesicle. Further glycosylation occurs during transport to the plasma membrane (reviewed by Lamb and Choppin, 1983). Transport occurs in smooth uncoated vesicles (Rodriguez-Boulan et al., 1984). In epithelial cells in monolayer cultures, HA is transported to apical surfaces (Rodriguez-Boulan and Sabatini, 1978), differing from VSV glycoprotein in this respect (Rindler et al., 1982). Glycosylation is not required in the determination of virus polarity for apical or basilar surfaces (Roth et al., 1979).

The carbohydrate composition of the HA is described in Chapter 3. Glycosylation, as with *N*-glycosidic glycoproteins in general, involves assembly of oligosaccharides by the dolichol pathway, then *en bloc* transfer to asparagine residues of the nascent polypeptide in the rough endoplasmic reticulum. During transport of the HA, there follows sequential removal of glucose (and most mannose) residues and attachment of *N*-acetylglucosamine, galactose, and fucose (Schwarz et al., 1977; Nakamura and Compans, 1979; Nakamura et al, 1980; Keil et al., 1984).

The composition and completeness of processing of carbohydrates on the HA are variable, depending on both virus and host cell. An important structural determinant for the oligosaccharide side chains is the structure of the glycoprotein itself (Keil et al., 1985). Sulfation of the HA occurs in both rough and smooth cytoplasmic membranes by attachment of sulfate to oligosaccharides, thus contributing to the negative charge on the virion surface (Nakamura and Compans, 1977).

Neuraminidase

Synthesis of neuraminidase (NA) probably follows the same pathways used in HA synthesis. Fewer data have been obtained, however, because of technical problems in resolving NA on polyacrylamide gel electrophoresis and the lesser amount of NA in the cell. NA, like HA, is found only in the cytoplasm and accumulates at apparently the same time, as is shown by immunofluorescence staining. Antibodies specific for HA block the intracellular staining of NA, suggesting that their synthesis occurs at the same or contiguous intracellular sites (Maeno and Kilbourne, 1970). Newly synthesized NA appears to be metabolically active within cells because intracytoplasmic membranes of infected cells contain much less sialic acid than those of uninfected cells (Nakamura and Compans, 1979).

Unlike most integral membrane proteins, the NA lacks a cleavable signal sequence. However, studies with an HA–NA chimera containing the N-terminal hydrophobic region of NA provide evidence that the N-terminal (anchor) sequence has a signal function and is probably involved in translocation of the NA (Bos et al., 1984). The NA polypeptide contains structural features capable of directing the transport of the protein to apical cell surfaces (Jones et al., 1985).

Viral Maturation and Assembly

The initial event in influenza virus assembly is the insertion of the membrane glycoproteins HA and NA into the cell's plasma membrane. This event must be preceded or accompanied by displacement of host proteins, as none is found in the virus particle with the occasional exception of small amounts of actin (Wang et al., 1976).

The route of HA and NA to the plasma membrane is through the membranes of the endoplasmic reticulum, at which time glycosylation of both proteins and proteolytic cleavage of the HA into HA_1 and HA_2 take place.

M and NP, and presumably the other viral structural proteins, reach the membrane directly by diffusion. M is seen as an electron-dense layer beneath the plasma membrane (Schulze, 1972). The synthesis of M may be rate limiting for virus particle formation (Lazarowitz et al., 1971). The formation of RNP occurs free in the cytoplasm (Compans and Caliguiri, 1973). The RNPs (nucleocapsids) appear to be influenced in their orientation by the presence of the overlying glycoproteins, as they are found only in locations beneath the sites into which these proteins have been previously inserted into the plasma membrane (Fig. 4-8) (Choppin et al., 1971; Lenard and Compans, 1974).

Assembly probably entails a transmembrane interaction between surface glycoproteins and M protein. The low specificity of the interaction is suggested by the formation of viral pseudotypes between different enveloped viruses (e.g., influenza and VSV) possessing M proteins (Simons and Garoff, 1980).

Virus Budding and Release

The actual budding or evagination process by which influenza virus particles push out from the plasma membrane (Fig. 4-9) is not understood. It has been proposed that with M-protein-containing viruses, polymerization of the M protein might be

> coupled to the binding of the nucleocapsid to the budding patch concomitant with the folding of the nucleocapsid into its final shape. To cause the outward bulging of the plasma membrane around the nucleocapsid, enough spike proteins would have to be included into the budding site to anchor the M protein layer to the membrane (Simons and Garoff, 1980).

In the course of efficient productive infection, when few defective virus particles are formed, virus particles do not accumulate on the cell surface and are

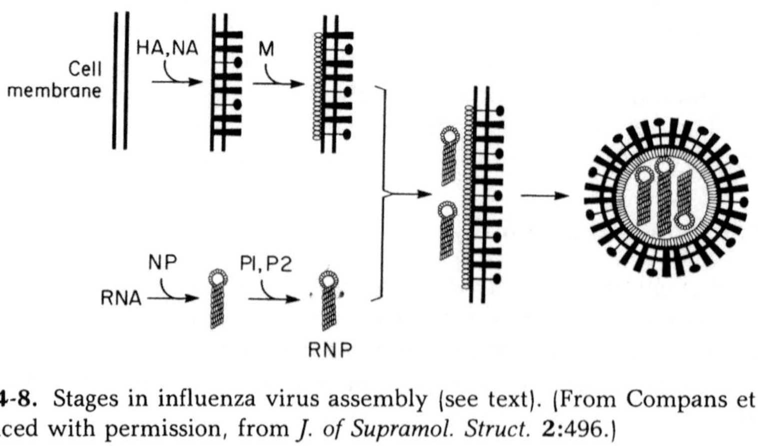

Figure 4-8. Stages in influenza virus assembly (see text). (From Compans et al., 1974. Reproduced with permission, from *J. of Supramol. Struct.* **2**:496.)

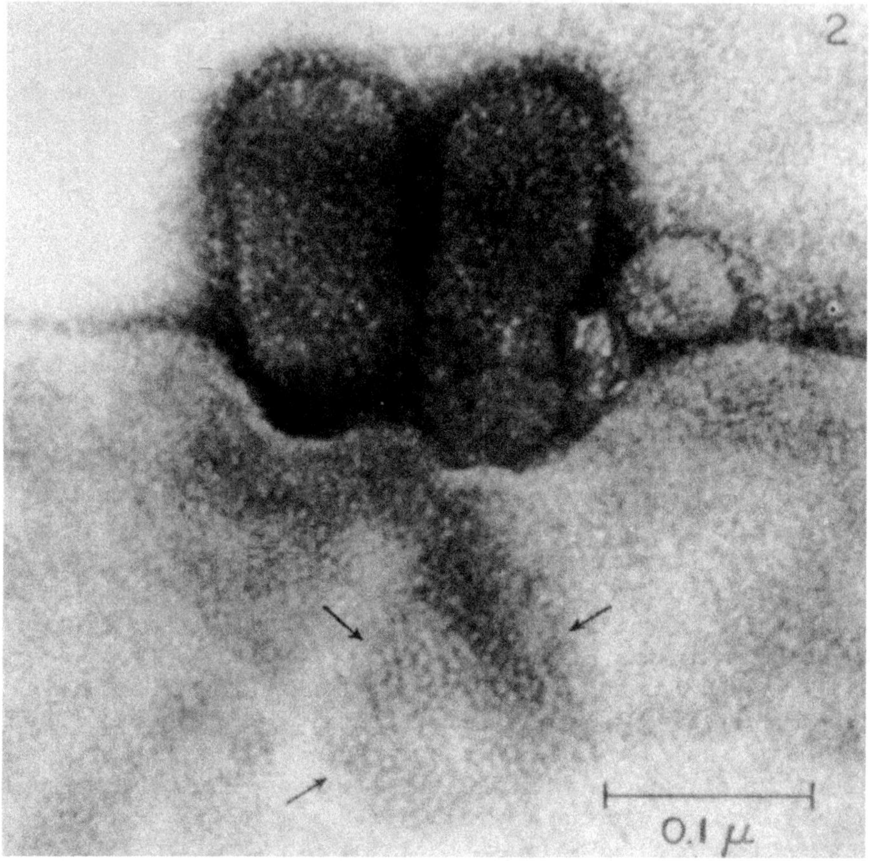

Figure 4-9. Budding of filamentous influenza virus particles from the cell plasma membrane. Arrows indicate the emergence of a particle toward the observer. (From Choppin, 1963. Reprinted with permission, from *Virology* **21**:279.)

promptly released. Virus replication in nonpermissive host cells may be attended by budding into intracytoplasmic vesicles (Caliguiri and Holmes, 1979).

The final event of detachment also is not well understood. Influenza virus particles are pleomorphic, ranging from spherical to filamentous (see Chapter 3), suggesting that they are pinched off from the cell surface according to different programs under viral genetic control (Kilbourne and Murphy, 1960). A linkage of spherical morphology with high virus yield (Kilbourne and Murphy, 1960) has been noted, and high yield has been associated with RNA 7 [M gene(s)] (Baez et al., 1980; Downie, 1984). Other evidence associates viral morphology with the M genes (R.P. Aaronson and E.D. Kilbourne, unpublished data).

Unlike most other enveloped viruses, influenza viruses do not acquire neuraminic acid during assembly and budding, presumably because of the incorporation of an active neuraminidase in the virus particles. NA-defective mutants do

contain neuraminic acid, and such particles are bound together in large aggregates at the cell surface by viral HA, for which neuraminic acid is, of course, the receptor (Palese et al., 1974; Palese and Compans, 1976) (Fig. 4-10). Thus, NA is important in the final release and detachment of virus. It may facilitate virus release not only by the prevention of particle aggregation but by destruction of cellular neuraminic acid (reviewed by Griffin et al., 1983).

ABORTIVE AND INEFFICIENT VIRUS REPLICATION

The productivity of influenza virus infection with respect to yield of infectious virus represents a gradient ranging from the translation of a limited number of viral proteins (Minor and Dimmock, 1976) to replication with release of a relatively high ratio of infective to defective virus particles (Choppin, 1969). Because it is difficult to demonstrate small amounts of infective virus under conditions in which replication is defective or incomplete and viral interference is high, the distinction between "abortive" and "inefficient" replication is somewhat arbitrary. In the present discussion the term abortive is reserved for situations in which virus is not released from the cell.

The efficiency of infective virus production is a product of virus–host cell interaction, will vary according to any virus–host cell pair, and will be influenced as well by the multiplicity of infection, by the presence of defective interfering virus in the inoculum, and by environmental factors. It is well to bear in mind that some of these factors, demonstrably important in cell cultures in vitro, may have little relevance to natural infection engendered in vivo by small inocula.

Because neuraminic acid receptors are ubiquitously distributed, a wide range of cells are potentially infectable by influenza viruses, but in only a few cell lines does productive infection occur (see beginning of this chapter). Blocks in replication can be assigned to the following categories:

1. Block in RNP transport from nucleus to cytoplasm.
2. Defective posttranslational cleavage of HA.
3. Defective M protein synthesis late in infection.
4. Defective budding of virus from the plasma membrane.

Expression of only a limited number of viral proteins has been observed only with heterokaryons of avian erythrocyte nuclei and BHK cells (Minor and Dimmock, 1976). In all other cases studied, all viral proteins are produced, but the rate of synthesis or distribution within the cell has been anomalous.

A number of studies involving both L and HeLa cells have shown (originally by immunofluorescense techniques) that transportation of NP or RNP from nucleus to cytoplasm may be impaired (Franklin and Breitenfeld, 1959; Ter Meulen and Love, 1967; Gandhi et al., 1971), leading to abortive replication. Whether or not this replication defect is associated with a reported absence of vRNA in L cells (Avery, 1975) is uncertain.

The requirement for proteolytic posttranslational cleavage of newly synthesized viral HA has already been considered (Chapter 3). This important replica-

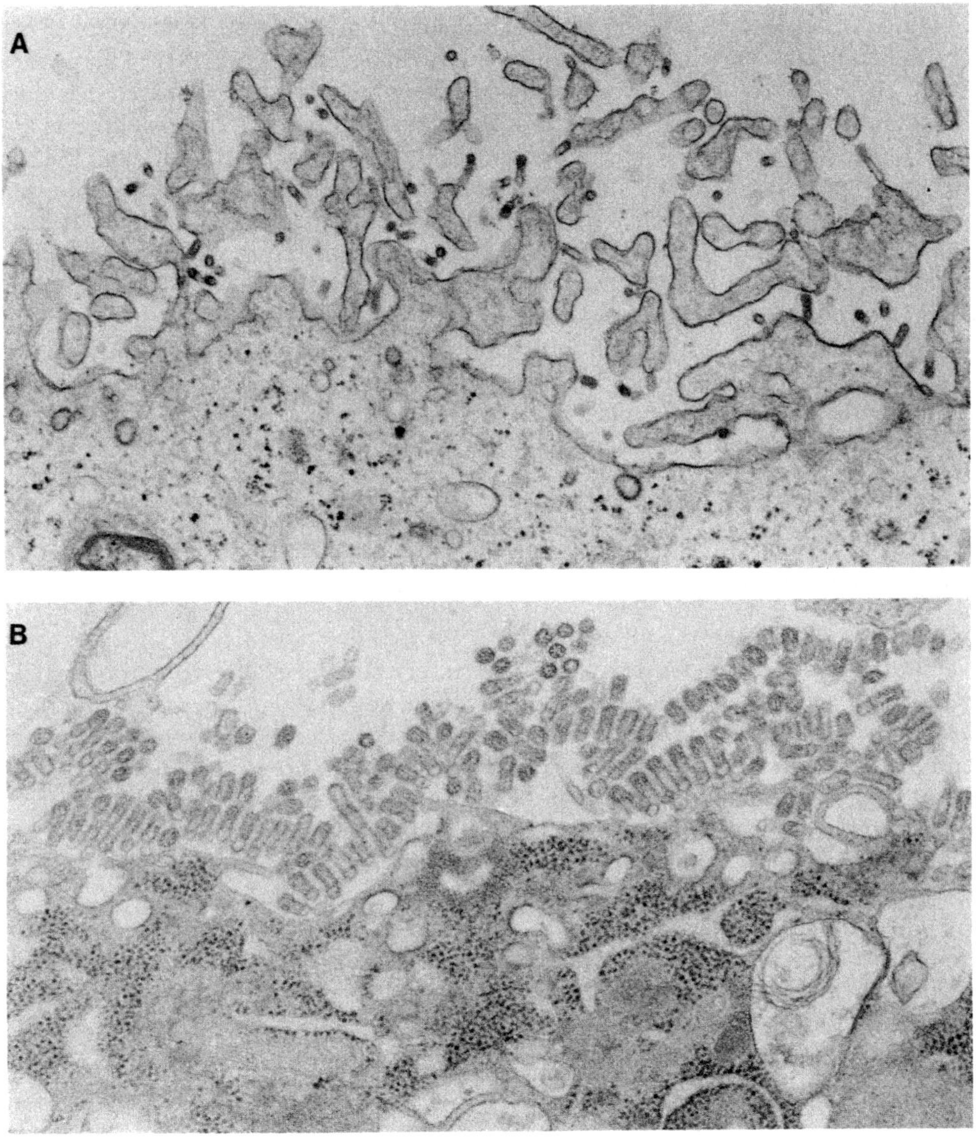

Figure 4-10. Replication of a ts NA-defective mutant at permissive (A) and nonpermissive (B) temperatures. At nonpermissive temperature, aggregation and accumulation of virus particles occur at the cell surface. (From Palese et al., 1974. Reproduced with permission, from *Virology* **61**:403.)

tive step is obviously dependent on the availability of endogenous host-cell proteases (reviewed by Scheid and Choppin, 1980).

Defects in M protein synthesis have been reported for a number of cell types in which abortive or restricted replication occurs, including L, HeLa, Vero, KB, and P 815 mastocytoma cells (Bosch et al., 1978; Valcavi et al., 1978; Lohmeyer et al., 1979; Hackett et al., 1980; Inglis and Brown, 1984). It is probable that the defect in late synthesis of M protein is associated with reduction in synthesis of RNA 7 mRNA from reduced template vRNA synthesis (Smith and Hay, 1982). In turn, this defect in expression of vRNA 7 in L cells might be linked to the capacity of these cells to splice vRNA 7 transcripts in uncontrolled fashion. It has been postulated that accumulation of M protein might be required to promote synthesis of vRNA 7 late in infection and that in L cells this positive regulation might be interrupted (Inglis and Brown, 1984).

Finally, under conditions in which no changes in viral protein synthesis or limitation of their intracellular distribution have been detected, deficiency in budding of virus particles at the plasma membrane has been described (Caliguiri and Holmes, 1979; Takeura et al., 1984).

ANOMALOUS VIRAL REPLICATION AND THE FORMATION OF DEFECTIVE VIRUS

The formation of genetically defective interfering virus following sequential passage of virus at high multiplicity of infection was first described for influenza viruses (von Magnus, 1951) but has since been recognized as a common phenomenon with a large number of viruses (Huang and Baltimore, 1970; Huang, 1973). Early studies, including those of von Magnus, described these defective products of viral replication as "incomplete"; their capacity to interfere with the replication of "standard," "infectious," or "complete" virus was appreciated soon after their discovery.

Even under ordinary conditions, the yield of multicycle replication of influenza viruses contains a heterogeneous mixture of infectious and noninfectious virus particles. The latter include (1) infective particles denatured by thermal inactivation during incubation of viral cultures, (2) genetically defective particles capable of interference (classical DI particles), and (3) other subgenomic particles lacking one or more genomic RNA segments. Those particles in the third category are noninterfering (Hirst and Pons, 1973) and have been termed defective noninterfering (DNI) particles (Carter and Mahy, 1982).

Because of difficulties in discriminating between DI and DNI particles, it is only relatively recently that the analogy of influenza viral DI particles to those of other viruses has been confirmed. Thus, like other DI particles, influenza virus DI particles (1) are produced and amplified under conditions of passage in which the ratio of virus particles to cell (multiplicity of infection) is high, (2) are defective and require concomitant inoculation of infective virus as a helper function for their replication, (3) cause interference with the replication of homologous standard infective virus, and (4) contain deletions in one or more genome RNA segments. The RNA segments containing deletions are responsible for interference (Nayak et al., 1978, 1985).

A distinction can be made between the abortive viral replication discussed previously and DI virus-mediated interference in that the latter is best effected in highly permissive host-cell systems that permit standard virus to replicate to high titer and thus help DI particle production. However, it has been shown that nonpermissive HeLa cells can produce some infective virus if the inoculum is relatively free of DI virus (Choppin and Pons, 1970).

Serial passage of influenza virus at high multiplicity in permissive host cells results in (1) a reduction in total virus yield and (2) a dramatic increase in the proportion of hemagglutinating but noninfective (DI) to infective influenza virus particles.

In their recent comprehensive review, Nayak et al. (1985) distinguished between the initial generation of DI particles and their subsequent amplification as two independent phenomena. Generation of DI RNAs (that contain internal deletions) does not depend on the multiplicity of infection. However, amplification of DI particles into which the defective RNA has been incorporated depends on the helper function of standard virus. To insure that coinfection of cells with both DI and standard virus occurs, a high multiplicity of infection is required.

The DI particles are not reproducibly distinguishable from standard virus by their morphological or physical characteristics. However, they appear to be more heterogeneous in morphology and variable in their content of the viral structural proteins HA, NA, M, and NP (reviewed by Nayak et al., 1985). It has been reported that they differ in lipid composition from standard influenza virus particles (Blough and Merlie, 1970).

The DI particles have reduced amounts of viral RNA, usually involving one or more of the polymerase (P) genes. In addition, they contain small RNA segments that are virus specific and negative stranded like viral RNA (Fig. 4-11). Oligonucleotide mapping demonstrates that these small RNAs are truncated P genes (Davis and Nayak, 1979; Nakajima et al., 1979). These small RNAs possess both 5′ and 3′ termini characteristic of influenza viral genomic RNA segments; their defect involves internal deletions (Nayak, 1980). Nucleotide sequencing of cDNA clones of these defective viruses has shown that they may contain single or multiple deletions, or they may be complex in having insertions, transpositions, base changes, or new sequences in addition to one or more internal deletions (Jennings et al., 1983). The existence of mosaic DI RNA, apparently produced through intersegmental recombination, has been described (Moss and Brownlee, 1981) but requires confirmation by direct RNA sequencing (Buonagurio et al., 1984). However, the majority of influenza virus DI RNAs studied have single internal deletions and principally involve the PB_2 gene (RNA 1). Only six of 41 subgenomic RNAs studied arose from non-P genes (Nayak et al., 1985).

Studies of the structure of DI RNAs have demonstrated that with continuing high-multiplicity virus passage, these RNAs may experience further deletions and evolve "to smaller forms or even disappear" (De and Nayak, 1980; Nayak et al., 1985; Sivasubramanian and Nayak, 1983).

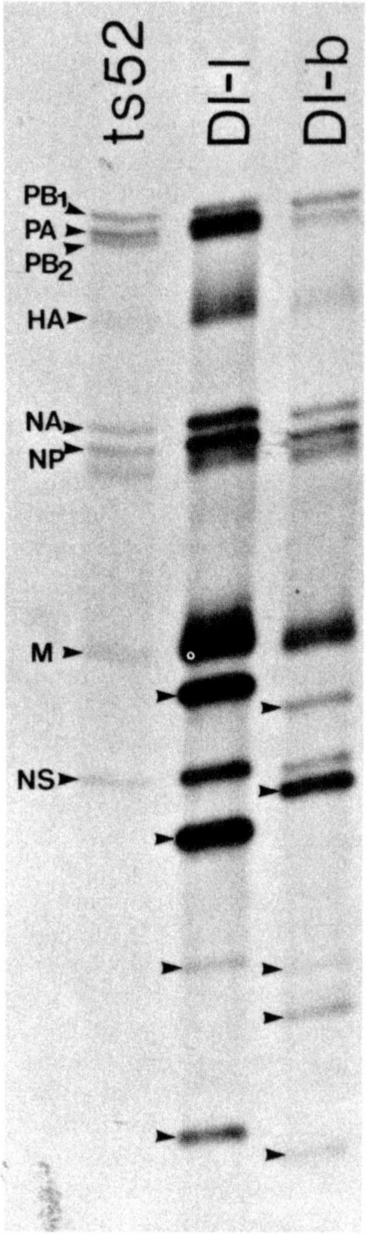

Figure 4-11. Electropherograms of RNA from influenza virus ts mutant ts52 and progeny (DI-1 and DI-6) passaged undiluted. Extra small (DI) RNA bands are shown by arrows. Coding assignments of other RNAs are indicated for ts52. (Modified by D. Nayak, personal communication.)

Obviously the generation of DI RNAs must involve a replication error, but the nature of this error is unknown. It seems most likely that the error involves interrupted transcription by the polymerase to produce deletions. Errors in RNA splicing or posttranscriptional processing appear to have been ruled out (Nayak et al., 1982).

The eventual predominance of DI virus in serial passage depends on its capacity to interfere with the replication of standard virus. Although DI particles alter mRNA transcription, and reduction in total RNA occurs (Pons and Hirst, 1969; Pons, 1980), the most striking effect is the decreased synthesis of P-gene transcripts in cells coinfected with standard and DI particles (Chambers et al., 1984). As a consequence, the synthesis of one or more polymerase proteins is specifically decreased in coinfected cells. Decrease in specific polymerase transcripts and P proteins is evident at the level of both primary and secondary transcription (Akkina et al., 1984a). Influenza DI particles can be transcribed into complementary mRNAs (Chanda et al., 1983), and novel peptides translated from them have been identified in DI virus-infected cells (Akkina et al., 1984b; Penn and Mahy, 1985). The role, if any, of these novel peptides in interference is unknown.

Although investigation of defective interfering viruses has been extraordinarily valuable in providing new insights into the mechanisms of viral replication, the role of DI viruses in the pathogenesis of disease remains conjectural.

INTERFERENCE

It has long been known that simultaneous inoculation of related or unrelated viruses can interfere with the replication of either or both agents (reviewed by Schlesinger, 1959). The interference that is mediated by interferon is, of course, broadly nonspecific, whereas that produced by DI particles (see preceding section) is by definition very specific. In between these extremes are the intrinsic interference described by Marcus and Carver (1967) that is specific for NDV but induced by a variety of unrelated RNA viruses and interference related to destruction of cellular receptors by viral neuraminidase (Cairns, 1955).

With influenza viruses, interference can be induced with infective, inactivated, or, of course, DI particles, as described above. Because homologous interference among influenza viruses of the same type is potentially associated with the genetic interaction of such strains, this phenomenon is discussed in the chapter on influenza virus genetics (Chapter 6). Concerning effects on viral replication, it is important to stress that the interference induced by influenza viruses inactivated by physical agents (Henle and Henle, 1943; Ziegler et al., 1944) is at least in part interferon mediated (Kilbourne et al., 1961). This interference (Kilbourne, 1957) and interferon synthesis can be suppressed by corticosteroids by mechanisms not yet understood (Kilbourne et al., 1961; Smart and Kilbourne, 1966).

Simultaneous high-multiplicity infection of cells with influenza A and B viruses results in heterologous interference (Ziegler and Horsfall, 1944; Gotlieb and Hirst, 1954; Tobita and Ohori, 1979; Mikheeva and Ghendon, 1982; Kaverin et al., 1983). This interference may be mediated at the level of primary transcription (Mikheeva and Ghendon, 1982; Aoki et al., 1984). Interference may favor one virus or be reciprocal, depending on the ratio of input viruses (Tobita and Ohori, 1979). Influenza B virus appears to be the stronger competitor (Kaverin et al., 1983).

Interference of one type A virus (FPV) with another (swine influenza virus) may preclude genetic interaction among these genetically compatible viruses, depending on the timing and multiplicity of infection (Rott et al., 1981).

REFERENCES

Akkina, R. K., Chambers, T. M., and Nayak, D. P., 1984a, Mechanism of interference by defective interfering particles of influenza virus: Differential reduction of synthesis of specific polymerase proteins in infected cells, *Virus Res.* **1**:687–702.

Akkina, R. K., Chambers, T. M., and Nayak, D. P., 1984b, Expression of defective-interfering influenza virus-specific transcripts and polypeptides in infected cells, *J. Virol* **51**:395–403.

Almond, J. W., and Felsenreich, V., 1982, Phosphorylation of the nucleoprotein of an avian influenza virus, *J. Gen. Virol.* **60**:295–305.

Aoki, H., Nishiyama, Y., Tsurumi, T., et al., 1984, Mechanism of interference between influenza A/WSN and B/Kanagawa viruses, *J. Gen. Virol.* **65**:1385–1393.

Avery, R. J., 1975, Abortive infection of L cells by influenza virus: Absence of virion RNA synthesis, *J. Virol.* **16**:311–314.

Baez, M., Palese, P., and Kilbourne, E. D., 1980, Gene composition of high-yielding influenza vaccine strains obtained by recombination, *J. Infect. Dis.* **141**:362–365.

Beaton, A. R., and Krug, R. M., 1984, Synthesis of the templates for influenza virus RNA replication *in vitro, Proc. Natl. Acad. Sci. U.S.A.* **81**:4682–4686.

Blough, H.A., and Merlie, J. P., 1970, The lipids of incomplete influenza virus, *Virology* **40**:685–692.

Bos, T. J., Davis, A. R., and Nayak, D. P., 1984, NH_2-terminal hydrophobic region of influenza virus neuraminidase provides the signal function in translocation, *Proc. Natl. Acad. Sci. U.S.A.* **81**:2327–2331.

Bosch, F. X., Hay, A. J., and Skehel, J. J., 1978, RNA and protein synthesis in a permissive and an abortive influenza virus infection, in: *Negative Strand Viruses and the Host Cell* (B. W. J. Mahy and R. D. Barry, eds.), Academic Press, London, pp. 465–474.

Breitenfeld, P. M., and Schafer, W., 1957, The formation of fowl plague virus antigens in infected cells as studied by fluorescent antibodies, *Virology* **4**:328–345.

Briedis, D. J., Conti, G., Munn, E. A., et al., 1981, Migration of influenza virus-specific polypeptides from cytoplasm to nucleus of infected cells, *Virology* **111**:154–164.

Buonagurio, D. A., Krystal, M., Palese, P., et al., 1984, Analysis of an influenza A virus mutant with a deletion in the NS segment, *J. Virol.* **49**:418–425.

Burnet, F. M., 1940, Influenza virus infections of the chick embryo by the amniotic route, *Aust. J. Exp. Biol. Med. Sci.* **18**:353–360.

Cairns, H. J. F., 1955, Multiplicity reactivation of influenza virus, *J. Immunol.* **75**:326–329.

Caliguiri, L. A., and Holmes, K. V., 1979, Host-dependent restriction of influenza virus maturation, *Virology* **92**:15–30.

Carroll, S. M., Higa, H. H., and Paulson, J. C., 1981, Different cell-surface receptor determinants of antigenically similar influenza virus hemagglutinins, *J. Biol. Chem.* **256**:8357–8363.

Carter, M. J., and Mahy, B. W. J., 1982, Incomplete avian influenza virus contains a defective non-interfering component, *Arch. Virol.* **71**:13–25.

Chambers, T. M., Akkina, R. K., and Nayak, D. P., 1984, *In vivo* transcription and translation of defective interfering particle-specific RNAs of influenza virus, in: *Segmented Negative Strand Viruses* (R. W. Compans and D. H. L. Bishop, eds.), Academic Press, San Diego, pp. 85–91.

Chanda, P. K., Chambers, T. M., and Nayak, D. P., 1983, In vitro transcription of defective interfering particles of influenza virus produces polyadenylic acid-containing complementary RNAs, *J Virol.* **45**:55–61.
Choppin, P. W., 1963, On the emergence of influenza virus particles from host cells, *Virology* **21**:278–281
Choppin, P. W , 1969, Replication of influenza virus in a continuous cell line High yield of infective virus from cells inoculated at high multiplicity, *Virology* **39**:130–134.
Choppin, P W., and Pons, M. W , 1970, The RNAs of infective and incomplete influenza virions grown in MDBK and HeLa cells, *Virology* **42**:603–610.
Choppin, P. W., and Scheid, A , 1980, The role of viral glycoproteins in adsorption, penetration, and pathogenicity of viruses, *Rev Infect Dis.* **2**:40–61.
Choppin, P W., Klenk, H -D., Compans, R. W , et al , 1971, The parainfluenza virus SV5 and its relationship to the cell membrane, in: *Perspectives in Virology,* Vol. VII (M. Pollard, ed), Academic Press, New York, pp 127–158.
Compans, R W., and Caliguiri, L A., 1973, Isolation and properties of an RNA polymerase from influenza virus-infected cells, *J Virol* **11**:441–448
Compans, R W , Meier-Ewart, H , and Palese, P , 1974, Assembly of lipid-containing viruses, *J Supramol Struct* **2**:496–511.
Davis, A. R , and Nayak, D P., 1979, Sequence relationships among defective interfering influenza viral RNAs, *Proc Natl. Acad. Sci U.S.A.* **76**:3092–3096
De, B K , and Nayak, D. P , 1980, Defective interfering influenza viruses and host cells Establishment and maintenance of persistent influenza virus infection in MDBK and HeLa cells, *J. Virol.* **36**:847–859
Dimmock, N. J., 1982, Initial stages in infection with animal viruses, *J. Gen. Virol.* **59**:1–22.
Downie, J. C., 1984, A genetic and monoclonal analysis of high-yielding reassortants of influenza A virus used for human vaccines, *J. Biol. Stand.* **12**:101–110.
Edwardson, J M , 1984, Effects of monensin on the processing and intracellular transport of influenza virus haemagglutinin in infected MDCK cells, *J Cell Sci* **65**:209–221.
Fidgen, K. J , and Tisdale, M., 1981, An 'on grid' electron microscopic method for studying the interaction and fusion of influenza A virus with human erythrocyte membranes, *J. Virol. Methods* **3**:271–276
Franklin, R. M., and Breitenfeld, P. M., 1959, The abortive infection of Earle's L-cells by fowl plague virus, *Virology* **8**:293–307.
Gandhi, S. S., Bell, H. B., and Burke, D. C., 1971, Abortive infection of L cells by fowl plague virus: Comparison of RNA and protein synthesis in infected chick and L cells, *J Gen. Virol.* **13**:423–432.
Garten, W., and Klenk, H.-D., 1983, Characterization of the carboxypeptidase involved in the proteolytic cleavage of the influenza haemagglutinin, *J. Gen. Virol.* **64**:2127–2137.
Garten, W., Bosch, F. X., Linder, D., et al., 1981, Proteolytic activation of the influenza hemagglutinin: The structure of the cleavage site and the enzymes involved in cleavage, *Virology* **115**:361–374.
Gotlieb, T., and Hirst, G. K., 1954. The experimental production of combination forms of virus III The formation of doubly antigenic particles from influenza A and B virus and study of the ability of individual particles to yield two separate strains, *J. Exp. Med.* **99**:307–320.
Greenspan, D., Krystal, M., Nakada, S., et al., 1985, Expression of influenza virus NS2 nonstructural protein in bacteria and localization of NS2 in infected eucaryotic cells, *J. Virol.* **54**:833–843.
Griffin, J. A., Basak, S., and Compans, R. W., 1983, Effects of hexose starvation and the role of sialic acid in influenza virus release, *Virology* **125**:324–334.
Hackett, C.J., Askonas, B. A., Webster, R. G., et al., 1980, Monoclonal antibodies to influenza matrix protein Detection of low levels of matrix protein on abortively infected cells, *J. Gen. Virol.* **47**:497–501.
Hay, A J , Lomniczi, B., Bellamy, A. R., et al., 1977, Transcription of the influenza virus genome, *Virology* **83**:337–355.
Henle, W., and Henle, G , 1943, Interference of inactive virus with the propagation of virus of influenza, *Science* **98**:87–89
Herz, C., Stavnezer, E., and Krug, R. M., 1981, Influenza virus, an RNA virus, synthesizes its messenger RNA in the nucleus of infected cells, *Cell* **26**:391–400.

Hirst, G. K., and Pons, M. W., 1973, Mechanism of influenza virus recombination. II. Virus aggregation and its effect on plague formation by so-called non-infection virus, *Virology* **56**:620–631.
Horsfall, F. L., Jr., 1955, Reproduction of influenza viruses. Quantitative investigations with particle enumeration procedures on the dynamics of influenza A and B virus reproduction, *J. Exp. Med.* **102**:441–473.
Hosaka, Y., Yasuda, Y., and Fukai, K., 1983, Hemolysis of liposomes containing influenza virus hemagglutinins, *J. Virol.* **46**:1014–1017.
Huang, A. S., 1973, Defective interfering viruses, *Annu. Rev. Microbiol.* **27**:101–117.
Huang, A. S., and Baltimore, D., 1970, Defective viral particles and viral disease processes, *Nature* **226**:325–327.
Huang, R. T. C., Rott, R., and Klenk, H.-D., 1981a, Influenza viruses cause hemolysis and fusion of cells, *Virology* **110**:243–247.
Huang, R. T. C., Wahn, K., Schmidt, M. F. G., et al., 1981b, Productive infection of chick embryo cells by influenza viruses tightly bound on substratum, *Med. Microbiol. Immunol.* **170**:91–98.
Inglis, S. C., and Brown, C. M., 1984, Differences in the control of virus mRNA splicing during permissive or abortive infection with influenza A (fowl plague) virus, *J. Gen. Virol.* **65**:153–164.
Inglis, S. C., and Mahy, B. W. J., 1979, Polypeptides specified by the influenza virus genome. 3. Control of synthesis in infected cells, *Virology* **95**:154–164.
Jackson, D. A., Caton, A. J., McCready, S. J., et al., 1982, Influenza virus RNA is synthesized at fixed sites in the nucleus, *Nature* **296**:366–368.
Janda, J. M., Davis, A. R., Nayak, D. P., et al., 1979, Diversity and generation of defective interfering influenza virus particles, *Virology* **95**:48–58.
Jennings, P. A., Finch, J. T., Winter, G., et al., 1983, Does the higher order structure of the influenza virus ribonucleoprotein guide sequence rearrangements in influenza viral RNA? *Cell* **34**:619–627.
Jones, L. V., Compans, R. W., Davis, A. R., et al., 1985, Surface expression of influenza virus neuraminidase, and aminoterminally anchored viral membrane glycoprotein, in polarized epithelial cells, *Mol. Cell. Biol.* **5**:2181–2189.
Kamata, T., and Watanabe, Y., 1977, Role for nucleocapsid protein phosphorylation in the transcription of influenza virus genome, *Nature* **267**:460–462.
Kato, A., Mizumoto, K., and Ishihama, A., 1985, Purification and enzymatic properties of an RNA polymerase—RNA complex from influenza virus, *Virus Res.* **3**:115–127.
Kaverin, N. V., Varich, N. L., Sklyanskaya, E. I., et al., 1983, Studies on heterotypic interference between influenza A and B viruses: A differential inhibition of the synthesis of viral proteins and RNAs, *J. Gen. Virol.* **64**:2139–2146.
Kawakami, K., and Ishihama, A., 1983, RNA polymerase of influenza virus. III. Isolation of RNA polymerase-RNA complexes from influenza virus PR8, *J. Biochem.* **93**:989–996.
Kawakami, K., Mizumoto, K., Ishihama, A., et al., 1985, Activation of influenza virus-associated RNA polymerase by Cap-1 structure (M^7GpppNm), *J. Biochem.* **97**:655–661.
Kawasaki, K., Sato, S., and Ohnishi, S.-I., 1983, Membrane fusion activity of reconstituted vesicles of influenza virus hemagglutinin glycoproteins, *Biochim. Biophys. Acta* **733**:286–290.
Keil, W., Niemann, H., Schwarz, R. T., et al., 1984, Carbohydrates of influenza virus. V. Oligosaccharides attached to individual glycosylation sites of the hemagglutinin of fowl plague virus, *Virology* **133**:77–91.
Keil, W., Geyer, R., Dabrowski, J., et al., 1985, Carbohydrates of influenza virus. Structural elucidation of the individual glycans of the FPV hemagglutinin by two-dimensional ^{1}H n.m.r. and methylation analysis, *EMBO J.* **4**:2711–2720.
Kida, H., Webster, R. G., and Yanagawa, R., 1983, Inhibition of virus-induced hemolysis with monoclonal antibodies to different antigenic areas on the hemagglutinin molecule of A/Seal/Massachusetts/1/80 (H7N7) influenza virus, *Arch. Virol.* **76**:91–99.
Kilbourne, E. D., 1957, The influence of cortisone on experimental infection. IV. Negation of interference as the mechanism by which cortisone induces increased virus yields, *J. Exp. Med.* **106**:863–881.
Kilbourne, E. D., 1983, The evolution of influenza viral genetics—a perspective, in: *Genetics of Influenza Viruses* (P. Palese and D. W. Kingsbury, eds.), Springer-Verlag, New York, pp. 1–20.
Kilbourne, E. D., and Murphy, J. S., 1960, Genetic studies of influenza viruses. I. Viral morphology

and growth capacity as exchangeable genetic traits. Rapid *in ovo* adaptation of early passage Asian strain isolates by combination with PR8, *J. Exp Med* **111**:387–406.

Kilbourne, E. D., Smart, K. M., and Pokorny, B A , 1961, Inhibition by cortisone of the synthesis and action of interferon, *Nature* **190**:650–651

Klenk, H.-D., Rott, R., Orlich, M., et al., 1975, Activation of influenza A viruses by trypsin treatment, *Virology* **68**:426–439

Klenk, H.-D., Rott, R., and Orlich, M., 1977, Further studies on the activation of influenza virus by proteolytic cleavage of the haemagglutinin, *J Gen. Virol.* **36**:151–161

Krug, R. M , 1971, Cytoplasmic and nucleoplasmic viral RNPs newly synthesized during the latent period of viral growth in MDCK cells, *Virology* **44**:125–136

Krug, R. M , 1972, Cytoplasmic and nucleoplasmic viral RNPs in influenza virus-infected MDCK cells, *Virology* **50**:103–113.

Krug, R M , 1983, Transcription and replication of influenza viruses, in *Genetics of Influenza Viruses* (P. Palese and D W Kingsbury, eds), Springer-Verlag, New York, pp. 70–98.

Krug, R M , and Etkind, P R , 1973, Cytoplasmic and nuclear virus specific proteins in influenza virus-infected MDCK cells, *Virology* **56**:334–348

Krug, R. M., and Soeiro, R., 1975, Studies on the intranuclear localization of influenza virus-specific proteins, *Virology* **64**:378–387.

Lamb, R A , and Choppin, P W , 1976, Synthesis of influenza virus proteins in infected cells Translation of viral polypeptides, including three P polypeptides, from RNA produced by primary transcription, *Virology* **74**:504–519.

Lamb, R A , and Choppin, P W , 1977, Synthesis of influenza virus polypeptides in cells resistant to alpha-amanitin· Evidence for the involvement of cellular RNA polymerase II in virus replication, *J. Virol.* **23**:816–819.

Lamb, R A., and Choppin, P. W., 1983, The gene structure and replication of influenza virus, *Annu Rev Biochem* **52**:467–506

Laver, W G , 1971, Separation of two polypeptide chains from the hemagglutinin subscript of influenza virus, *Virology* **45**:275–288.

Laver, W. G , Colman, P. M., Webster, R. G., et al., 1984, Influenza virus neuraminidase with hemagglutinin activity, *Virology* **137**:314–323.

Lazarowitz, S. G., and Choppin, P. W., 1975, Enhancement of the infectivity of influenza A and B viruses by proteolytic cleavage of the hemagglutinin polypeptide, *Virology* **68**:440–454

Lazarowitz, S. G., Compans, R. W., and Choppin, P. W., 1971, Influenza virus structural and nonstructural proteins in infected cells and their plasma membranes, *Virology* **46**:830–843.

Lazarowitz, S. G., Goldberg, A. R., and Choppin, P. W., 1973, Proteolytic cleavage by plasmin of the hemagglutinin polypeptide of influenza virus. Host cell activation of serum plasminogen, *Virology* **56**:172–180.

Lenard, J., and Compans, R. W., 1974, The membrane structure of lipid-containing viruses. *Biochim Biophys. Acta* **344**:51–94.

Lenard, J., Bailey, C. A., and Miller, D. K., 1982, pH dependence of influenza A virus-induced haemolysis is determined by the haemagglutinin gene, *J. Gen. Virol* **62**:353–355

Lin, B.-C., and Lai, C.-J., 1983, The influenza virus nucleoprotein synthesized from cloned DNA in a simian virus 40 vector is detected in the nucleus, *J. Virol.* **45**:434–438.

Lohmeyer, J., Talens, L. T., and Klenk, H.-D., 1979, Biosynthesis of the influenza virus envelope in abortive infection, *J Gen Virol* **42**:73–88.

Maeno, K., and Kilbourne, E D , 1970, Developmental sequence and intracellular sites of synthesis of three structural protein antigens of influenza A_2 virus, *J. Virol.* **5**:153–164.

Mahy, B. W. J , 1983, Mutants of influenza virus, in: *Genetics of Influenza Viruses* (P Palese and D. W. Kingsbury, eds.), Springer-Verlag, New York, pp. 192–254.

Mahy, B. W. J., Barrett, T , Nichol, S. T., et al , 1981, Analysis of the function of the influenza virus RNA segment by use of temperature-sensitive mutants of fowl plague virus, in *The Replication of Negative Strand Viruses—Developments in Cell Biology*, Vol 7 (D. H. L Bishop and R W. Compans, eds.), Elsevier/North Holland, Amsterdam, pp. 379–387

Marcus, P I , and Carver, D. H., 1967, Intrinsic interference: A new type of viral interference, *J Virol.* **1**:334–343

Marsh, M., Bolzav, E., and Helenius, A., 1983, Penetration of Semliki Forest virus from acidic prolysosomal vacuoles, *Cell* **32**:931–940.

Matlin, K. S., Reggio, H., Helenius, A., et al., 1981, Infectious entry pathway of influenza virus in a canine kidney cell line, *J. Cell Biol.* **91**:601–613.

Maxfield, F. R., 1982, Weak bases and nonophores rapidly and reversibly raise the pH of endocytic vesicles in cultured mouse fibroblasts, *J. Cell Biol.* **95**:676–681.

Meier-Ewert, H., and Compans, R. W., 1974, Time course of synthesis and assembly of influenza virus proteins, *J. Virol.* **14**:1083–1091.

Mikheeva, A., and Ghendon, Y. Z., 1982, Intrinsic interference between influenza A and B viruses, *Arch. Virol.* **73**:287–294.

Minor, P. D., and Dimmock, N. J., 1976, The multiplication of influenza virus in enucleated BHK cells fused with chicken erythrocytes, *Virology* **69**:336–338.

Minor, P. D., Hart, J. G., and Dimmock, N. J, 1979, Influence of the host cell on proteins synthesized by different strains of influenza virus, *Virology* **97**:482–487.

Moss, B. A., and Brownlee, G. G., 1981, Sequence of DNA complementary to a small RNA segment of influenza virus A/NT/60/68, *Nucleic Acids Res.* **9**:1941–1947.

Naeve, C. W., Hinshaw, V. S., and Webster, R. G., 1984, Mutations in the hemagglutinin receptor-binding site can change the biological properties of an influenza virus, *J. Virol.* **51**:567–569.

Nakajima, K., Ueda, M., and Sugiura, A., 1979, Origin of small RNA in von Magnus particles of influenza virus, *J. Virol.* **29**:1142–1148.

Nakamura, K., and Compans, R. W., 1977, The cellular site of sulfation of influenza virus glycoproteins, *Virology* **79**:381–392.

Nakamura, K., and Compans, R. W., 1979, Biosynthesis of the oligosaccharides of influenza viral glycoproteins, *Virology* **93**:31–47.

Nakamura, K., Bhown, A. S., and Compans, R. W., 1980, Glycosylation sites of influenza viral glycoproteins. Tryptic glycopeptides from the A/WSN (H1N1) hemagglutinin glycoprotein, *Virology* **107**:208–221.

Nayak, D. P., 1980, Defection interfering influenza viruses, *Annu. Rev. Microbiol.* **34**:619–644.

Nayak, D. P., Tobita, K., Janda, J. M., et al., 1978, Homologous interference mediated by defective interfering influenza virus derived from a temperature-sensitive mutant of influenza virus, *J. Virol.* **28**:375–386.

Nayak, D. P., Sivasubramanian, N., Davis, A. R., et al., 1982, Complete sequence analyses show that two defective interfering influenza viral RNAs contain a single interval deletion of a polymerase gene, *Proc. Natl. Acad. Sci. U.S.A.* **79**:2216–2220.

Nayak, D. P., Chambers, T. M., and Akkina, R. K., 1985, Defective-interfering (DI) RNAs of influenza viruses: Origin, structure, expression, and interference, *Curr. Top. Microbiol. Immunol.* **114**:103–151.

Palese, P., and Compans, R. W., 1976, Inhibition of influenza virus replication in tissue culture by 2-deoxy-2,3-dehydro-*N*-trifluoroacetylneuraminic acid: Mechanism of action, *J. Gen. Virol.* **33**:159–163.

Palese, P., Tobita, K., Ueda, M., et al., 1974, Characterization of temperature sensitive influenza virus mutants defective in neuraminidase, *Virology* **61**:397–410.

Penn, C. R., and Mahy, B. W. H., 1985, Novel polypeptides encoded by influenza virus subgenomic (DI type) virion RNAs, *Virus Res.* **3**:311–321.

Petri, T., Patterson, S., and Dimmock, N. J., 1982, Polymorphism of the NSI proteins of type A influenza virus, *J. Gen. Virol.* **61**:217–231.

Plotch, S. J., Buoloy, M., Ulmanen, I., et al., 1981, A unique cap(m^7GpppXm)-dependent influenza viron endonuclease cleaves capped RNAs to generate the primers that initiate viral RNA transcription, *Cell* **23**:847–858.

Pons, M. W., 1980, The genome of incomplete influenza virus, *Virology* **100**:43–52.

Pons, M. W., and Hirst, G. K., 1969, The single- and double-stranded RNAs and the proteins of incomplete influenza virus, *Virology* **38**:68–72.

Privalsky, M. L., and Penhoet, E. E., 1978, Influenza virus proteins: Identity, synthesis, and modification analyzed by two-dimensional gel electrophoresis, *Proc. Natl. Acad. Sci. U.S.A.* **75**:3625–3629.

Rees, P. J., and Dimmock, N. J., 1982, Kinetics of synthesis of influenza virus ribonucleoprotein structures, *J. Gen. Virol.* **59**:403-408.

Richardson, C. D , Scheid, A., and Choppin, P. W., 1980, Specific inhibition of paramyxovirus and myxovirus replication by oligopeptides with amino acid sequences similar to those at the N-termini of the F_1 or HA_2 viral polypeptides, *Virology* **105**:205-222.

Rindler, M. J., Ivanov, I. E., Rodriguez-Boulan, E., et al., 1982, Biogenesis of epithelial cell plasma membranes, *Ciba Found. Symp.* **92**:184-202.

Rodriguez-Boulan, E., and Sabatini, D. D., 1978, Asymmetric budding of viruses in epithelial monolayer A model system for study of epithelial polarity, *Proc Natl Acad Sci U.S A* **75**:5071-5075.

Rodriguez-Boulan, E., Paskiet, K T , Salas, P J. I , et al., 1984, Intracellular transport of influenza virus hemagglutinin to the apical surface of Madin-Darby canine kidney cells, *J Cell Biol* **98**:308-319

Rogers, G N , and Paulson, J C , 1983, Receptor determinants of human and animal influenza virus isolates Differences in receptor specificity of the H3 hemagglutinin based on species of origin, *Virology* **127**:361-373

Roger, G N , Paulson, J C., Daniels, R. S., et al., 1983, Single amino acid substitutions in influenza hemagglutinin change receptor binding specificity, *Nature* **304**:76-78.

Romanos, M A., and Hay, A J., 1984, Identification of the influenza virus transcriptase by affinity-labeling with pyridoxal 5′-phosphate, *Virology* **132**:110-117

Roth, M G , Fitzpatrick, J. P., and Compans, R. W., 1979, Polarity of influenza and vesicular stomatitis virus maturation in MDCK cells· Lack of a requirement for glycosylation of viral glycoproteins, *Proc. Natl. Acad. Sci U S.A.* **76**:6430-6434.

Rott, R., Orlich, M., and Scholtissek, C , 1981, Intrinsic interference between swine influenza and fowl plague virus, *Arch. Virol* **69**:25-32.

Sato, S. B , Kawasaki, K , and Ohnishi, S.-I., 1983, Hemolytic activity of influenza virus hemagglutinin glycoproteins activated in mildly acidic environments, *Proc. Natl Acad Sci. U S.A.* **80**:3153-3157.

Scheid, A , and Choppin, P. W., 1980, The role of viral glycoproteins in adsorption, penetration, and pathogenicity of viruses, *Rev. Infect. Dis.* **2**:40-61

Schlesinger, R. W., 1959, Interference between animal viruses, in: *The Viruses* (F. M. Burnet and W. M. Stanley, eds.), Academic Press, New York, London, pp. 157-194.

Schmidt, M F. G., and Lambrecht, B , 1985, On the structure of the acyl linkage and the function of fatty acyl chains in the influenza virus haemagglutinin and the glycoproteins of Semliki Forest virus, *J. Gen. Virol.* **66**:2635-2647.

Scholtissek, C , and Klenk, H.-D., 1975, Influenza virus replication, in: *The Influenza Viruses and Influenza* (E. D. Kilbourne, ed.), Academic Press, New York, pp. 215-242

Schulze, I. T , 1972, The structure of influenza virus. II. A model based on the morphology and composition of subviral particles, *Virology* **47**:181-196.

Schwarz, R. T., Schmidt, M. F. G., Anwer, W., et al., 1977, Carbohydrates of influenza virus. I. Glycopeptides derived from viral glycopeptides after labeling with radioactive sugars, *J Virol* **23**:217-226.

Shaw, M. W., and Choppin, P. W., 1984, Studies on the synthesis of the influenza B and virus NB glycoprotein, *Virology* **139**:178-184.

Shaw, M W , and Lamb, R A , 1984, A specific sub-set of host-cell mRNAs prime influenza virus mRNA synthesis, *Virus Res* **1**:455-467.

Shaw, M. W., Choppin, P. W , and Lamb, R A., 1983, A previously unrecognized influenza B virus glycoprotein from a bicistronic mRNA that also encodes the viral neuraminidase, *Proc Natl. Acad. Sci. U.S.A.* **80**:4879-4883.

Shibata, M., Maeno, K., Tsurumi, T., et al., 1982, Role of viral glycoproteins in haemolysis by influenza B virus, *J Gen. Virol.* **59**:183-186

Simons, K., and Garoff, H , 1980, The budding mechanisms of enveloped animal viruses, *J Gen Virol* **50**:1-21.

Sivasubramanian, N., and Nayak, D. P., 1983, Defective interfering influenza RNAs of polymerase

3 gene contain single as well as multiple internal deletions, *Virology* **124**:232–237.
Skehel, J. J., 1972, Polypeptide synthesis in influenza virus-infected cells, *Virology* **49**:23–36.
Skehel, J. J., 1973, Early polypeptide synthesis in influenza virus-infected cells, *Virology* **56**:394–399.
Skehel, J. J., and Waterfield, M. D., 1975, Studies of the primary structure of the influenza virus hemagglutinin, *Proc. Natl. Acad. Sci. U.S.A.* **72**:93–97.
Skehel, J. J., Bayley, P. M., Brown, E. B., et al., 1982, Changes in the conformation of influenza virus hemagglutinin at the pH optimum of virus-mediated membrane fusion, *Proc. Natl. Acad. Sci. U.S.A.* **79**:968–972.
Smart, K. M., and Kilbourne, E. D., 1966, The influence of cortisone on experimental viral infection. VI. Inhibition by hydrocortisone of interferon synthesis in the chick embryo, *J. Exp. Med.* **123**:299–307.
Smith, G. L., and Hay, A. J., 1982, Replication of the influenza virus genome, *Virology* **118**:96–108.
Spooner, L. L. R., and Barry, R. D., 1977, Participation of DNA-dependent RNA polymerase II in replication of influenza viruses, *Nature* **268**:650–652.
Takeura, S., Aoki, H., Tsurumi, T., et al., 1984, Abortive infection of L cells by influenza B virus: Defect in bud formation, *Microbiol. Immunol.* **28**:427–437.
Ter Meulen, V., and Love, R., 1967, Virological studies of four HeLa cell lines infected with two strains of influenza virus, *J. Virol.* **1**:626–639.
Tobita, K., and Ohori, K., 1979, Heterotypic interference between influenza viruses A/Aichi/2/68 and B/Massachusetts/1/71, *Acta Virol.* **23**:263–266.
Tycko, B., and Maxfield, F. R., 1982, Rapid acidification of endocytic vesicles containing α_2-macroglobulin, *Cell* **28**:643–651.
Ulmanen, I., Broni, B., and Krug, R. M., 1983, Influenza virus temperature-sensitive cap (m^7GpppNm)-dependent endonuclease, *J. Virol.* **45**:27–35.
Valcavi, P., Conti, G., and Schito, G. C., 1978, Molecular synthesis during abortive infection of KB cells by influenza virus, in: *Negative Strand Viruses and the Host Cell* (B. W. J. Mahy and R. D. Barry, eds.), Academic Press, London, New York, pp. 475–482.
von Magnus, P., 1951, Propagation of the PR8 strain of influenza virus in chick embryos. II. The formation of "incomplete" virus following inoculation of large doses of seed virus, *Acta Pathol. Microbiol. Scand.* **28**:278–293.
Wang, E., Wolf, B. A., Lamb, R. A., et al., 1976, The presence of actin in enveloped viruses, in *Cell Motility* (R. Goldman, T. Pollard, and J. Rosenbaum, eds.), Cold Spring Harbor Press, Cold Spring Harbor, New York, p. 589.
White, J., Helenius, A., and Gething, M.-J., 1982, Haemagglutinin of influenza virus expressed from a cloned gene promotes membrane fusion, *Nature* **300**:658–659.
Yewdell, J. W., Gerhard, W., and Bachi, T., 1983, Monoclonal antihemagglutinin antibodies detect irreversible antigenic alterations that coincide with the acid activation of influenza virus A/PR/834-mediated hemolysis, *J. Virol.* **48**:239–248.
Yoshida, T., Shaw, M. W., Young, J. F., et al., 1981, Characterization of the RNA associated with influenza A cytoplasmic inclusions and the interaction of NS protein with RNA, *Virology* **110**:87–97.
Yoshimura, A., and Ohnishi, S.-I., 1984, Uncoating of influenza virus in endosomes, *J. Virol.* **51**:497–504.
Yoshimura, A., Kuroda, K., Kawasaki, K., et al., 1982, Infectious cell entry mechanism of influenza virus, *J. Virol.* **43**:284–293.
Young, J. F., Desselberger, U., Palese, P., et al., 1983, Efficient expression of influenza virus NS1 nonstructural proteins in *Escherichia coli. Proc. Natl. Acad. Sci. U.S.A.* **80**:6105–6109.
Zebedee, S. L., Richardson, C. D., and Lamb, R. A., 1985, Characterization of the influenza virus M_2 integral membrane protein and expression at the infected-cell surface from cloned cDNA, *J. Virol.* **56**:502–511.
Zhirnov, O. P., and Bukrinskaya, A. G., 1981, Two forms of influenza virus nucleoprotein in infected cells and virions, *Virology* **109**:174–179.

Zhirnov, O. P., and Bukrinskaya, A. G , 1984, Nucleoproteins of animal influenza viruses, in contrast to those of human strains, are not cleaved in infected cells, *J. Gen. Virol.* **65**:1127–1134.

Ziegler, J. E., Jr., and Horsfall, F. L., Jr., 1944, Interference between the influenza viruses. I. The effect of active virus upon the multiplication of influenza viruses in the chick embryo, *J. Exp. Med.* **79**:361–377.

Ziegler, J. E , Jr., Lavin, G. I., and Horsfall, F. L., Jr., 1944, Interference between the influenza viruses. II The effect of virus rendered non-infective by ultraviolet radiation upon the multiplication of influenza viruses in the chick embryo, *J. Exp. Med.* **79**:379–400.

5

Cytopathogenesis and Cytopathology of Influenza Virus Infection of Cells in Culture

INTRODUCTION

A critical appraisal of available evidence suggests that all target cells in which influenza virus replication occurs are doomed to destruction. Studies of apparently noncytopathic viruses in which foci of infection were localized with a chromogenic neuraminidase substrate demonstrated that in all instances when neuraminidase-positive foci could be demonstrated, cell necrosis (cytopathic effect, CPE) was subsequently detected by microscopy (Palese et al., 1973; E.D. Kilbourne, unpublished data). Assuming, then, the absence of temperate or latent infection, differences in virulence, pathogenicity, and persistence among influenza viruses must be explained as a resultant of virus and host-cell factors that determine the probability of host cell escape from infection under defined conditions. These conditions may favor the emergence of DI particles or the induction of interference or interferon that may influence the pattern of microinfection (plaque formation) or the persistence of virus in serially propagated cell cultures. In persistent infection of laboratory cell cultures, influenza virus mutants may evolve with altered replication, neuraminidase activity, and interfering properties. However, even these mutants were demonstrably cytopathic (Frielle et al., 1984).

PRIMARY INTRINSIC CYTOTOXICITY OF INFLUENZA VIRUSES

Effects in Intact Animals

Granted that influenza viruses destroy cells, can they do so without replicating in them? That is, do they have intrinsic cytotoxicity? A priori, one would think so because of the occasional toxic effects observed with influenza vaccines rendered noninfective with formalin or other agents. In the past, such reactions have been attributable in part to vaccine components other than the virus, including bacterial endotoxins. However, contemporary vaccines are highly purified yet still elicit some local soreness and, in the young, febrile reactions. Furthermore, even purified influenza virus free of endotoxin has been shown to be pyrogenic in rabbits, a host not known to support influenza virus replication (Bennett et al., 1949; Wagner et al., 1949; Siegert et al., 1976). This effect has been related to the release of endogenous pyrogen from leukocytes (Kanoh and Kawasaki, 1966; Kosel and Kohlhage, 1970; Táborský, 1972). Recently, production of interleukin 1 and interleukin 1 inhibitor has been demonstrated in human macrophages exposed to influenza virus (Roberts et al., 1986).

Administration of infective influenza virus in large amounts by nonrespiratory routes can produce toxic effects leading to brain, liver, or ocular damage and often death (Schlesinger, 1950; Henle and Henle, 1946a,b; Sugg, 1949; Ginsberg, 1954; Oh, 1965). But such effects ,including CPE in "nonpermissive" cells in culture (Henle et al., 1955), probably are associated with some virus replication even though it may not be demonstrably productive of infective virus. But toxic effects of inactivated vaccine cannot be similarly explained. The possibility that some form of anomalous replication may occur, perhaps by multiplicity reactivation, following the introduction of inactivated virus seems remote. Therefore, vaccine-induced and other effects of inactivated virus (see the following sections) may indeed represent primary viral toxicity.

In Vitro Effects of Influenza Viruses on Polymorphonuclear Leukocytes

Peripheral blood leukocytes are not ordinarily considered to be targets for influenza virus infection, although recent studies suggest that the virus can infect lymphocytes with production of viral antigens "without direct cytopathic effects" (Casali et al., 1984; Wilson et al., 1976). However, synthetic function of the infected lymphocytes was impaired, and subsequent proliferative response precluded (Casali et al., 1984).

Most studies have been of the effect of influenza viruses on polymorphonuclear leukocytes (PMNL), in which virus replication has not been demonstrated (Ginsberg and Blackmon, 1956) or in which some effects can be induced with inactivated virus or isolated viral glycoproteins (Table 5-1). Such studies are of particular interest with reference to the question of primary intrinsic toxicity of influenza viruses. Table 5-1 summarizes representative studies of the effects of influenza viruses on leukocytes in in vitro systems divorced from complicating factors inherent in studies in intact animals or man.

Following the provocative experiments of Merchant and Morgan (1950) and Fisher and Ginsberg (1956a,b) that demonstrated impairment of metabolic function and phagocytosis by influenza viruses (Table 5-1), little was done for the next 20 years despite the obvious relevance of virus-induced leukocyte damage to the pathogenesis of influenza and its complications. Recently, the problem of influenza virus effects on polymorphonuclear leukocytes (PMNLs) has received renewed attention by investigators skilled primarily in cytological methods. Therefore, although the evidence of impaired leukocyte function in response to virus is persuasive, the literature summarized in Table 5-1 provides scanty information on purity of virus preparations, viral quantitation, degree of viral inactivation, or specificity of the effects described. However, the following conclusions seem justified on the basis of this evidence.

Exposure in vitro of PMNLs to infective influenza A and B viruses impairs anaerobic glycolysis, stimulates a "respiratory burst" as measured by chemiluminescence, and inhibits random locomotion of PMNLs and their response to chemotactic stimuli.

Perhaps secondary to these initial effects on the cell, infective and ultraviolet (UV)-"inactivated" influenza viruses depress phagocytosis and the respiratory burst and the superoxide production that accompanies this process. In addition, lysosome–phagosome fusion after phagocytosis is depressed, lysosomal enzyme release is inhibited, and killing of secondarily ingested organisms is impaired (Abramson et al., 1982c).

With a single exception (Abramson et al., 1982b), heating of virus at 50–80°C abolished its effects on PMNL metabolism and phagocytic function. Although not actually demonstrated in the reported studies, it is probable that these temperatures inactivated both viral neuraminidase activity and infectivity. The more selective inactivation procedure of UV or formalin treatment would be expected to reduce viral infectivity markedly but would probably leave neuraminidase function unimpaired. Therefore, it is significant that UV-irradiated viruses are substitutive for infective virus in effects on respiratory burst (Mills et al., 1981; Henricks et al., 1982; Abramson et al., 1982b) and phagocytosis (Henricks et al., 1982; Abramson et al., 1982b). Purified influenza virus neuraminidase failed to stimulate chemiluminescence in mouse spleen cells (Peterhans, 1980), but HANA liposomes stimulated chemiluminescence of human PMNLs but had no effect on phagocytosis. On the other hand, partial removal of neuraminic acid from PMNLs by bacterial or viral neuraminidase enhances phagocytosis and generation of superoxide and chemiluminescence in response to opsonized staphylococci (Henricks et al., 1982). Thus, the role of the viral envelope proteins in viral effects on PMNLs remains unresolved. The NA activity is probably not the cause of influenza-virus-induced neutrophil dysfunction (Abramson et al., 1985). Intact virus particles are demonstrable within PMNLs (Busse et al., 1979) within 5 min of virus–cell contact (Abramson et al., 1982b). But evidence of abortive replication (e.g., viral protein synthesis) has not been reported.

In summary, the influenza virus–PMNL system is a promising model for the study of intrinsic viral cytotoxicity without replication. However, further studies are required to ascertain (1) whether or not gene expression of live virus or virus presumably inactivated by UV irradiation occurs and (2) which, if any, of the effects on PMNL function are induced through the glycoproteins of input viruses.

Viral Effects on Other Blood Leukocytes

Influenza virus inhibits the ability of macrophages to enhance lymphocyte transformation (Roberts, 1982) and also induces formation of interleukin 1 and its inhibitor (Roberts et al., 1986). Effects of virus on peripheral blood monocytes resemble effects on PMNLs with respect to stimulation of respiratory burst

Table 5-1. In Vitro Effects of Influenza Virus on Polymorphonuclear Leukocytes

Reference	Virus	Cell (PMNL)[a]	Effect
		Effects of live virus	
Merchant and Morgan (1950)	A/PR/8/34	Guinea pig	↓Phagocytosis of anthrax bacilli
Ginsberg and Blackmon (1956)	A/PR/8/34 B/Lee/40	Guinea pig	Virus adsorbed to cells but did not elute or → infective virus ↓Phagocytosis
Fisher and Ginsberg (1956a)	A/PR/8/34 B/Lee/40	Guinea pig	↓Anaerobic glycolysis
Fisher and Ginsberg (1956b)	B/Lee/40	Guinea pig	↓Phagocytosis of yeast cells
Larson and Blades (1976); Larson et al. (1977)	B/Eng/13/65	Human	↓Chemotaxis ↓Phagocytosis of staphylococci
Ruutu (1977)	A/Finland/7/69 B/Finland/1/69	Rat	↓Chemotaxis ↓Random locomotion
Busse et al. (1979)	A/reassortant B/reassortant	Human	↓Inhibition of lysosomal enzyme release Induced by pharmacological agents
Mills et al. (1981)	A/Texas/77	Human	↑Chemiluminescence ↑Superoxide ("respiratory burst")
Henricks et al. (1982)	A/Texas/77	Human	↑Phagocytosis ↑Chemiluminescence ↑Superoxide
Abramson et al. (1982a)	A/Texas/77	Human	↑Initial chemiluminescence ↓Chemiluminescence response to secondary phagocytosis ↓Bactericidal activity
Abramson et al. (1982b)	Various A strains	Human	↑Respiratory burst ↓Phagocytosis
Abramson et al. (1982c)	A/reassortants A/wild type strains	Human	↓Lysosome-phagosome fusion after phagocytosis of staphylococci

			↓Lysosomal enzyme release ↓Killing
Abramson et al. (1984)	A/PR/8/34 A/Texas/77	Human	↓Chemotaxis ↓Lysosomal enzyme degranulation ↓Adherence ↑Bipolar cells
Masihi et al. (1984)	A/PR/8/34 B/Singapore/222/79	Mouse	↓Chemiluminescence in response to zymosan
	Effects of inactivated virus or components		
Merchant and Morgan (1950)	A/PR/8/34 (△ 50°C)	Guinea pig	Effect on phagocytosis abolished
Fisher and Ginsberg (1956a)	B/Lee/40 (△ 56°C)	Guinea pig	Effect on glycolysis abolished
Larson et al (1977)	B/Eng/13/65 (△ 50°C)	Human	Effects on chemotaxis and phagocytosis abolished
Mills et al. (1981)	A/Texas/77 (△ 80°C)	Human	Effects on respiratory burst abolished
	(UV-irradiated)		Effects preserved
Henricks et al. (1982)	A/Texas/77 (△ 80°C)	Human	Effects on phagocytosis and respiratory burst abolished
	(UV-irriated)		Effects preserved
Abramson et al. (1982b)	A/X-47 (△ 56°C) (UV-irradiated)	Human	Effects on respiratory burst and phagocytosis preserved
	A/X-31 HANA liposomes		↑Chemiluminescence No effect on phagocytosis
Masihi et al (1984)	A/PR/8/34 UV-irradiated	Mouse	↓Zymosan-induced chemiluminescence
	Inactivated vaccines		↓Zymosan-induced chemiluminescence

[a]PMNL, polymorphonuclear leukocyte

and reduction of bactericidal activity (Abramson et al., 1982a; Gardner and Lawton, 1982).

In human peripheral blood lymphocytes, limited viral protein synthesis has been reported. This abortive replication occurred only in mitogen-transformed cells (Brownson et al., 1979). In another study, viral protein synthesis occurred in resting lymphocytes. Infection terminated cell proliferation and IgG production but not natural killer cell activity (Casali et al., 1984). Influenza virus antigen has been reported to be present on circulating lymphocytes during human infection (Wilson et al., 1976).

Viral Effects on Other Cells

Both living and formalin-inactivated influenza A virus (A/Hong Kong/68) induced chromosome aberrations in mouse spermatogonia (Thadani and Polasa, 1979). Abortive in vitro replication of influenza A viruses in human macrophages without immediate cytopathic effect has been reported (Rodgers and Mims, 1982).

CELL SYSTEMS SUPPORTIVE OF PRODUCTIVE REPLICATION AND PLAQUE FORMATION BY INFLUENZA VIRUSES

As mentioned in the preceding section, the number and variety of cell types that will support replication of influenza virus productive of infective virus still await definition in light of the need for HA cleavage by host-cell or exogenously supplied proteases to maximize the efficiency of the infectious process. Nevertheless, it seems useful to summarize here past experience with host-cell systems in relation to their variety and susceptibility to cytopathic effects as a function of the cytocidal effects of representative influenza viruses. The listing of such systems in Table 5-2 does not pretend to completeness but does include those primary cell cultures and cell lines most widely used in recent years.

In reviewing this table (and the studies it summarizes), it is important to note the following.

With few exceptions, the viruses represented here are so-called "standard" strains that have been passaged in a number of laboratory hosts. Some have acquired unusual virulence properties; e.g., WSN and NWS are neurovirulent in mice. Therefore, they are not representative of isolates derived directly from human or other hosts infected in nature.

No useful correlates can be made of host species of virus origin and host species of cell origin with respect to cellular susceptibility (for an exception, see Flehmig et al., 1976). Mammalian viruses prosper in avian (chick embryo) cells, and viruses of avian origin can replicate in certain mammalian cells (e.g., human conjunctival clone 1-5C-4 cells) (Kilbourne, 1969; Gerber et al., 1973).

Plaquing efficiency defined in relation to chick embryo infectivity (EID_{50}) varies widely and obviously is influenced by both virus and host cell. No single host cell supports replication of all influenza strains, and no virus is pantropic.

As with other virus–host cell systems, the organ of cell origin of cells susceptible to virus in vitro is often irrelevant to the in vivo organ target of natural infection.

CYTOPATHIC EFFECTS OF INFLUENZA VIRUSES IN SPECIALIZED CELLS

The WSN and NWS neurotropic variants of A/WS/33 (H1N1) virus appear to be more broadly cytopathogenic than nonneurotropic mutants. WSN produces cytopathic effects in differentiated human skeletal muscle cells (Armstrong et al., 1978) and in a neuroblastoma cell line (Nakajima and Sugiura, 1980). WSN is also pathogenic in ependymal organ cultures (Kohn et al., 1981) and in organotypic cultures of embryonic mouse hypothalamus (Gamboa et al., 1974). NWS replicates productively with cytopathic effect in human diploid cell lines in which other influenza viruses cannot be serially propagated (Kilbourne et al., 1964; Kutinova et al., 1967) except with the addition of protease (Herrero-Uribe et al., 1983).

THE NATURE OF INFLUENZA VIRUS CYTOPATHIC EFFECTS

Cytopathology

Given the wide variety of influenza virus-cell systems that have for the most part been employed in studies focused primarily on viral replication, it is perhaps not surprising that few systematic studies have been made of influenza virus cytopathogenesis. Optimally, such studies demand the use of cloned, uniformly susceptible cells and cloned, genetically defined viruses. The conditions of cell cultivation must be rigorously maintained and standardized, and the role of endogenously generated inhibitors such as interferon must be appreciated. These conditions cannot be met with primary or secondary cell cultures in which as few as one of 1000 cells may yield infectious virus (Tucková and Vonka, 1970). Even in the case of relatively homogeneous continuous cell lines, constant surveillance and occasional recloning are required to maintain a population of uniformly susceptible cells (E.D. Kilbourne, unpublished experiments). With a single virus there was a striking difference in the proportion of cells with virus-associated changes in four HeLa cell lines (Ter Meulen and Love, 1967).

Regardless of the cell type in which infection occurs, the following cytopathic effects of influenza virus are usually observed: (see Fig. 5-1) rounding of cells, cytoplasmic retraction, granularity and vacuolization, nucleolar enlargement, and clumping of chromatin (Armstrong et al., 1978). An early change in the nucleus is the coarsening and increased prominence of chromatin granules and nucleolus-associated chromatin (Niven et al., 1962).

By electron microscopy, changes can be seen as early as 4 hr after infection, including nucleolar distention with the acquisition of "dense spots" (Compans and Dimmock, 1969) and the formation of fibrillar and electron-dense structures in the nucleus (Ciampor and Turcan, 1972; Anisimová et al, 1973; Miranda et al., 1978). Comparison of the time course of virus formation and the appearance of nucleolar alterations suggested that the dense masses observed were not precursors of virus particles (Compans and Dimmock, 1969).

In both the nucleus and cytoplasm, elongated fibrillar structures 22 to 33 nm in diameter may be observed in some cells (Archetti et al., 1970; Ciampor and

Table 5-2. Cell Systems Supportive of Productive Replication and Plaque Formation by Influenza Viruses

Cell system	Type of culture	Influenza viruses	Plaquing efficiency[a] (%)	Reference
Chick embryo fibroblasts	Primary (9-day)	A/WSN (H1)	1.0	Granoff (1955)
Chick embryo lung	Primary (16-day)	A/PR8; Mel (H1)	0.2	Granoff (1955)
Chick embryo lung	Primary (14-day)		NS[b]	Ledinko (1955)
Chick embryo kidney–skin–muscle	Primary (18- to 20-day)	A/WS (H1)	10.0	Chang et al. (1966)
Chick embryo kidney	Primary (18- to 20-day)	A/PR8 (H1)	4–100.0	Wright and Sagik (1958)
Chick embryo fibroblasts	Primary (11-day)	A/Jap.305 (H2)	NS	Peter and Goedemans (1966)
Chick embryo fibroblasts	Primary (5-day)	A/ (H1)(H2), B	NS	Mizutani and Nagase (1967)
Monkey kidney	Primary	A/575[c] (H1)	NS	Henry and Youngner (1957)
Monkey kidney	Primary	A/RI/5[c] (H2)	28–70	Choppin (1962, 1963)
Monkey kidney	Primary	A/Singapore[c] (H2)	6	Vonka (1965)
Monkey kidney	Primary	A/Conn/1/65 (H2)		Walder and Hsiung (1967)
Calf kidney	Primary	A/Swine (H1)	10	Zimmermann and Schäfer (1959)
Calf kidney	Primary	A (H1), B	10–300	Lehmann-Grube (1963)
Hamster kidney	Primary	A (H1)(H2)	0.1–10	Grossberg (1964)
Human conjunctival (clone 1-5C-4)	Aneuploid cell line	A/Avian, Sw (H1)(H2), B	2–70	Sugiura and Kilbourne (1965)

Human bone carcinoma (G2)	Aneuploid (?) cell line	A (H1)(H2)	18–>100	Hatano and Morita (1967)
Human embryo lung (LEP-12)	Diploid line	A/NWS (H1)	≅1 0	Kutinova et al (1967)
BHK-21	Aneuploid line	A/WSN (H1)	100	Zitcer et al (1967)
		A/NWS (H1)	NS	Milliken (1967)
Chick kidney	Primary	A/Ann Arbor (H1)	NS	Maassab (1968)
		A/PR8, A/FM1	NS	Babiker and Rott (1968)
		A/Sing (H1, H2), B/Lee		
Canine kidney (MDCK)	Aneuploid line	4 A and 3 B strains	>100	Gaush and Smith (1968)
Human cervical carcinoma (HeLa) cells	Aneuploid lines	Adapted A/FPV (H7) (A/Ty/Eng.)	NS	Gerber et al. (1973)
Various monkey kidney cell lines with trypsin	Aneuploid lines	A and B prototypes	NS	Ørstavik (1981)
Ferret kidney	Primary	A/Bangkok (H3)	NS	Pons et al (1981)
Swine kidney (ESK)	Aneuploid line	B/Kanagawa	NS	Shibata et al. (1982)
		A/Prague (H7)	100	Yamagishi et al (1984)
		A/Miami (H3)		

[a]In relation to chick embryo infectivity (EID_{50}) as 100%
[b]NS, not stated
[c]Monkey-kidney-adapted virus used

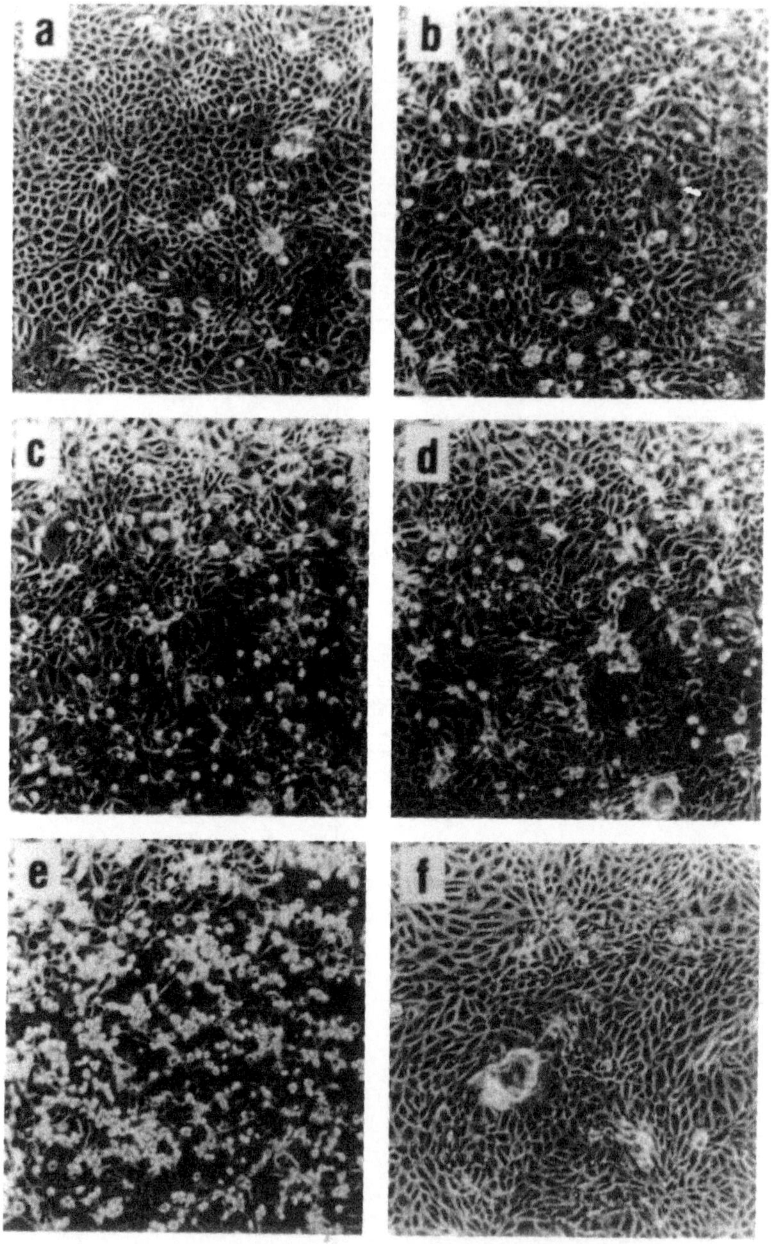

Figure 5-1. Cytopathic effect of influenza B virus and a highly cytolytic mutant in MDCK cells. (a,b,c) Wild-type virus 9, 12, and 15 hr after infection. (d,e) Mutant, 9 and 12 hr after infection. (f) Uninfected cell control. (From Tanaka et al., 1984. Reproduced with permission, from *Virology* **135**:518.)

Turcan, 1972). As early as 4 hr after infection, aggregation of ribosomes around dense fibrous material is seen in the cytoplasm (Compans and Dimmock, 1969). The frequency with which the preceding cytopathic changes are observed appears to be dependent on the cell–virus system studied. Changes observed late in infection (after 24 hr) include rupture of cytoplasmic nucleolar and nuclear membranes, clumping of chromatin, and cytoplasmic vacuolization, all of which are nonspecific effects of cytocidal viral infection in general.

Nuclear and Cytoplasmic Inclusions

So-called inclusions or "inclusion bodies" have been a hallmark of viral infections since the early days of pathology. Until recently there was little information about the nature of these inclusions and their relation to the viral replication cycle or to viral gene products, although it has been appreciated that they tend to appear late in infection.

In studies of influenza B virus infection of monkey kidney (MK) cells, basophilic cytoplasmic inclusions were noted late in infection. These inclusions stained positively for RNA, and electron microscopy showed them to be associated with large aggregates of ribosomelike particles (Porebska et al., 1968). Inclusions have also been found in the cytoplasm of influenza A virus-infected cells, in which they appear by electron microscopy throughout the cytoplasm as discrete darkly stained bodies with a crystalloid fine structure; they are closely associated with ribosomes (Kopp et al., 1968; Saito et al., 1970). It is now known that these prominent cytoplasmic paracrystalline inclusions (Fig. 5-2) represent the nonstructural viral protein NS bound to cellular (predominantly ribosomal) RNA (Yoshida et al., 1981; Morrongiello and Dales, 1977; Shaw and Compans, 1978). The nature of tubular inclusions sometimes seen in the nucleus is not yet clear.

The Kinetics of CPE: Influenza Virus Plaque Formation in Cell Cultures as a Model of Microinfection

In monolayer cell cultures maintained under agar, it can be shown that influenza-virus-induced plaques usually are initiated by single virions except when complementation of noninfective aggregated particles occurs (Hirst and Pons, 1973). Adsorption of virus to cells occurs rapidly, more than 50% being adsorbed within 15 min in the case of the well-studied NWS virus–clone 1-5C-4 human conjunctival cell system. Viral entry measured by adsorption irreversible with specific antibody requires more time (90 to 180 min) (Fig. 5-3) (Sugiura and Kilbourne, 1965). After this point, the kinetics of virus replication and the concomitant and sequential cytopathic effect are influenced by the susceptibility and permissiveness of the cell, the properties of the virus, and the environmental conditions (e.g., temperature and pH) maintained during viral replication. Figure 5-4 illustrates typical replication curves in which infective and hemagglutinating (total) virus are plotted separately. In the system illustrated, A/NWS or its reas-

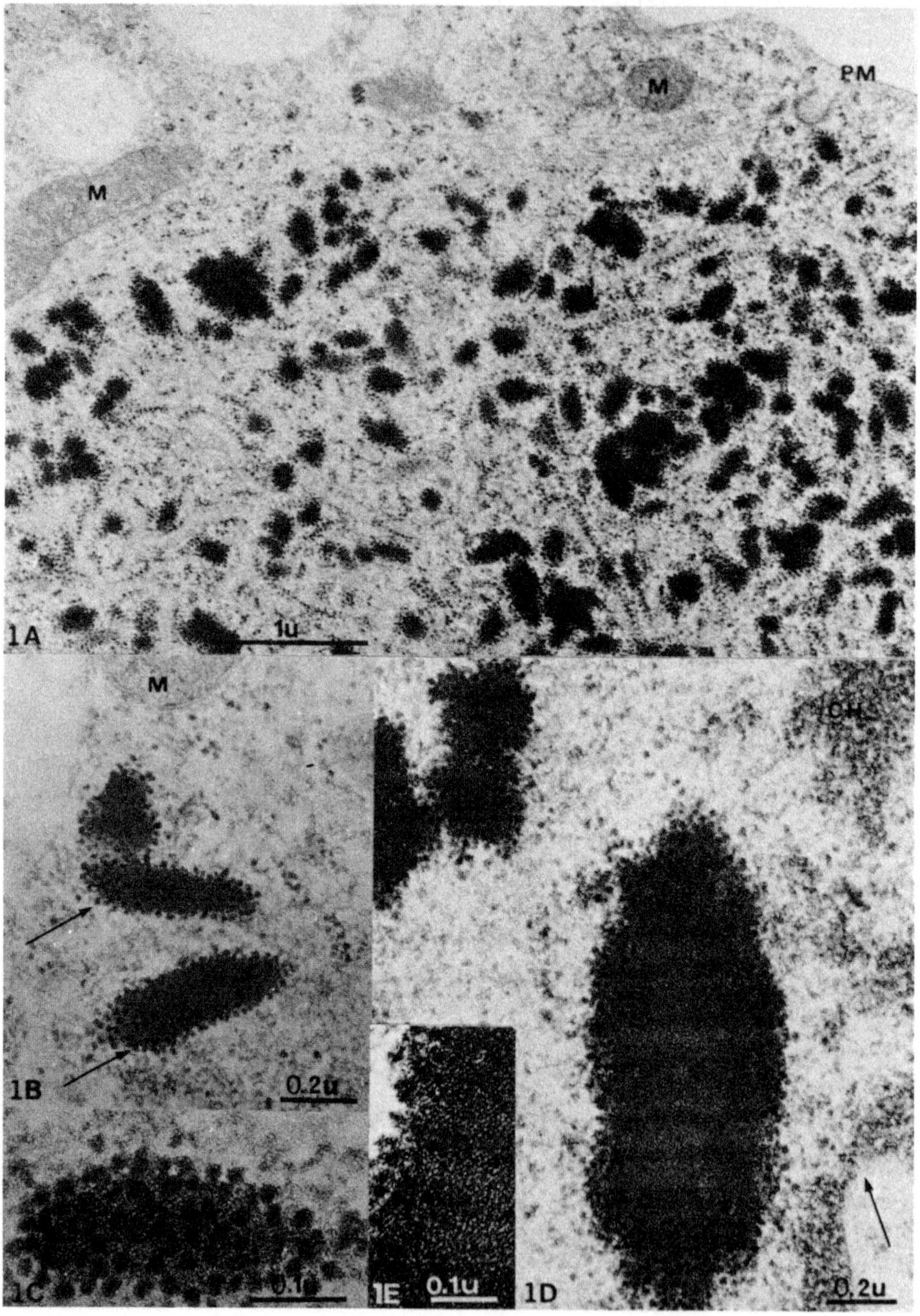
M
M
PM
1A
1u
M
CH
1B
0.2u
1C
0.1u
1E
0.1u
1D
0.2u

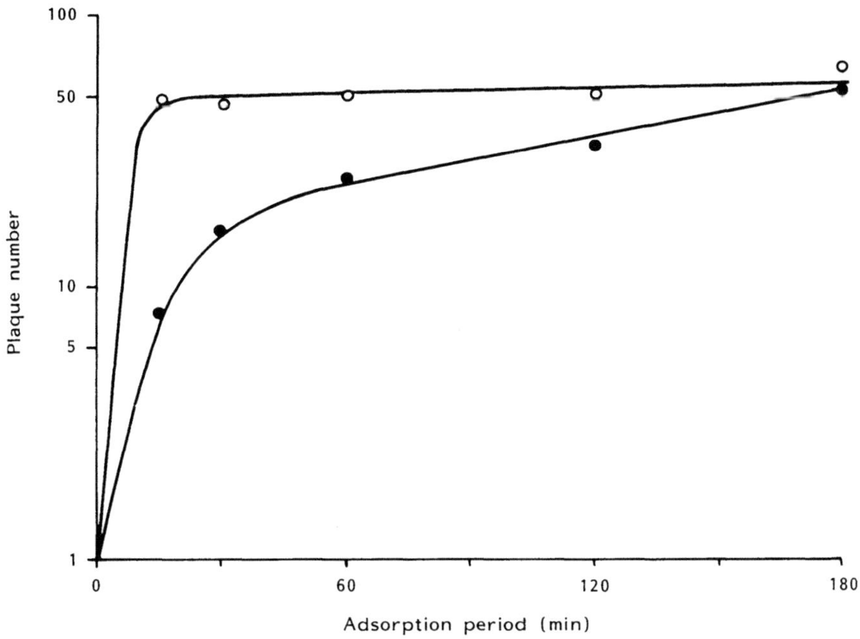

Figure 5-3. Viral adsorption not reversible with antiserum. After the indicated adsorption periods, plates were washed twice with PBS and overlaid with agar medium (open circles), or plates were treated with antiserum, washed three times with PBS, and then overlaid with agar medium (filled circles). Each circle represents the average of plaque counts on four replicate plates. (From Sugiura and Kilbourne, 1965. Reprinted with permission, from *Virology* **26**:484.)

sortant X-7 in clone 1-5C-4 cells (Sugiura and Kilbourne, 1965; Jahiel and Kilbourne, 1966), cell damage sufficient to detach cells is first evident in about 24 hr, and microscopically visible plaques appear at 48 hr (Jahiel and Kilbourne, 1966). With successive rounds of replication, plaque growth kinetics are linear with respect to plaque radius, as illustrated in Fig. 5-5.

In clone 1-5C-4 cells, NWS produces large clear plaques with sharply defined borders (Fig. 5-6a; Table 5-3); the same virus in another aneuploid cell line, BHK-

Figure 5-2. Selected areas of thinly sectioned infected cells sampled 19 hr post-infection, chosen to illustrate the appearance in intracellular crystals. (1A) Low power shows a large number of cytoplasmic crystals and associated monoribosomes. ×24,000. (1B) As 1A but at higher resolution. The arrows point toward attached ribosomes. ×59,000. (1C) As 1B, but the high-resolution image reveals the periodicity in the crystalline lattice. ×145,000. (1D) An intranuclear crystal. The arrow points toward the nuclear periphery. ×58,000. (1E) A segment of 1D illustrates the same periodicity as in 1C. However, the peripheral granules are smaller, implying that they are incomplete ribosomes. ×104,000. CH, chromatin; M, mitochondrion; N, nucleus; PM, plasma membrane. (From Morrongiello and Dales, 1977. Reproduced with permission of S. Karger AG, Basel, from *Intervirology* **8**:283.)

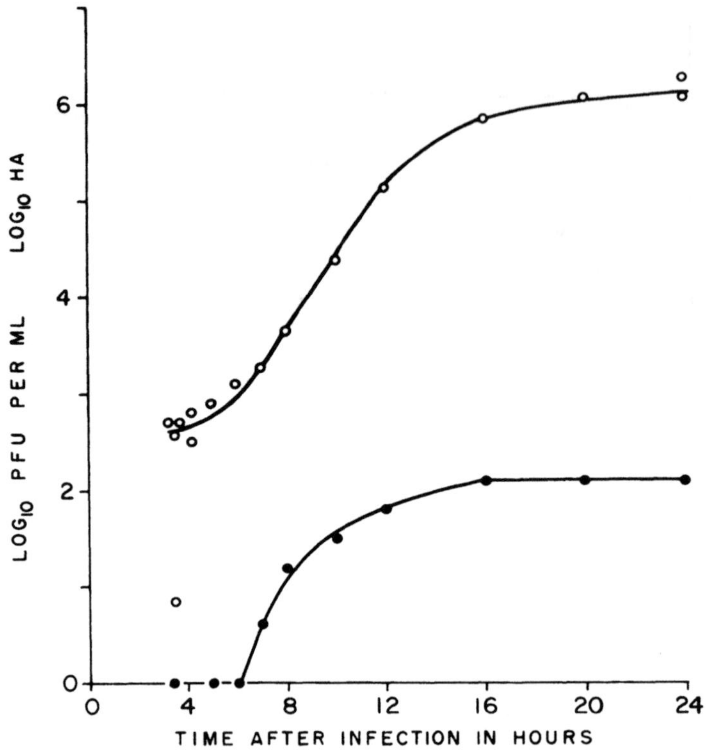

Figure 5-4. Single-step growth curve of influenza A virus in human conjunctival cells in monolayer culture after high-multiplicity infection. Infective virus titers are shown by filled circles, and HA titers by closed circles. (From Sugiura and Kilbourne, 1965. Reproduced with permission, from *Virology* **26:**484.)

21, produces medium-sized, turbid plaques. Whereas WSN produces clear plaques in both clone 1-5C-4 cells and chick embryo fibroblasts (CEF), its lineal antecedent, WS, produces clear plaques only in clone 1-5C-4 (Table 5-3). Thus, characterization of the cytopathogenic potential of any virus cannot be based on its effects on a single cell type.

Clear plaques are best demarcated in living cultures by the introduction of neutral red dye, which is taken up by the lysosomes of normal viable cells, thus outlining the necrotic, nonstaining clear plaque centers. At their earliest microscopic stage, NWS plaques contain nonstaining cells, indicating that cell necrosis occurs rapidly following replication of this virus strain. Early in infection with certain strains of virus, increased uptake of neutral red, initially restricted to cell lysosomes, reflects increased permeability of these organelles (Allison and Mallucci, 1965). Later, dying cells will show decreased neutral red uptake as plaques develop.

A cytopathic characteristic of some influenza virus strains is their production of red-bordered plaques in cells stained with neutral red (Fig. 5-6b). This

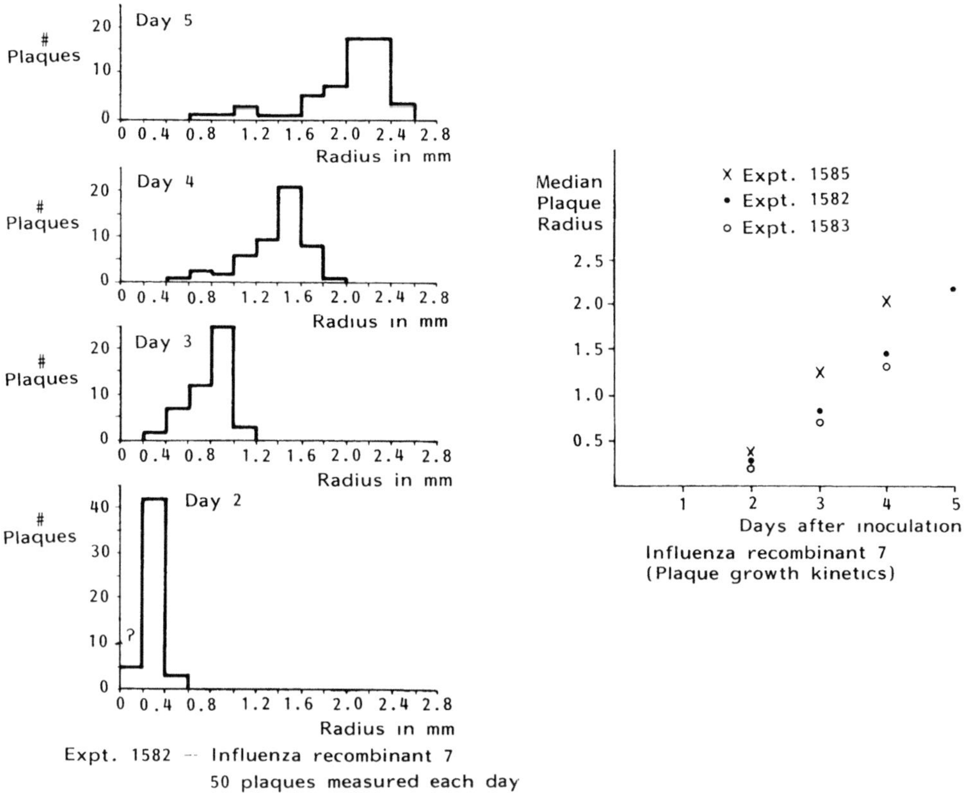

Figure 5-5. Influenza virus plaque growth kinetics are linear with respect to plaque radius. (R.I. Jahiel and E.D. Kilbourne, unpublished data.)

cytopathic property reflects an increased uptake of neutral red by infected but still viable cells at the plaque margins. With one such virus, A/CAM/56, this cytopathic property was transferable by genetic reassortment with A/NWS (Sugiura and Kilbourne, 1966). $NWSr^+$ virus is a reassortant of NWS antigenic phenotype that produces plaques with clear centers demarcated by damaged but viable cells at the plaque margins. Although the overall diameter of $NWSr^+$ plaques is less than that of NWS plaques at similar periods after infection, and plaque size generally is a function of viral replication rate, red-bordered plaques cannot be explained only on the basis of delayed plaque evolution. From the outset, differences in plaque phenotype exist and must reflect a more temperate relationship of virus and cell. Furthermore, medium-sized clear plaques are produced by other NWS-CAM reassortants in this system in which red borders cannot be discerned. Definitive understanding of this interesting cytopathic effect will require inquiry into the details of intracellular stages of viral replication with special attention to the possible role of interferon and lysosomal enzymes in what appears to be a unique model for the study of cytopathogenesis.

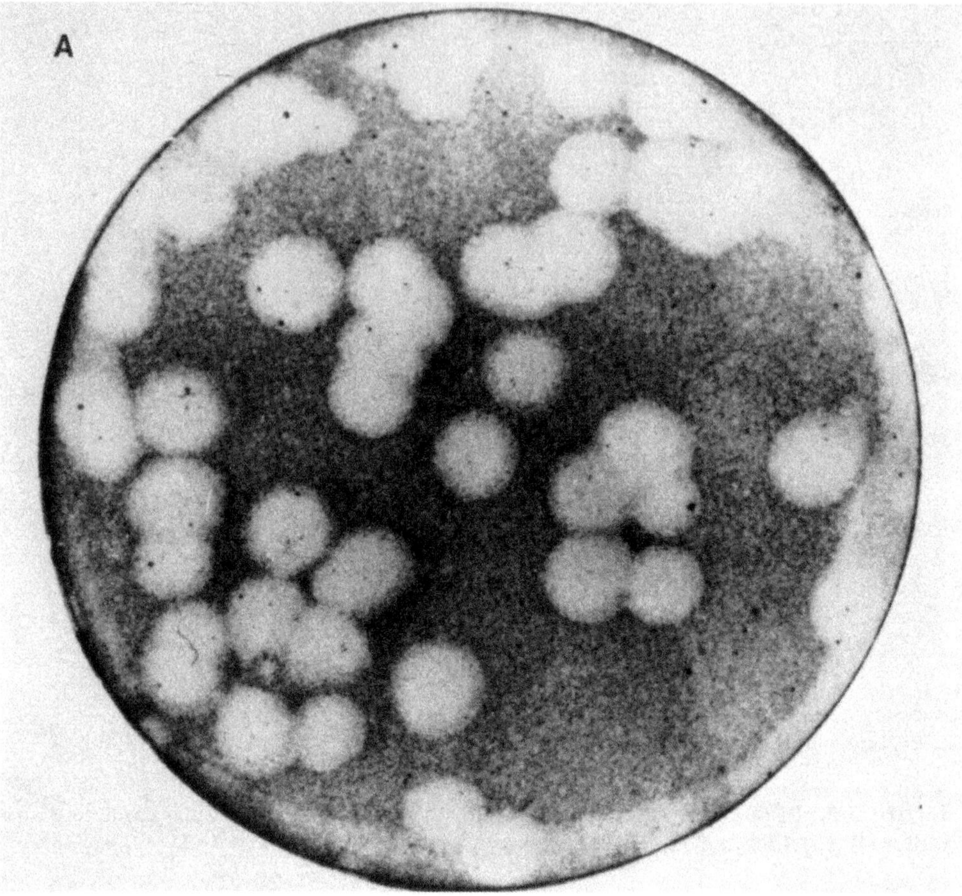

Figure 5-6. (A) Large clear plaque produced by A/NWS virus in clone 1-5C-4 cells (neutral red stain). (B) Red-bordered plaque produced in same cells by A/NWS r+, a reassortant of A/NWS with A/CAM, a virus bearing the "red-border" trait. (From Kilbourne, 1969. Reproduced with permission, from *Fundamental Techniques in Virology,* K. Habel and N. Salzman (eds.), pp. 154-155, Academic Press, New York.)

Table 5-3. Plaque Morphology is a Function of Virus Strain and Cell Type

Reference	Virus	Cell	Plaque morphology
Sugiura and Kilbourne (1965)	A/NWS	Clone 1-5c-4	Large, clear
Milliken (1967)	A/NWS	BHK-21	Medium-sized, turbid
Sugiura and Kilbourne (1965)	A/WS	Clone 1-5c-4	Large, clear
Simpson and Hirst (1961)	A/WS	CEF[a]	"Faint"
Sugiura and Kilbourne (1965)	A/WSN	Clone 1-5c-4	Large, clear
Simpson and Hirst (1961)	A/WSN	CEF	Large, clear

[a]Chick embryo fibroblast culture

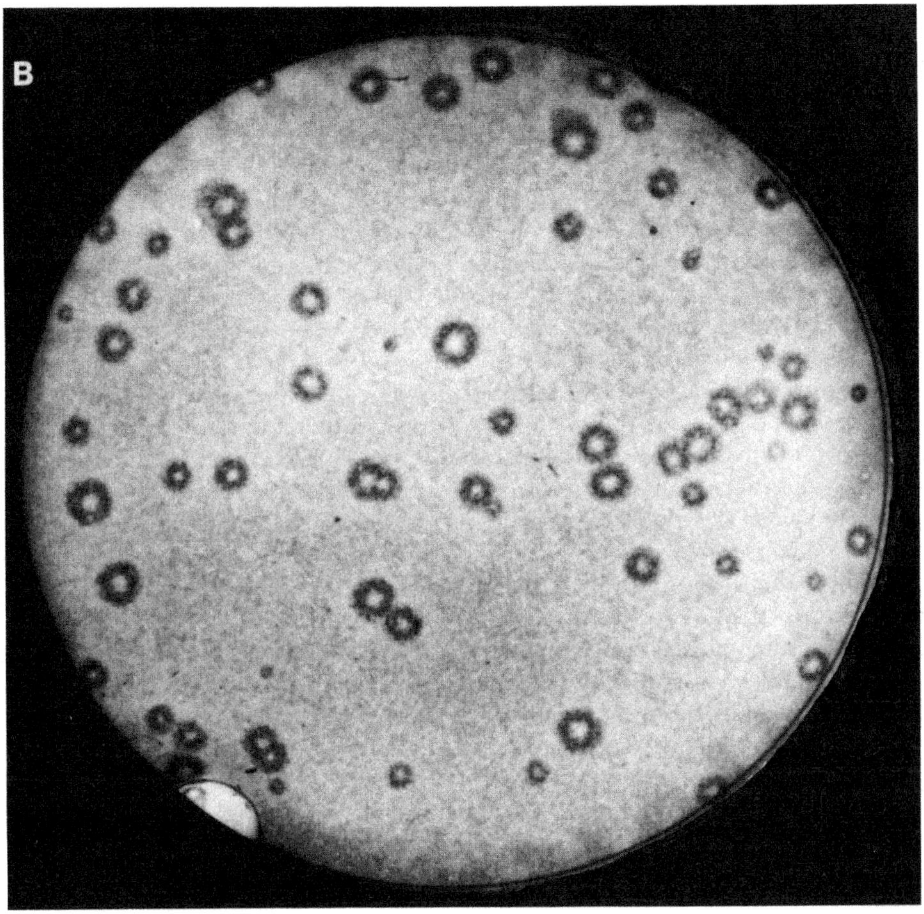

Figure 5-6. *(Continued)*

The Possible Role of Lysosomal Enzymes in Influenza Virus CPE

When the effects of virulent and noncytopathic influenza viruses in chick embryo cell culture were compared, Allison and Mallucci (1965) found histochemical changes within 4 to 5 hr after infection with A/FPV inoculated at high multiplicity. Lysosomes, ordinarily impermeable to an acid phosphatase substrate in unfixed cells, became permeable. Later, when necrosis and detachment of cells were demonstrable in 6 to 8 hr, lysosomal enzymes were diffusely distributed throughout the cell. At this stage (comparable to the state of cells in lytic plaque centers), reduced lysosomal uptake of neutral red was seen. It remains unclear, however, to what extent lysosomal enzyme release is a cause or consequence of influenza-virus-induced cell damage.

The Molecular Basis of CPE

A consequence of influenza virus infection is the shut-off of host-cell protein synthesis.This event offers a reasonable explanation for virus-induced host-cell damage and eventual necrosis, although in the case of poliovirus infection, the cellular degeneration seen occurs more rapidly than can be explained by inhibition of cellular RNA and protein (Bablanian et al., 1965).

Viruses can effect inhibition of host-cell protein synthesis through effects on nuclear RNA polymerase II transcription, preferential transcription of viral rather than host mRNAs (Ehrenfeld, 1982), or selective degradation of host-cell mRNAs (reviewed by Katze and Krug, 1984). Present evidence is conflicting on which mechanism operates in influenza virus infection. Inglis (1982) has reported degradation of host messenger RNA measured by blot hybridization. Katze and Krug (1984), however, measuring host mRNA translational activity in vitro found no inactivation of host mRNA function for 3 hr post-infection and concluded that selective translation of viral mRNA must occur. In the same study, it was shown that newly synthesized polymerase II transcripts were degraded in the nucleus, perhaps as a consequence of cleavage of their 5′ ends by the influenza virus cap-dependent endonuclease. Thus, 1 to 2 hr after infection, the appearance of newly processed host mRNAs in the cytoplasm was markedly depressed.

These studies suggest that profound effects on normal host-cell synthetic processes occur very soon after infection. The extent to which these effects are the basis of subsequent cell death remains to be determined. Possibly relevant to these recent findings are studies published almost two decades ago that demonstrated by cytochemical staining the disappearance of "ribonucleoprotein" from nuclei soon after infection of HeLa cells with A/PR8 virus (Ter Meulen and Love, 1967).

REFERENCES

Abramson, J. S., Mills, E. L., Giebink, G. S., et al., 1982a, Depression of monocyte and polymorphonuclear leukocyte oxidative metabolism and bactericidal capacity by influenza A virus, *Infect. Immun.* **35:**350–355.

Abramson, J. S., Lyles, D. S., Heller, K. A., et al., 1982b, Influenza A virus-induced polymorphonuclear leukocyte dysfunction, *Infect. Immun.* **37:**794–799.

Abramson, J. S., Lewis, J. C., Lyles, D. S., et al., 1982c, Inhibition of neutrophil lysome–phagosome fusion associated with influenza virus infection in vitro. Role in depressed bactericidal activity, *J. Clin. Invest.* **69:**1393–1397.

Abramson, J. S., Parce, J. W., Lewis, J. C., et al., 1984, Characterization of the effect of influenza virus on polymorphonuclear leukocyte membrane responses, *Blood* **64:**131–138.

Abramson, J. S., Wiegand, G. L., and Lyles, D. S., et al., 1985, Neuraminidase activity is not the cause of influenza virus-induced neutrophil dysfunction, *J. Clin. Microbiol.* **22:**129–131.

Allison, A. C., and Mallucci, L., 1965, Histochemical studies of lysosomes and lysosomal enzymes in virus-infected cell cultures, *J. Exp. Med.* **121:**463–476.

Anisimová, E., Tucková, E., and Vonka, V., 1973, Morphological changes in BHK-21 cells infected with S/N (H_2N_1) influenza virus, *Arch. Ges. Virusforsch.* **43:**221–234.

Archetti, I., Bereczky, E., Rosati-Valente, F., et al., 1970, Elongated structures present in cells infected with influenza viruses, *Arch. Ges. Virusforsch.* **29:**275–286.

Armstrong, C L., Miranda, A F , Hsu, K. C , et al., 1978, Susceptibility of human skeletal muscle culture to influenza virus infection I Cytopathology and immunofluorescence, *J Neurol Sci.* **35**:43-57.

Babiker, H. A., and Rott, R., 1968, Plaque formation by influenza viruses in monolayers of chicken kidney cells, *J. Gen. Virol.* **3**:285-287.

Bablanian, R., Eggers, H. J., and Tamm, I., 1965, Studies on the mechanism of poliovirus-induced cell damage I The relation between poliovirus-induced metabolic and morphological alterations in cultured cells, *Virology* **26**:100-113

Bennett, I L , Wagner, R R , and LeQuire, V S , 1949, Pyrogenicity of influenza virus in rabbits, *Proc Soc Exp. Biol Med* **71**:132-133

Brownson, J M , Mahy, B W , and Hazleman, B L , 1979, Interaction of influenza A virus with human peripheral blood lymphocytes, *Infect Immun* **25**:749-756

Busse, W. W , Cooper, W , Warshauer, D. M , et al , 1979, Impairment of isoproterenol, H2 histamine, and prostaglandin E response of human granulocytes after incubation in vitro with live influenza vaccines, *Am Rev Respir Dis* **119**:561-569

Casali, P , Rice, G P A , and Oldstone, M B A., 1984, Viruses disrupt functions of human lymphocytes Effects of measles virus and influenza virus on lymphocyte-mediated killing and antibody production, *J. Exp Med* **159**:1322-1337

Chang, H -J , Wang, C.-L , and In, C -J , 1966, Studies on plaque formation by WS strain in mixed monolayer culture of chick embryo kidney and skin-muscle cells, *Acta Microbiol. Sin.* **12**:158-162

Choppin, P W , 1962, Plaque formation by influenza A2 virus in monkey kidney cells, *Virology* **18**:332-334.

Choppin, P. W., 1963, Multiplication of two kinds of influenza A2 virus particles in monkey kidney cells, *Virology* **21**:342-352

Ciampor, F , and Turcan, P., 1972, Electron microscopy of tissue culture cells infected with myxoviruses. II Nucleo-cytoplasmic changes in fowl plaque virus-infected cells, *Acta Virol* **16**:177-182

Compans, R W., and Dimmock, N. J., 1969, An electron microscopic study of single-cycle infection of chick embryo fibroblasts by influenza virus, *Virology* **39**:499-515.

Ehrenfeld, E., 1982, Poliovirus-induced inhibition of host-cell protein synthesis, *Cell* **28**:435-436

Fisher, T. N., and Ginsberg, H. S , 1956a, The reaction of influenza viruses with guinea pig polymorphonuclear leucocytes. II. The reduction of white blood cell glycolysis of influenza viruses and receptor-destroying enzyme (RDE), *Virology* **2**:637-655

Fisher, T N., and Ginsberg, H. S., 1956b, The reaction of influenza viruses with guinea pig polymorphonuclear leucocytes. III. Studies on the mechanism by which influenza viruses inhibit phagocytosis, *Virology* **2**:656-664.

Flehmig, B., Vallbracht, A., and Gerth, H.-J., 1976, Influenza virus Association of mouse-lung virulence with plaque formation in mouse kidney cells, *Intervirology* **7**:201-210.

Frielle, D. W., Huang, D D., and Youngner, J. S., 1984, Persistent infection with influenza A virus Evolution of virus mutants, *Virology* **138**:103-117.

Gamboa, E. T., Harter, D. H., Wolf, A., et al., 1974, Replication of neurotropic influenza virus in organotypic cultures of embryonic mouse hypothalamus, *J Neuropathol. Exp. Neurol.* **33**:571-581

Gardner, I. D., and Lawton, J. W., 1982, Depressed human monocyte function after influenza infection in vitro, *J. Reticuloendothel. Soc.* **32**:443-448.

Gaush, C. R , and Smith, T. F., 1968, Replication and plaque assay of influenza virus in an established line of canine kidney cells, *Appl Microbiol.* **16**:588-594

Gerber, A., Sauter, C., and Lindenmann, J., 1973, Fowl plague virus adapted to human epithelial tumor cells and human myeloblasts in vitro. I. Characteristics and replication in monolayer cultures, *Arch. Ges. Virusforsch.* **40**:137-151.

Ginsberg, H. S., 1954, Formation of non-infectious influenza virus in mouse lungs· Its dependence upon extensive pulmonary consolidation initiated by the viral inoculum, *J. Exp. Med* **100**:581-603

Ginsberg, H S., and Blackmon, J. R , 1956, Reactions of influenza viruses with guinea pig polymorphonuclear leucocytes. I Virus-cell interactions, *Virology* **2**:618-636

Granoff, A., 1955, Plaque formation with influenza strains, *Virology* **1**:252.

Grossberg, S. W., 1964, Human influenza A viruses: Rapid plaque assay in hamster kidney cells, *Science* **144**:1246-1247.

Hatano, M., and Morita, O., 1967, Multiplication and plaque assay of influenza viruses in a continuous cell line (G2) of human origin, *Arch. Ges. Virusforsch.* **20**:305–313.

Henle, G., and Henle, W., 1946a, Studies on the toxicity of influenza viruses. I. The effect of intracerebral injection of influenza viruses, *J. Exp. Med.* **84**:623–637.

Henle, W., and Henle, G., 1946b, Studies on the toxicity of influenza viruses. II. The effect of intraabdominal and intravenous injection of influenza viruses, *J. Exp. Med.* **84**:639–660.

Henle, G., Girardi, A., and Henle, W., 1955, A non-transmissible cytopathogenic effect of influenza virus in tissue culture accompanied by formation of non-infectious hemagglutinins, *J. Exp. Med.* **101**:25–41.

Henricks, P. A. J., van Erne-van der Tol, M. E., and Verhoef, J., 1982, Partial removal of sialic acid enhances phagocytosis and the generation of superoxide and chemiluminescence by polymorphonuclear leukocytes, *J. Immunol.* **129**:745–750.

Henry, C., and Youngner, J. S., 1957, Influence of normal animal sera on influenza viruses in cultures of trypsin-dispersed monkey kidney cells, *J. Immunol* **78**:273–281.

Herrero-Uribe, L., Mann, G. F., Zuckerman, A. J., et al., 1983, Replication of influenza A and B viruses in human diploid cells, *J. Gen. Virol.* **64**:471–475.

Hirst, G., and Pons, M., 1973, Mechanism of influenza recombination. II. Virus aggregation and its effect on plaque formation by so-called noninfective virus, *Virology* **56**:620–631.

Inglis, S. C., 1982, Inhibition of host protein synthesis and degradation of cellular mRNAs during infection by influenza and herpes simplex virus, *Mol. Cell. Biol.* **2**:1644–1648.

Jahiel, R. I., and Kilbourne, E. D., 1966, Reduction in plaque size and reduction in plaque number as differing indices of influenza virus-antibody reactions, *J. Bacteriol.* **92**:1521–1534.

Kanoh, S., and Kawasaki, H., 1966, Studies on myxovirus pyrogen (I) interaction of myxovirus and rabbit polymorphonuclear leucocytes, *Biken J.* **9**:177–184.

Katze, M. G., and Krug, R. M., 1984, Metabolism and expression of RNA polymerase II transcripts in influenza virus-infected cells, *Mol. Cell. Biol.* **4**:2198–2206.

Kilbourne, E. D., 1969, Plaque formation by influenza viruses, in: *Fundamental Techniques in Virology* (K. Habel and N. Salzman, eds.), Academic Press, New York, pp. 146–160.

Kilbourne, E. D., Sugiura, A., and Wong, S. C., 1964, Serial multiplication of an influenza A virus (NWS) in certain human diploid cell strains, *Proc. Soc. Exp. Biol. Med.* **116**:225–228.

Kohn, D. F., Chinookoswong, N., and Magill, L. S., 1981, Pathogenicity of influenza A virus in ependymal organ culture, *Teratology* **24**:201–213.

Kopp, J. V., Kempf, J., and Kroeger, A. V., 1968, Cytoplasmic inclusions observed by electron microscopy late in influenza virus infection of chicken embryo fibroblasts, *Virology* **36**:681–683.

Kosel, B., and Kohlhage, H., 1970, Effects of actinomycin on *in vivo* and *in vitro* formation of endogenous rabbit pyrogen induced with myxoviruses or stimulated with endotoxin, *Arch. Ges. Virusforsch.* **32**:209–220.

Kutinová, L., Vonka, V., and Rezácová, D., 1967, Plaque formation by NWS influenza virus in a human diploid cell strain, *Acta Virol.* **11**:372–378.

Larson, H. E., and Blades, R., 1976, Impairment of human polymorphonuclear leucocyte function by influenza virus, *Lancet* **1**:283.

Larson, H. E., Parry, R. P., Gilchrist, C., et al., 1977, Influenza virus and staphylococci *in vitro*: Some interactions with polymorphonuclear leucocytes and epithelial cells, *Br. J. Exp. Pathol.* **58**:281–288.

Ledinko, N., 1955, Production of plaques with influenza viruses, *Nature* **175**:999.

Lehmann-Grube, F., 1963, A sensitive plaque assay for influenza viruses, *Virology* **21**:520–522.

Maassab, H. F., 1968, Plaque formation of influenza virus at 25°C, *Nature* **219**:645–646.

Masihi, K. N., Lange, W., and Müller, S., 1984, Depression of chemiluminescence response in mouse spleen cells by infective and inactivated influenza virus, *Clin. Immunol. Immunopathol.* **33**:23–30.

Merchant, D. J., and Morgan, H. R., 1950, Inhibition of the phagocytic action of leucocytes by mumps and influenza viruses, *Proc. Soc. Exp. Biol. Med.* **74**:651–653.

Milliken, S., 1967, A delay in maturation as a cause of small plaque size with the NM strain of influenza A virus, *J. Gen. Virol.* **1**:189–198.

Mills, E. L., Debets-Ossenkopp, Y., Verbrugh, H.A., et al., 1981, Initiation of the respiratory burst of human neutrophils by influenza virus, *Infect. Immun.* **32**:1200–1205.

Miranda, A. F., Gamboa, E. T., Armstrong, et al., 1978, Susceptibility of human skeletal muscle culture to influenza virus infection, *J. Neurol. Sci.* **36**:63–81.

Mizutani, H , and Nagase, M , 1967, Plaque formation with influenza viruses in explants of chick fibroblasts from embryos of different ages, *Arch. Ges. Virusforsch.* **20**:278–282.

Morrongiello, M. P , and Dales, S , 1977, Characterization of cytoplasmic inclusions formed during influenza/WSN virus infection of chick embryo fibrolast cells, *Intervirology* **8**:281–293.

Nakajima, S , and Sugiura, A., 1980, Neurovirulence of influenza virus in mice. II. Mechanism of virulence as studied in a neuroblastoma cell line, *Virology* **101**:450–457.

Niven, J S. F., Armstrong, J. A., Balfour, B M., et al., 1962, Cellular changes accompanying the growth of influenza virus in bovine kidney cell cultures, *J Pathol. Bacteriol.* **84**:1–18.

Oh, J. O , 1965, The effect of bacterial endotoxins on toxic corneal lesions produced by influenza and Newcastle disease viruses, *Am J. Pathol.* **46**:177–195.

Ørstavik, I., 1981, Susceptibility of continuous lines of monkey kidney cells to influenza and parainfluenza viruses in the presence of trypsin, *Acta Pathol Microbiol Scand [B]* **89**:179–183

Palese, P., Bucher, D., and Kilbourne, E D., 1973, Applications of a synthetic neuraminidase substrate, *Appl Microbiol* **25**:195–201

Peterhans, E , 1980, Chemiluminescence An early event in the interaction of Sendai and influenza viruses with mouse spleen cells. I. The role of the envelope glycoproteins in the stimulation of chemiluminescence, *Virology* **105**:445–455

Peters, A , and Goedemans, W T , 1966, A Sensitive plaque assay for influenza virus A Japan in chick embryo fibroblast cultures, *Arch. Ges Virusforsch* **19**:190–196.

Pons, M W , Rochovansky, O M , and Brody, J. M , 1981, Primary ferret kidney cell cultures as a substrate for the growth of clinical isolates of influenza virus, *Clin Res.* **29**:769A.

Porebska, A , Pereira, H G., and Armstrong, J. A , 1968, Cytoplasmic inclusions and nuclear changes in cells infected with influenza B virus, *J. Med Microbiol* **1**:145–151.

Roberts, N J., Jr., 1982, Different effects of influenza virus, respiratory syncytial virus, and Sendai virus on human lymphocytes and macrophages, *Infect. Immun* **35**:1142–1146.

Roberts, N J , Jr., Prill, A H., and Mann, T. N., 1986, Interleukin 1 and interleukin 1 inhibitor production by human macrophages exposed to influenza virus or respiratory syncytial virus, *J. Exp. Med* **163**:511–519.

Rodgers, B C., and Mims, C. A., 1982, Influenza virus replication in human alveolar macrophages, *J Med Virol.* **9**:177–184.

Ruutu, P , 1977, Depression of rat neutrophil exudation and motility by influenza virus, *Scand J Immunol.* **6**:1113–1120.

Saito, Y., Yoshioka, I., Igarashi, Y., et al., 1970, Nuclear inclusions observed by electron microscope in cynomolgus monkey kidney cells infected with influenza virus, *Virology* **40**:408–409.

Schlesinger, R W , 1950, Incomplete growth cycle of influenza virus in mouse brain, *Proc. Soc. Exp. Biol. Med* **74**:541–548.

Shaw, M. W., and Compans, R. W., 1978, Isolation and characterization of cytoplasmic inclusions from influenza A virus-infected cells, *J. Virol.* **25**:608–615.

Shibata, M., Maeno, K., Isomura, S., et al., 1982, Plaque formation by influenza B virus in a porcine kidney cell line, *Microbiol. Immunol.* **26**:441–444.

Siegert, R , Philipp-Dormston, W. K., Radsak, K., et al., 1976, Mechanism of fever induction in rabbits, *Infect. Immun.* **14**:1130–1137.

Simpson, R. W., and Hirst, G. K., 1961, Genetic recombination among influenza viruses. I. Cross reactivation of plaque-forming capacity as a method for selecting recombinants from the progeny of crosses between influenza A strains, *Virology* **15**:436–451.

Sugg, J. Y., 1949, An influenza virus pneumonia of mice that is nontransferable by serial passage, *J Bacteriol.* **57**:399–403.

Sugiura, A., and Kilbourne, E.D., 1965, Genetic studies of influenza viruses. II. Plaque formation in influenza viruses in a clone of a variant human heteroploid cell line, *Virology* **26**:478–488

Sugiura, A., and Kilbourne, E D., 1966, Genetic studies of influenza viruses. III. Production of plaque type recombinants with A_0 and A_1 strains, *Virology* **29**:84–91.

Táborský, I., 1972, Effect of cycloheximide on in vitro formation of rabbit granulocyte pyrogen induced with influenza A/PR8 virus or endotoxin, *Acta Virol* **16**:376–381.

Tanaka, T., Urabe, M , Goto, H., et al., 1984, Isolation and preliminary characterization of a highly cytolytic influenza B virus variant with an aberrant NS gene, *Virology* **135**:515–523.

Ter Meulen, V., and Love, R., 1967, Virological, immunochemical, and cytochemical studies of four

HeLa cell lines infected with two strains of influenza virus, *J. Virol.* **1**:626–639.

Thadani, M. A., and Polasa, H., 1979, Cytogenetic effects of inactivated influenza virus on male germ cells of mice, *Hum. Genet.* **51**:253–258.

Tucková, E., and Vonka, V., 1970, Interactions between an influenza A-2 virus and rhesus monkey kidney cells, *Arch. Ges. Virusforsch.* **30**:181–186.

Vonka, V., 1965, Plaque technique with an influenza A-2 virus in rubber stoppered bottle cultures from monkey kidney cells, *Arch. Ges. Virusforsch.* **15**:514–524.

Wagner, R. R., Bennett, I. L., Jr., and LeQuire, V. S., 1949, The production of fever by influenzal viruses. I. Factors influencing the febrile response to single injections of virus, *J. Exp. Med.* **90**:321–334.

Walder, R., and Hsiung, G. D., 1967, Biological and antigenic characteristics of an A2 influenza virus isolated in 1965, *Public Health Rep.* **82**:689–696.

Wilson, A. B., Planterose, D. N., Nagington, J., et al., 1976, Influenza A antigens on human lymphocyte in vitro and probably in vivo, *Nature* **259**:582.

Wright, B. S., and Sagik, B. P., 1958, Plaque formation in monolayers of chicken embryo kidney cells, *Virology* **5**:573–574.

Yamagishi, H., Ide, S., Eiki, T., et al., 1984, Plaque assay of equine influenza virus, *Vet. Microbiol.* **9**:187–192.

Yoshida, T., Shaw, M. W., Young, J. F., et al, 1981, Characterization of the RNA associated with influenza A cytoplasmic inclusions and the interaction of NS protein with RNA, *Virology* **110**:87–97.

Zimmermann, T., and Schäfer, W., 1959, Erzeugung von "Plaques" durch Schweine-Influenza-Virus und andere Viren der Myxovirusgruppe in Gewebekulturen von Kälbernieren-Zellen, *Z. Naturforsch.* **14b**:213–214.

Zitcer, E. M., Bruening, G., and Agrawal, H. O., 1967, The multiplication of an influenza virus strain in a continuous line of mammalian cells, *Arch. Ges. Virusforsch.* **20**:137–141.

6

Influenza Virus Genetics, Viral Adaptation, and Evolution

INTRODUCTION

Influenza virus genetics comprises the genetics of a highly mutable RNA virus and the unusual genetics of a segmented-genome virus capable of genetic interaction with homologous strains. Genetic reassortment occurs in nature and has been used in the laboratory in the fabrication of vaccine and reagent viruses with desirable characteristics.

The primary structure of all influenza virus genes is now known, and most gene products have been identified. All genes of influenza A virus have been cloned in prokaryotic or eukaryotic vectors, and some of these genes have been expressed. This recent explosion of knowledge addresses appropriately the basic problem of viral variation that underlies the riddle of influenza and its epidemiology.

THE NATURE OF THE VIRAL GENOME

The influenza virus genes and their gene products are described in Chapters 3 and 4. Current information about them is summarized in Table 6-1, which reviews for influenza A virus the relative size of the vRNAs, their size variability, if any, and the coding potential of their genes. Gene size is variable for HA and NA and for NS genes; the others seem invariant. This fact may reflect either the tight structural constraints on the genes of the transcription complex or merely the more limited sampling of their RNA sequences. It is worth noting that the RNAs of relatively few viral subtypes have yet been completely sequenced.

Table 6-1. The Influenza A Virus Genome: A Consensus of Nucleotide Sequences of cDNA Transcripts from vRNA

vRNA segments	Gene product(s)	Nucleotides[a]	Comparative RNA lengths	Variation in vRNA size	Published data[b] based on nucleotide sequencing: Number of cDNAs sequenced	Number of virus strains	Viral subtypes
1	PB2	2,341	ααααααααααααααααααααααα	None	3	3	H1, H7
2	PB1	2,341	ααααααααααααααααααααααα	None	3	3	H1, H3
3	PA	2,233	αααααααααααααααααααααα	None	3	3	H1, H3, H7
4	HA	1,778	αααααααααααααααααα	1,742–1,778	33	25	H1, H2, H3, H7
5	NP	1,565	αααααααααααααααα	None	4	4	H1, H3, H7
6	NA	1,413	αααααααααααααα	1,409–1,467	15	15[d]	(N1, N2)
7	M_1, M_2, ?M_3	1,027	αααααααααα	None	5	5	H1, H3, H7
8	NS_1, NS_2	890	ααααααααα	—[c]	9	9	H1, H3, H7, H12
Total		13,588		13,548–13,642			

[a]In prototype A/PR8 strain.
[b]Published data of: Winter and Fields, 1980, 1982, Bishop et al., 1982a, b, Kaptein and Nayak, 1982, Sivasubramanian and Nayak, 1982; Roditi and Robertson, 1984; Robertson et al., 1984; Winter et al., 1981a, b; Hiti et al., 1981, Gething et al., 1980; Fang et al., 1981, Verhoeyen et al , 1980; Both and Sleigh, 1980, 1981; Sleigh et al., 1980, Min Jou et al., 1980, Porter et al , 1979; Both et al , 1983b, Newton et al , 1983, Tomley and Roditi, 1984; Lentz et al , 1984, Fields et al , 1981, Hiti and Nayak, 1982, Markoff and Lai, 1982, Bentley and Brownlee, 1982, Van Rompuy et al., 1981, 1982, Elleman et al., 1982; Steuler et al , 1984, Concannon et al , 1984, Allen et al., 1980; Lamb and Lai, 1980, 1981, 1982; McCauley et al , 1982, Ortin et al., 1983; Porter et al., 1980; Baez et al , 1980, 1981; Buonagurio et al , 1984; Krystal et al , 1983
[c]Size variation occurs on basis of partial and complete sequence data (Parvin et al , 1983, Krystal et al , 1983)
[d]Nine field strains and six variants of one strain

The relationship of genes to each other in the virion remains to be determined. It is usually assumed that eight RNA pieces are packaged per virion in equimolar amounts and without heterozygosity. This may be true as a statistical mean, but one knows that viral populations are heterogeneous and contain defective particles with RNAs of subgenomic size (Fig. 4-11), some of which can be expressed (see Chapter 4).

GENETIC SYSTEMS

In selecting viruses for genetic studies in the laboratory, it is probably unnecessary that they retain all or any of the characteristics that distinguish them as disease-producing agents in nature. Formal genetic analysis of viral strains demands a reductionist approach in the simplest of systems for the establishment of principles that will have general application in the understanding of virus infection, with appropriate extrapolation to virus disease.

Nevertheless, it is well to be aware of the nature and derivation of the viruses under study and to bear in mind that each is a highly selected, amplified sample of virus as it comes from the natural host (see Fig. 6-1).

As is stressed in Chapter 2, the precise identification of virus strains is critical because of the variability of laboratory stocks. The reproduction of experi-

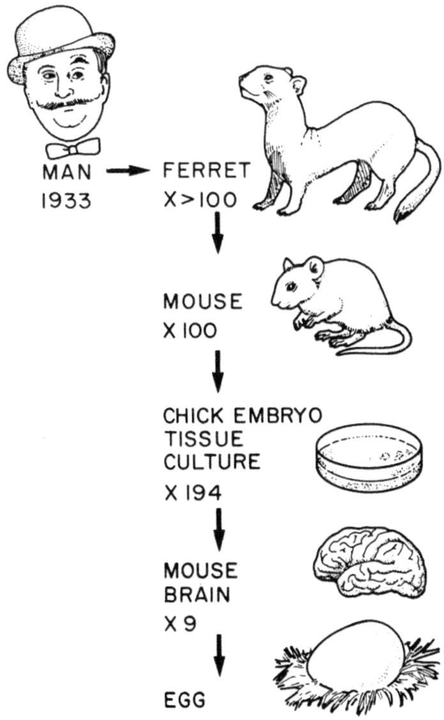

Figure 6-1. Summary of the genealogy of the standard WSN laboratory strain since its isolation from man in 1933. [From Kilbourne, 1984. Reproduced with permission, from *The Molecular Virology and Epidemiology of Influenza,* C. Stuart-Harris and C. W. Potter (eds.), Academic Press, London.]

ments in different laboratories may require the exchange of viruses and/or cells. Variation may reflect only passage levels or conditions of passage. Genetic variation was demonstrable in sequentially passaged influenza virus (WSN) stocks in all of three culture systems simultaneously studied in a single laboratory (Brand and Palese, 1980). Study of contemporary samples of the old laboratory prototype strains A/WSN/33 and A/PR/8/34 isolated only 1 year apart reveals significant variation in nucleotide sequence of their RNA 1 segments, suggesting that "laboratory drift" has occurred on subsequent passage of these viruses (Jones et al., 1983). Long ago, Burnet and Bull (1943) reported emergence on chick embryo passage of "D" or derived virus adapted to replication in the allantoic sac from original "O" virus capable only of amniotic replication. Other biological changes in D virus appeared pleiotropically, including increased capacity to agglutinate chicken RBC and capacity to cause mouse lung lesions. Recently, it has been demonstrated that selection of antigenic variants may be dependent on the host cell in which newly isolated virus is first cultivated (Schild et al., 1983). A 1980 H3N2 virus isolated in MDCK cells was temperature sensitive in chick embryos but not in MDCK cells (E. D. Kilbourne, unpublished data).

Viral Heterogeneity

Influenza viruses are most commonly propagated and preserved as yields from entodermal cells of the chick embryo allantoic sac (allantoic fluid virus). Such seed virus stocks comprise mixtures of infective, thermally inactivated, and DI particles, all of which are measurable as hemagglutinating virus. Therefore, inoculation of virus at high virus/cell multiplicity may lead simultaneously to autointerference, complementation, and multiplicity reactivation (see Chapter 4). Multiplicity reactivation is a genetic event that may be related to the presence of individually noninfective but genetically functional viral particles in the inoculum. Depending on the multiplicity of infection, viral yields may comprise 34–90% noninfective virus particles (Horsfall, 1954; Isaacs and Donald, 1955). If selection pressures are applied (e.g., passage with antibody or inhibitor), uncloned seed virus may be revealed as polymorphic, comprising two or more phenotypes (Mowshowitz and Kilbourne, 1975; Kilbourne, 1978).

Host Cell Heterogeneity

If the species of host cell influences the replication and selection of viral genetic variants (as it can; see above), then homogeneity of cell type would seem to be desirable in genetic systems. Therefore, primary or secondary cell cultures hypothetically are less desirable for the conduct of genetic studies. Even continuous cell lines must be cloned from time to time to assure the maintenance of a population of cells of maximum viral susceptibility (Sugiura and Kilbourne, 1965).

There is no published evidence on the influence of intraspecific host cell variants on the selection of influenza virus mutants.

VIRAL MUTATION AND MUTANTS

Mutation Frequency

Although the basic problem of influenza is the frequency with which epidemiologically significant viral mutants emerge in nature, there is only limited evidence that influenza viruses are more highly mutable than other viruses with RNA genomes. RNA viruses, in general, demonstrate high mutation frequencies, in part because they lack proofreading exonucleases to remove misincorporated bases from newly synthesized RNA strands (reviewed by Holland et al., 1982). Estimates of mutation frequency, whether measured by the appearance of ts mutants or antigenic mutants selected by monoclonal antibody, average about 10^{-5} per replication (see following section). Holland et al. (1982) have calculated overall genome mutation frequencies of 10% or more based on reversion rates of 10^{-4} to 10^{-5} for a 10-kilobase virus and assuming an equal probability of base substitution at all positions. Such mutation frequencies exceed those of DNA genomes by 100,000- to 1,000,000-fold. Parvin et al. (1986) have studied the comparative mutation rates of the influenza A virus NS gene and the VP1 gene of polio virus and have calculated mutation rates of 1.5×10^5 and 2.1×10^6 mutations/nucleotide per infection cycle for influenza virus and polio virus, respectively. These findings, comparing the genes of only two viruses, suggest that a higher mutation rate of influenza virus may promote its rapid evolution in nature.

Frequency of Influenza Virus Antigenic Variation

Just as mutation frequency of the influenza virus genome may not exceed that of other RNA viruses, the rate of antigenic variation is not uniquely high. The frequency of antigenic variants in cloned populations of Sendai, VSV, and influenza A viruses detected with monoclonal antibodies was similar ($10^{-4.5}$ to $10^{-4\ 7}$) for these three dissimilar RNA viruses (Portner et al., 1980). However, marked differences in the frequencies of influenza virus antigenic variants selected with different monoclonal antibodies are occasionally noted (Lubeck et al., 1980; Nakajima and Kendal, 1981; Webster et al., 1981). Other studies have encountered less extreme variation. Antigenic variants of the HA, selected with anti-HA monoclonal antibodies, were found to be present at a frequency of about 10^{-5} in stock virus of either H1 (Yewdell et al., 1979) or H3 (Webster et al., 1980) subtype. A similar frequency was found with NA variants selected from a cloned H2N2 virus (Webster et al., 1982), although in nature antigenic changes in the NA may occur less frequently than in HA (Webster et al., 1979).

Gerhard et al. (1983) have used 11 different monoclonal antibodies for determination of the frequency of single-epitope mutants in four independently cloned PR8 (H1) virus seeds. Single-epitope mutants occurred in all of five different antigenic sites at comparable frequency, averaging $10^{-5\ 67}$. This rate was roughly comparable to that noted in other studies (see above). In relating the mutation frequency of observed phenotypes to a deduced genomic (HA gene) mutation rate, an estimated rate of $10^{-7.7}$ was calculated (Gerhard et al., 1983), in conflict with the very high (approximately 10^{-1}) rate proposed by Holland et al. (1982) for RNA viruses in general. As acknowledged by both groups of authors, such estimates of mutation frequency are imprecise and may be influenced by such factors as (1) phenotypic mixing of mutant and parental virus during selection by antibody (Sugiura, 1975; Gerhard et al., 1983), (2) wide range of individual mutation frequencies, or (3) concomitant alteration of antigenic site and structure affecting viral growth rate or stability (Erickson and Kilbourne, 1980; Kilbourne, 1978; Gerhard, 1983; and see biologically selected antigenic variants below).

In any event, the antigenic mutability of influenza viruses certainly does not exceed that of such viruses as parainfluenza (Portner et al., 1980; Yewdell and Gerhard, 1982), rabies virus (Wiktor and Koprowski, 1980), VSV (Portner et al., 1980), and measles virus (Birrer et al., 1981).

Molecular Basis of Influenza Virus Mutations

The base sequences of all eight influenza virus genomic RNAs are now known from direct RNA sequencing or have been inferred from the primary structures of cDNA copies of vRNA. Comparison of complete or partial nucleotide sequences of different wild-type viruses or their mutants has shown that mutations reflect a variety of molecular changes in genome RNA, including insertions, deletions, transitions, and transmissions.

Laboratory-Derived Mutants and Their Contribution to Genetic Analysis of Influenza Viruses

Phenotypic Markers

Influenza virus genetic markers are of two sorts: (1) those demonstrable through the in vitro biological activity of the external glycoproteins and (2) those demonstrable only through replication of the virus (Kilbourne, 1963; Sugiura, 1975). The former include antigenicity as measured by HA- and NA-specific antibody, stability of HA or NA activity, HA/NA ratios (reviewed by Mowshowitz and Kilbourne, 1975), or HA receptor affinity/specificity (Choppin and Tamm, 1960; Carroll et al., 1981). Viral morphology (spherical or filamentous) is a genetically determined viral phenotype not linked to HA or NA genes but probably determined by the M protein gene (E. D. Kilbourne, unpublished data).

Early studies of influenza virus genetics—indeed, those that demonstrated high-frequency recombination and the probability of a segmented genome—utilized replication-dependent markers such as virulence for animals or differences in plaque formation in cell culture (see Chapter 5) (Kilbourne, 1963, 1983). Often these markers were polygenic and susceptible to host-dependent variation in their expression. Therefore, they are of little use today except as they may be investigated in studies of virulence (see below).

Operational Mutants: Conditional Lethal Mutants Used in Definitive Genetic Studies of Influenza Viruses

Temperature-Sensitive (ts) Mutants. The use of systems in which elevated temperature is restrictive to the replication of mutants under study has permitted the isolation of some 388 mutants of eight different wild-type parental strains that carry defects in one or another of the eight genome RNA segments. Collectively, these mutants, studied in at least ten different laboratories during the past 18 years, define eight complementation–recombination groups. [The details of these studies have been comprehensively and succinctly summarized by Mahy (1983) and are not repeated here.] As reviewed by Sugiura (1975), ts mutants possess the following advantages in genetics studies. (1) The genome of a ts mutant may be identical to that of wild-type virus except at a single locus. Problems of gene incompability or polygenic control of a marker are thus eliminated. (2) Mutations can occur in all genes. (3) Selection of ts mutants that are "nonleaky" and have low back mutation (reversion) rates can afford powerful resolution, i.e., up to 10^6-fold differences in replication between permissive and nonpermissive temperatures.

A requirement in studies of ts mutants is for host cells that are highly permissive for wild-type virus. Accordingly, European investigators have made use of several fowl plague viruses (FPV) in chick embryo fibroblast (CEF) in which efficiency of infection is high even though they represent primary cultures. American investigators, not permitted the laboratory use of FPV, have adapted such permissive virus–cell systems as WSN in MDBK cells (Sugiura et al., 1972) or BK cells with other viruses (Spring et al., 1975).

Most studies have employed stocks of virus mutagenized with 5-fluorouracil or nitrosoguanidine, although some spontaneous mutants have been employed (Mahy, 1983). In typical experiments in which pairs of mutants were studied for their ability to complement defective functions, duplicate monolayer cultures were doubly or singly infected at high multiplicity, and then cultures were incubated at either 33°C (permissive temperature) or 39.5°C (nonpermissive temperature) for a sufficient period to allow single-cycle replication (10 hr). Harvest of virus was then titrated at both permissive and nonpermissive temperatures. Results were analyzed using a variation of the following formula for measurement of recombination (reassortment) frequency and efficiency of complementation (Sugiura et al., 1972). In this formula, mutant viruses are designated A and B.

Here, similarly, $[A_{39.5}]_{33}$ is the titer of a single virus yield grown at 39.5°C but assayed at the permissive temperature (33°C) (Sugiura et al., 1972; Mackenzie, 1970; Simpson and Hirst, 1968).

By this method, pairwise crosses of ts mutants demonstrated that reassortment and complementation occurred either at high efficiency or not at all—results consistent with the exchange of physically discrete genetic units (Sugiura et al., 1972).

$$\frac{[AB_{33}]_{39.5} - ([A_{33}]_{39.5} + [B_{33}]_{39.5})}{[AB_{33}]_{39.5}}$$

(In this formula $[AB_{33}]_{39.5}$ represents the titer of mixed yield grown at 33°C but assayed at 39.5°C, etc.)
Efficiency of complementation was calculated as

$$\frac{[AB_{39.5}]_{33} - ([AB_{39.5}]_{39.5} + [A_{39.5}]_{33} + [B_{39.5}]_{33})}{[A_{39.5}]_{33} + [B_{39.5}]_{33}}$$

Studies of ts mutants of influenza viruses have (1) combined with physicochemical evidence to confirm the segmented nature of the viral genome, (2) revealed the phenomenon of "pseudoreversion" or suppressor mutation (Murphy et al., 1980; Tolpin et al., 1981), (3) demonstrated the potential for intrasegmental complementation (Massicot et al., 1982), and (4) contributed importantly to understanding of viral replication as gene coding assignments have been determined. For example, a ts NA^- mutant fails to exclude neuraminic acid from its envelope at the nonpermissive temperature and accumulates at the cell surface in viral aggregates (Palese et al., 1974; Chapter 4). The use of ts and ca (cold adapted) mutants as live virus vaccines is discussed in Chapter 11.

Host Range (hr) Mutants. Far less extensively utilized in genetic studies have been host range or host-restricted (hr) mutants. From the limited data available, it is clear that any of several genes may enable virus to replicate in a host in which wild-type virus is restricted. As examples, two separate studies have implicated RNA 1 in the capacity of either Dobson (Almond, 1977) or Rostock (Scholtissek and Murphy, 1978) strains of FPV to replicate in BHK or MDCK cells, respectively. The NA gene of WSN facilitates cleavage of virion HA and enables virus to replicate in MDBK cells (Schulman and Palese, 1977).

Temperature-Dependent Host Range (td-hr) Mutants. Such mutants are doubly conditional because their host cell range is dependent on temperature as well as the host cell restriction. A/Udorn ts mutants bearing lesions in different genes produce plaques in rhesus monkey kidney (RMK) cells at 40°C, but are growth restricted at the same temperature in MDCK cells. Mutants with a reverse pattern of restriction were also found (Shimizu, 1982a, 1983). Host range mutants provide not only another system of conditional lethal mutants but potential insight into the mechanisms of viral pathogenesis.

Drug-Resistant Mutants. Studies on the emergence of amantadine-resistant mutants in cell culture have been well summarized by Mahy (1983). Further discussion of this topic is presented in Chapters 7 and 11.

VIRAL VARIATION

Antigenic Variants

Improvements in the precision of immunologic methods in recent years, and especially the use of monoclonal antibodies in selection and analysis of variants, have shown that antigenic variation in viruses is a frequent occurrence. With no other virus, however, does antigenic variation have the pervasive epidemiologic impact occasioned by antigenic changes in human influenza viruses. The nature of this impact is discussed in subsequent chapters, but antigenic variation is discussed here as a genetic event resulting either from cumulative point mutations or from genetic reassortment. Antigenic variants can arise either from selection by antibody or fortuitously through covariation in viruses selected for biologically important attributes. The frequency of antigenic variation is discussed under Mutation Frequency.

Sites of Antigenic Variation: The HA Polypeptide

Not unexpectedly, significant antigenic variation affects principally the two external glycoproteins of the virus, the HA and NA. But virion structural proteins less accessible to antibody and not involved in virus neutralization (e.g., NP) have also shown antigenic variation (see below), and even the nonstructural protein NS_1 is antigenically variable (Shaw et al., 1982).

Antigenic analysis of the HA protein has received major emphasis for appropriate reasons: (1) HA is the major glycoprotein of the virus and is involved in attachment and entry of the virus into the cell; (2) antibody directed against the HA neutralizes viral infectivity and therefore is highly selective for variants with changed antigenic sites or epitopes; (3) primary base sequences are known for the HA genes of many human strains, and the three-dimensional structure of an H3 prototype is well established.

As information has accrued from a variety of overlapping and interlocking approaches, much has been learned about HA-mediated antigenic variation. This information comes from studies of sequential antigenic changes in both laboratory (immunoselected) and field strains of virus correlated with determination of the primary amino acid sequences of the mutant HAs as deduced for the most part from studies of nucleotide sequences. (For a detailed presentation of the evidence, see Webster et al., 1983). These studies have demonstrated conservation of large areas of the HA molecule but have also revealed amino acid changes that tend to be clustered in restricted sites (Wiley et al., 1981) (Fig. 6-2). Most such changes involve the more exposed HA_1 portion of the molecule—the probable sites of antibody attachment. However, site D (Fig. 6-2) is at the monomer interface and is probably inaccessible to antibody binding. The dynamic nature of the HA molecule is revealed by its changing reactivity with monoclonal antibodies at different pHs (Webster et al., 1983; Daniels et al., 1983). Antibodies to site B

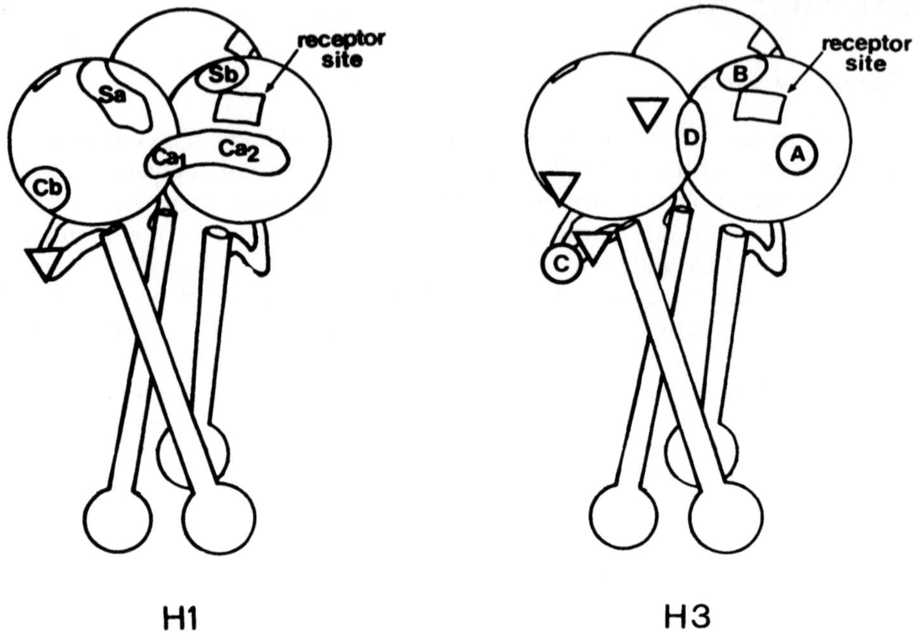

Figure 6-2. Comparative location of mapped antigenic and glycosylation sites on HA trimers of H1 and H3 influenza viruses. Letters show antigen, and triangles, carbohydrate attachment sites. (From Caton et al., 1982. Reproduced with permission, from *Cell* **31**:425. Copyright © by M.I.T.)

can be divided into two groups on the basis of their reactions with HA and conformational change characteristic of pH 5.0 (Daniels et al., 1983).

Even the limited information now available suggests that the location, number, and even specificity of antigenic sites are variable despite the evidence that HA structure in general appears to be conserved, even in different viral types (A and B) (see Chapter 3). On the other hand, the definitive structure of only a single HA subtype (H3) is known, so that comparisons among A subtypes must be made on the basis of assumptions about common three-dimensional structure. Furthermore, antigenic mapping of the PR8 (H1) HA, although partially confirmatory of antigenic site locations on H3, has revealed differences (Fig. 6-2), some dependent on differences in glycosylation.

Monoclonal antibodies to H1 (PR8) delineate both discrete (strain specific) and cross-reactive antigenic sites found in various epidemic strains (Caton et al., 1982). Glycosylation differences have been noted also among later H3 field isolates (Skehel et al., 1983). It seems unlikely that the surface of the HA molecule may indeed be an antigenic continuum (reviewed by Underwood, 1984), although recent studies of the H2 HA suggest that it does not have the discrete nonoverlapping regions found in early H3 influenza viruses (Yamada et al., 1984).

The Molecular Basis of HA Antigenic Variation

Studies of viral antigenic variants selected by neutralization by monoclonal antibody have shown that single amino acid changes can totally impair the ability of antibody effective in neutralizing the parental virus to bind to the selected variant. Absorption of antiserum raised to a selected variant by parental virus demonstrated residual antibodies specific for the variant, indicating that a new antigenic site must be present on the mutated HA (Webster et al., 1983).

Change in HA antigenicity apparently does not require large conformational changes in the molecule. Recent study of the three-dimensional structure of a selected antigenic variant of the H3 virus (X-31) of which the original three-dimensional structure had been determined showed that the addition of a single amino acid side chain, which resulted in no structural distortion of the molecule, is sufficient structural alteration for the virus to escape neutralization by a monoclonal antibody (Knossow et al., 1984). Independent studies of the H1 (PR8) HA have led to the conclusion that antigenic alteration in mutant HAs results from replacement of actual contact residues, i.e., amino acids with which antibody combining sites interact directly, or from very localized conformational changes but not from induction of long-range conformational changes (Gerhard et al., 1983).

Although it has been proposed that multiple amino acid substitutions in the HA might be necessary for a virus to acquire epidemic potential (Wiley et al., 1981), in A/USSR/77 viruses a single mutation not only changed the epitope but caused a significant antigenic variation detectable by postinfection ferret antiserum. This change contributed to antigenic drift and was considered to be epidemiologically significant (Nakajima et al., 1983). Minimal antigenic changes alone might be insufficient to promote viral dissemination but might require coincident cryptic "virulence" changes in other genes as well.

Antigenic Variation in the NA

The NA appears to possess at least four overlapping antigenic sites (Jackson and Webster, 1983), but a single site may be immunodominant (Lentz et al., 1984). Selection of variants with monoclonal antibodies has demonstrated single amino acid sequence changes in selected variants, some of which are also demonstrable in field strains. Antigenic drift in NA occurs in nature as it does in HA, representing the accumulation of single amino acid changes (Blok and Air, 1982; Ortin et al., 1980; Martinez et al., 1983). In field strains studied between 1957 and 1979, variation of amino acid sequence in the N2 NA occurred at an average rate of 5.5×10^{-3} changes per year and per amino acid (i.e., 73 amino acid changes) (Martinez et al., 1983).

Antigenic Variation in Other Viral Proteins

Although antigenic variation has been found in other influenza A virus-coded proteins, they seem to be minimal in degree, as would be expected for proteins less under selective pressure by antibody. By definition, all influenza A viruses are classified together on the basis of common M and NP protein antigens. Antigenic variation in the NPs of influenza A viruses seems to proceed independently of changes in the viral surface antigens; point mutations and genetic reassortment may account for NP variability (Van Wyke et al., 1980). Some evidence suggests the possibility of antigenic drift occurring with the NPs of human influenza A viruses (Schild et al., 1979). Three different overlapping antigenic regions are present on NS_1 of the H1N1 strain A/WSN/33. These antigenic determinants were common to NS_1 proteins found in a wide range of mammalian and avian influenza viruses. However, certain avian strains were demonstrably different (Brown et al., 1983).

Antigenic variation in internal proteins may be unrelated to immunoselection (see below).

Antigenic differences in the M protein were found in three of 26 influenza A viruses tested with monoclonal antibodies, although one of the three epitopes defined appeared to be invariant (Van Wyke et al., 1984).

Biologically Selected (Nonimmunoselected) Variants

Nonimmunoselected HA Variants

Antibody-induced selection of influenza virus variants has long been a major preoccupation of influenza virus research (Archetti and Horsfall, 1950; Gerber et al., 1955; Magill, 1955; Laver and Webster, 1968; Fazekas de St. Groth, 1978; Haaheim and Schild, 1976). However, it has also been appreciated by some that viral mutants fortuitously changed at antigenic sites might arise through biological selection (e.g., during mouse lung passage) (Horsfall, 1952; Hirst, 1947; Sugg, 1949). In other words, bioselection of a favorable mutation not mediated by antibody might still lead to antigenic variation, suggesting a linkage of certain biological and antigenic properties.

Variants Selected by Replication Characteristics. More recently, studies of two naturally occurring swine influenza virus mutants have demonstrated that these mutants differ pleiotropically in their replication capacity in laboratory and natural hosts and in their antigenicity defined either by a heterotypic polyclonal antiserum or by monoclonal antibodies (Kilbourne, 1978; Kilbourne et al., 1983). Both mutants are readily isolated from uncloned field isolates obtained from humans, swine, or turkeys (Kilbourne et al., 1979; E. D. Kilbourne, unpublished data). The mutants can be biologically or immunologically selected in the laboratory, and reversion occurs concordantly for biological and antigenic phenotype. Comparison of the nucleotide sequences of HA1 of these mutants has shown that a single amino acid change pleiotropically affects both viral antigenicity and replication. This change is at amino acid 155 of HA1, which in H1 virus corresponds to residue 158 in the H3 model. This change is near the tip of the HA monomer adjacent to the receptor binding site and therefore could credibly affect both viral antigenicity and replication (Both et al., 1983) (see also Genetic Dimorphism, below).

Single amino acid substitutions in the distal receptor pocket of the HA can change receptor binding specificity (Rogers et al., 1983) concordant with antigenic change in the molecule (Daniels et al., 1984). Such variants are selected not by antibody but by passage of virus (H3N2) with normal horse serum containing glycoprotein inhibitor. Adsorptive (Yewdell et al., 1986) and HA fusion mutants have also been recognized (Rott et al., 1984; Daniels et al., 1985; Doms et al., 1986).

An HA mutant of A/WSN virus with an increased host-cell affinity was shown to be changed in antibody reactivity, being less sensitive to neutralization than the parental virus. This finding suggested that preferential neutralization of parental virus could contribute to the selective advantage of the mutant already afforded by its increased affinity for cellular receptors (Nohinek et al., 1985). This affinity reflects elimination of an HA glycosylation site on the mutant. Furthermore, the host-determined structure of the oligosaccharide at that site affects the affinity of the parent virus for cellular receptors (Deom et al., 1986).

Host-Determined Antigenic Variation. Cultivation of influenza B viruses in chick embryos selects for HA mutants that differ antigenically from virus from the same source grown in mammalian (MDCK) cell cultures (Schild et al., 1983), reflecting the loss of a carbohydrate side chain (Robertson et al., 1985). This finding represents further evidence of nonimmune selection of antigenic variants. In its reproducibility, it extends earlier unsystematic studies that suggested host determination of antigenic variation resulting from mouse lung passage. (Hirst, 1947; Sugg, 1949; Horsfall, 1952), a phenomenon recently confirmed with monoclonal antibody studies (Gitelman et al., 1984).

In contrast to Hirst's (1947) finding of apparent random antigenic change with adaptation, Gitelman and colleagues (1984) found the same change in antigenic pattern with adaptation to the mouse of two different H1N1 viruses.

HA Glycosylation Mutants. A single amino acid substitution (Asp^{63}-Asn^{63}) generated a carbohydrate attachment site that in turn inhibited reaction of virus with the selecting monoclonal antibody (Skehel et al., 1984). This mutation was also found in the field strain A/Victoria/3/75. Such glycosylation could sterically hinder antibody access to the underlying molecular surface, precluding antigenic selection by stabilizing potential variability in that region (Skehel et al., 1984; Both et al., 1983; see also discussion of molecular epidemiology in Chapter 10).

Pleiotropism and Covariation

The preceding examples of covariation of antigenic and other viral properties have emphasized that a single amino acid change can be manifest in multiple phenotypes (pleiotropism), a phenomenon with important biological implications. But the covariation of traits implicit in pleiotropism must be rigorously demonstrated, particularly in the highly mutable RNA viruses in which cryptic simultaneous mutations at other sites might coincidentally affect viral phenotype. Therefore, definitive proof demands evidence (1) that reversion of the two or more linked phenotypes occurs concurrently, (2) that selection of phenotype by independent procedures (e.g., antigenic selection or plaque size difference) also demonstrates linked phenotypes, and (3) that change in phenotypes is reproducibly related to a single amino acid change.

The most obvious biological significance of pleiotropism with influenza viruses lies in the linkage of functional change in a protein to protein antigenicity. Thus, selection of a functional variant favored in replication might affect a site within or adjacent to an epitope, thereby concordantly selecting for an antigenic variant (e.g., see Kilbourne, 1978). Conversely, in another situation, immunoselection might yield a mutant conferring the virus with altered replication or virulence potential if the new epitope generated effected a significant structural change in the virus.

It is predictable that with the increasing use of monoclonal antibodies, which are exquisitely sensitive tools for detection of antigenic variation, more and more instances of covariation in single-step mutants will be found. Although single amino acid substitutions probably can result on occasion in epidemiologically important variation, many such minor antigenic changes will be important only as convenient markers for a pleiotropic phenotype (see also molecular epidemiology, Chapter 10).

The Genetics of Major Antigenic Change in Influenza A Viruses

On the basis of antigenic analysis (principally by hemagglutination inhibition and neutralization tests), 13 subtypes of HA and nine subtypes of NA are recognized in the classification of influenza A viruses (Chapter 2). Sequence data, in general, have been confirmatory of antigenic subtype definition and, indeed, led to the reclassification of viruses in 1980. However, antigenic cross reactions among HA subtypes indicate the sharing of some epitopes—not unexpected in proteins with common structure and sequence homology of 25–80%. In a sense, influenza A viruses of different HA,NA subtypes can be considered "antigenic variants" within a common family of viruses probably derived from a common evolutionary ancestor (Fig. 6-3). Indeed, there is an arbitrary aspect to the definition of subtypes (Smorodintsev et al., 1982). However, in epidemiologic and immunologic behavior, differences among subtypes outweigh their similarities, and it is useful to consider them as different but closely related viruses.

Except with the most closely related HA subtypes (H2 and H5), it is extremely unlikely that transition of one subtype to another through sequential point mutations could occur within a short period of time.

Evidence is now strong that reassortment of HA and NA genes occurs in nature, creating "antigenic shifts" leading to the emergence in the human population of viruses with apparently novel antigens. It is somewhat misleading, conceptually, to view such events as a variety of antigenic variation of a preexisting virus, whereas, in fact, the emerging virus is a truly new virus—the progeny of genetic reassortment of two or possibly more viruses. In any case, "antigenic shift" (which virus is shifting?) must represent either the sudden appearance of a newly created (e.g., H3N2) or an old emergent (e.g., H1N1) virus by nonmutational mechanisms (see Genetic Reassortment, below).

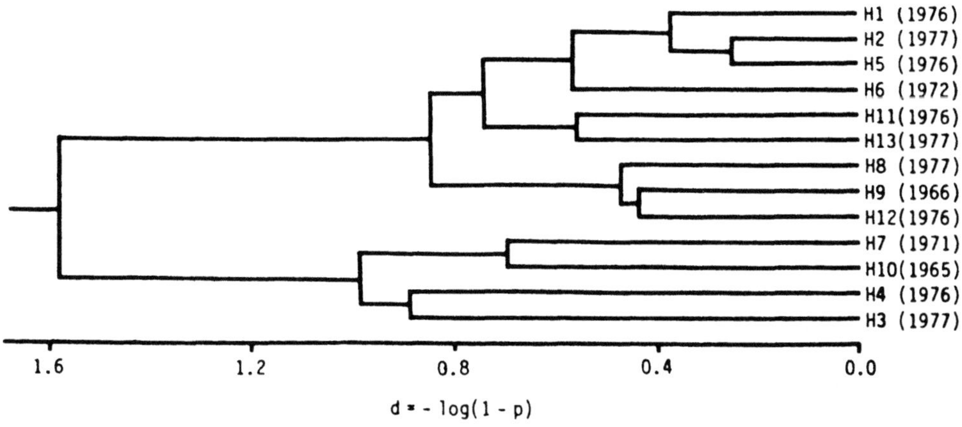

Figure 6-3. Phylogenetic tree of 13 HA subtypes of influenza A virus based on amino acid sequence. (From Saitou and Nei, 1986. Reproduced with permission, from *Mol. Biol. Evol.* 3:57–74. Copyright © 1986 by The University of Chicago Press.)

GENETIC AND NONGENETIC VIRAL INTERACTIONS

Phenotypic Mixing

Coincident infection of cells with certain pairs of related or unrelated viruses can result in the formation of viral pseudotypes—virus particles containing surface antigens of one or both viruses and a single genome. That is, particles can contain either a mosaic of antigens derived from both parental viruses or only antigen(s) heterologous to the genome (Fig. 6-4). Viral interactions of this type are nongenetic, in that genes are not mixed and phenotypically mixed particles breed true according to the genome they bear. The phenomenon is of interest chiefly for the insight that it provides into the intracellular interaction of viral proteins and the process of viral assembly.

Mosaic viral pseudotypes are recognized as particles that are doubly neutralizable, i.e., neutralized with antibody to either parent. Phenotypically mixed viruses have been noted with bacterial, plant, and animal viruses, the latter including both enveloped and nonenveloped viruses (reviewed by Zavada, 1976).

Influenza Virus–Influenza Virus Pseudotypes

Phenotypic mixing of influenza viruses has been described for influenza A strains and between influenza A and B viruses (Hirst and Gotlieb, 1953; Gotlieb and Hirst, 1954). In genetic reassortment experiments with influenza viruses capable of genetic interaction, as much as 90% of virus yield may consist of pseudotypes (Sugiura and Kilbourne, 1966). In the case of dual infection with influenza A and B viruses, not only does mixture of surface glycoproteins occur, but mixture of nucleocapsid proteins. However, in the course of virus assembly, pheno-

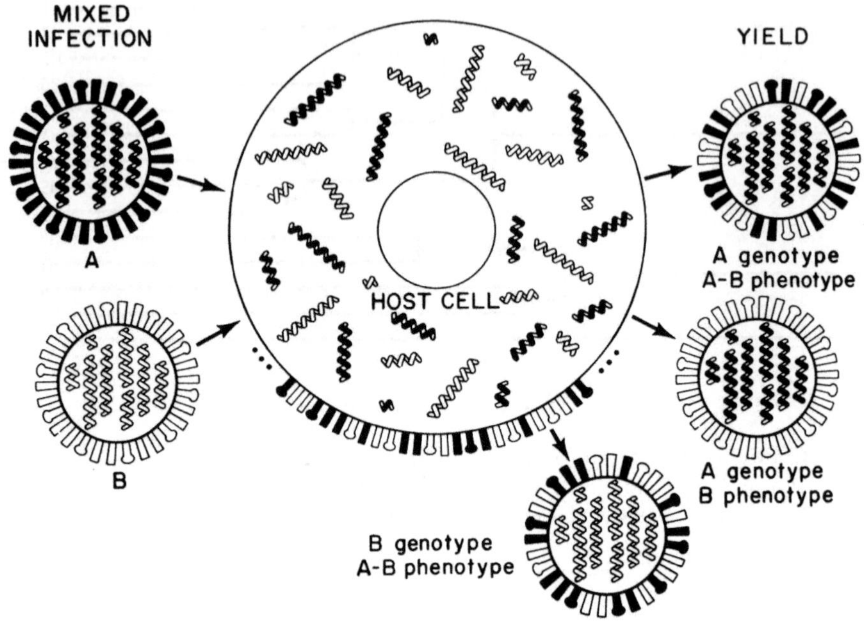

Figure 6-4. Diagram of phenotypic mixing in which influenza virus phenotypes are defined by surface proteins (see text).

typically mixed nucleocapsids are preferentially excluded from virions (Sklyanskaya et al., 1985).

Heterologous Pseudotypes

Phenotypic mixing of influenza viruses with the paramyxovirus NDV (Granoff and Hirst, 1954) and with an oncornavirus (Zavada and Rosenbergova, 1968) has been reported. Phenotypic mixing of influenza A virus (FPV) with vesicular stomatitis virus (VSV) is unipolar in that all pseudotypic particles contain the VSV genome (Zavada and Rosenbergova, 1972). Phenotypic mixing of influenza virus and VSV is of special interest as a model system because (1) these dissimilar enveloped viruses differ in morphology, providing a structural marker, and (2) in certain cells, assembly and budding occur at opposite (apical versus basolateral) membrane surfaces (see Chapter 4). Given the latter circumstance, the formation of pseudotypes is puzzling. However, Roth and Compans (1981) have shown that although the sorting mechanism responsible for directing the glycoproteins of VSV and influenza viruses to separate cell surfaces continues to operate during double infection, pseudotypes are formed late in infection coincident with CPE and the disruption of junctional complexes.

Genetic Reassortment

Within each type (A, B, or C), all strains of influenza virus appear to be genetically compatible. Therefore, concordant infection of a cell with any pair of influenza A viruses, for example, has the potential for yielding reassortment viruses of mixed genotype (Fig. 6-5). The high frequency of this genetic interaction is inconsistent with classical recombination and prompted Burnet (1956) to suggest that the genome of influenza virus comprised "a variable number of genetic determinants." Hirst (1962) postulated "an exchange of large pieces" (of the genome) in "recombination" occurring at a high rate.

Most evidence suggests that influenza virus genetic reassortment is essentially a random process. RNA segments reassort independently in three-factor crosses (Nakajima and Sugiura, 1977), and the frequency of reassortment at any given locus can be as high (50%) as that expected for independently segregating genes (Feng et al., 1983).

However, evidence of gene linkage has been found during influenza virus reassortment. Lubeck et al. (1979) found linkage of influenza virus genes involved in RNA replication, P1–P2 and P2–P3, in yields of single-cycle infection with PR8 and HK viruses. It is not surprising that the pattern of RNA reassortment might vary in relation to different viral pairs and different host cells. Apparent gene linkages might reflect differences in viral protein–protein compatibility or packaging problems based on RNA–RNA recognition or possibly differences in regulation of transcription. Skewing of parental donation can be accomplished by adjustment of infecting dose or partial inactivation of one donor. Frequency of allelic genes following reassortment differs from gene to gene (Feng et al., 1983).

Assembly and packaging of virion RNAs are not yet understood. Because all genes are required for infective virus, assembly requires (1) incorporation, by

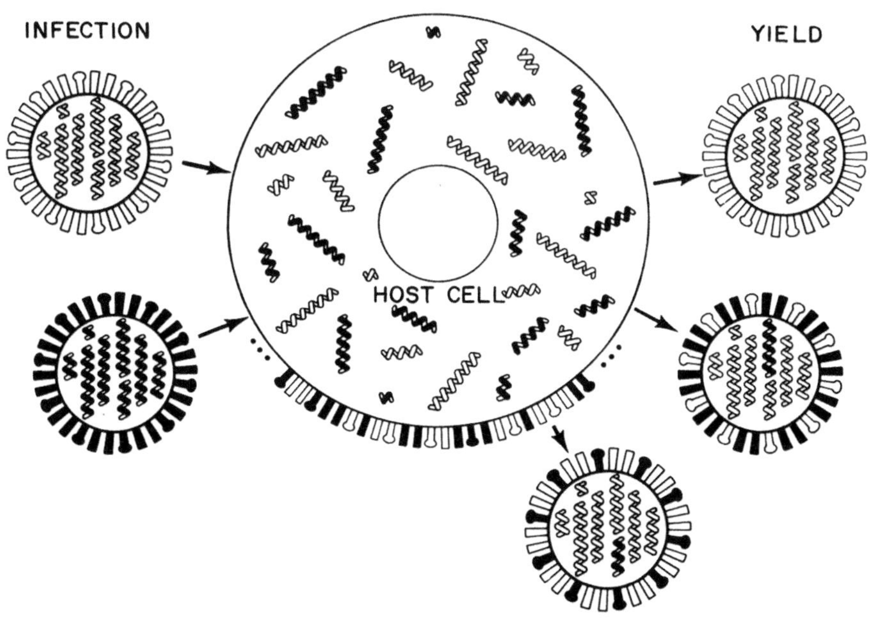

Figure 6-5. Intratypic genetic reassortment of influenza viruses, illustrating diagramatically three of 256 possible reassortant progeny from infection with two parental viruses.

chance, of the requisite number and type of genes in a virus particle as a random event of relatively low probability or (2) some mechanism of restrictive packaging and ordered gene selection (Kilbourne, 1981). With influenza viruses, the ratio of infective to viral particles determined by electron microscopy is inconsistent with random packaging; one of 20 rather than an expected one out of 416 particles is infective (Nakajima and Sugiura, 1977).

The lack of genotypic mixing of nonhomologous (A and B) influenza viruses has been assumed to reflect differences in polymerase specificity rather than packaging specificity. However, recent evidence suggests that influenza A-B hybrid nucleocapsids arising during reassortment attempts and containing mixed genomic RNA segments are preferentially excluded from the virion (Sklyanskaya et al., 1985).

If coordinated RNA packaging does occur, its mechanism must explain the ordering of RNA pieces with similar 3′ and 5′ terminal sequences (Desselberger et al., 1980).

Complementation–Reassortment

The classical genetic technique of complementation analysis has been extensively used in studies of conditional lethal mutants of influenza viruses (see above). Complementation appears always to be accompanied by recombination, an all-or-none phenomenon consistent with a reassortment mechanism. Initial studies indicated that complementation–reassortment groups, therefore, corresponded to the number of RNA segments. However, continuing studies of the ability of pairs of ts mutants to produce infective virus during coinfection at restrictive temperatures (complementation) demonstrated that the number of complementation groups exceeded the number of RNA segments. This finding was explained by the assumption that intrasegmental complementation had occurred (reviewed by Murphy and Webster, 1985; Massicot et al., 1982).

Another recently appreciated problem in complementation analysis is the occurrence of synergistic mutations in several RNA segments that only in combination can express the ts phenotype. Such mutants cannot complement other ts mutants defective in one gene (Scholtissek, 1984).

Multiplicity Reactivation and Gene Rescue

Noninfective virus is genetically functional in reassortment with infective parental (rescuing) virus. Indeed, non-plaque-forming virus was found to be the principal source of virus of mixed genotype during reassortment (Hirst and Pons, 1973). Thus, genes carried by individually noninfective particles can be rescued through complementation–recombination by genetic interaction with infective and with other noninfective particles.

Genetic complementation, of course, does not require interaction between different viruses but can occur endogenously among genes of a single virus inactivated at different loci. High-multiplicity infection with UV-irradiated or thermally inactivated virus leads to multiplication rates in excess of that expected from the residual infectivity of such preparations (Henle and Liu, 1951; Kilbourne, 1955) (Fig. 6-6). This effect may be difficult to demonstrate because of viral autointerference in partially inactivated preparations. In chick embryos, administration of 11-oxycorticoids negates these interfering effects (Kilbourne, 1957).

VIRULENCE AND ATTENUATION AS GENETIC PHENOMENA

Viral virulence is a composite of those properties of the virus that enable it to replicate in and damage cells. If viruses cannot attach to and enter cells, they cannot replicate; therefore, mutations that affect these functions are polar in determining virulence or attenuation and can be critical in restricting virus host range. But because all genes are required for viral replication, all are to some degree involved in virulence. Thus, the polygenic nature of influenza virus virulence has long been appreciated (Fraser, 1959; Kilbourne, 1963, 1969; Mayer et

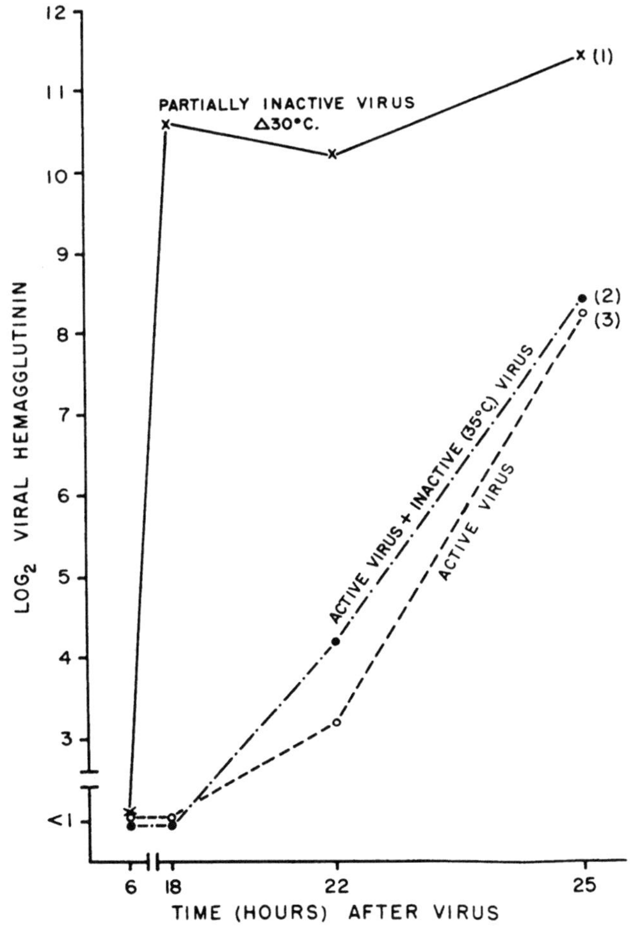

Figure 6-6. Multiplicity reactivation of thermally inactivated influenza virus. All three viral inocula contained the same number of titratable infective virions, but growth curves 1 and 2 demonstrate reactivation through complementation of particles rendered individually noninfective. Autointerference was reduced by inoculation of cortisone-injected eggs. (From Kilbourne, 1957. Reproduced from *J. Exp. Med.* **106:**873, by copyright permission of the Rockefeller University Press.)

al., 1973; Rott et al., 1976). However, although there are no single "virulence genes" and none that confers pantropicity, single genes may be critical in defined systems in altering a specific phenotype associated with virulence (Almond, 1977; Israël, 1980a,b). In some instances, an effect observed in cell culture can be correlated with in vivo pathogenicity (Bosch et al., 1981; Flehmig et al., 1976). But more often, the full expression of virulence requires the ancillary function of additional genes (Kilbourne, 1978; Sugiura and Ueda, 1980).

Identification of Single Genes Influencing Virulence and Attenuation

Evidence that mutations in single influenza virus genes can influence virulence is summarized in Table 6-2. It is notable and not surprising that mutations affecting genes for the HA and NA are of cardinal importance in determining virulence. These proteins are critical to interactions of the virus with host-cell receptors and, in the case of the HA, to virus–cell fusion and viral entry. The NA can facilitate the latter process by contributing to HA cleavage (Schulman and Palese, 1977). Substitution of only the HA gene of PR8 with the L mutant HA gene of A/NJ/11/76 swine influenza virus can endow that virus with the capacity to replicate in swine (Kilbourne et al., 1979); two mutations in the receptor binding site of the HA of a human influenza virus (amino acid residues 226 and 228) permitted replication of the virus in ducks. A single critical point mutation in the HA gene of avirulent A/Chicken/Pennsylvania/1/83 (H5N2) appears to be all that is required for the acquisition of virulence. (Webster et al., 1986). These mutations resulted in a receptor binding site sequence identical to the known avian influenza virus sequences (Naeve et al., 1984).

Each of the three polymerase genes has been shown to be contributory to virulence, particularly in determining host range.

RNA 7, coding for viral M proteins, is essential in determining maximal viral yields in the chick embryo (Baez et al., 1980). Which RNA 7 gene product is responsible has not been determined.

A deletion in gene 8 affecting NS_1 has resulted in attenuation of a cold-adapted reassortant virus for MDCK cells and ferrets (Maassab and De Borde, 1983; Buonagurio et al., 1984); an influenza B virus RNA 8 mutant arising in the course of influenza A and B reassortment attempts destroys cells rapidly (Tanaka et al., 1984).

Analysis of Virulence through Genetic Reassortment

The identification of single genes that determine virulence represents, perhaps, an artifact of experimental design—the fortuitous use of the proper virus in an in vitro or in vivo system in which the virulence phenotype is not complex and is readily observed (e.g., plaque formation or tropicity for special target organs). For the most part, virulence can be shown to result from appropriate gene combinations, particularly as they affect the more complex events initiated by infection of intact animals. In animals, the virulence endpoint of manifest disease is a demanding one, requiring maximal viral replication and the destruction of sufficient target cells to evoke the characteristic pathological response that will permit recognition of the phenotype (Table 6-3). Studies that quantitate viral replication in target organs as well as pathological response are necessary to define gradations of virulence. Results of two such studies are presented in Table 6-4 (Schulman, 1983; Sugiura and Ueda, 1980).

Study I of mouse lung virulence demonstrates with a pair of viruses, PR8 and HK, the cardinal importance of the HA. Substitution of only the HK HA in reassortant R1 (bearing all other genes from PR8) reduced virulence to a level intermediate between the two parental viruses. On the other hand, the substitution of only the PR8 HA in the HK genome (R2) endowed the virus with enhanced virulence. The full restoration of virulence to back reassortant R3 reconstituted the PR8 parental genotype, indicating the additive importance of other PR8 genes in mouse lung virulence. Such isogenic reassortants are essential controls for interpretation of experiments of this type in order to assess the possibility of significant mutation in any of the genes during reassortment. Reassortants with identical gene combinations may vary phenotypically, reflecting such mutations (Schulman and Palese, 1978; Kilbourne, 1978; Naeve et al., 1983).

Study II of mouse neurovirulence similarly demonstrates the essential contribution of the WSN (W) NA but also identifies the contributory role of RNA 7 and 8 gene products in mediation of maximal virulence. The influence of all three genes in viral replication was also observed in a neuroblastoma cell line (Nakajima and Sugiura, 1980).

Taken together, these studies (1) confirm the polygenic nature of virulence, (2) indicate that one gene may be of cardinal importance, and (3) fulfill earlier predictions that most reassortants of virulent and avirulent parents should be intermediate in virulence (Kilbourne, 1969).

However, study III demonstrates that reassortment of two highly virulent viruses, FPV (F) and A/Turkey/England/63 (T), can result in the production of progeny viruses less virulent than either parental virus through apparent gene incompatibility (e.g., R1 and R2). Nonpathogenic (nonlethal) reassortants were found to possess mixed genes in the polymerase complex RNAs 1, 2, 3, and 5; (R1 and R2 are examples). R12 (virulent) is but one example of several mixed-gene reassortants in this study in which polymerase complex genes were derived from a single parent.

An even more compelling argument for the importance of gene combination rather than the existence of "virulence genes" per se is the production of reassortants that exceed either parent in virulence (Scholtissek et al., 1979). This possibility had been suggested earlier (Kilbourne, 1969) on the basis of the fact that new phenotypes may arise during reassortment (Laver and Kilbourne, 1966).

Many examples can be cited of specific gene combinations critical for virulence for specific virus pairs and specific host systems. Both HA and NA are required for avian enterotropism (Hinshaw et al., 1983); RNA segments 3 and 5 are required for virulence in reassortants of WSN and a turkey virus (Webster and Bean, 1978); neurovirulence of FPV reassortants derived from virus N, HK, and PR8 crosses required the donation of RNA segments 4 from FPV and 2 from the other strain (Bonin and Scholtissek, 1983). Neurovirulence is negated, however,

Table 6-2. Molecular Determinants of Influenza Viral Virulence: Relation to Single Gene Changes

Viral				
RNA	Protein	Virus[a]	Phenotype influenced	Reference
1	PB2	FPV(D)	Plaquing in BHK cells	Almond (1977)
	PB2	FPV(R)	Plaquing in MDCK cells	Scholtissek and Murphy (1978)
	PB2	FPV(D)	Td–hr phenotype	Israël (1980a)
2	PB1	FPV(R)	Plaquing in MDCK cells	Scholtissek and Murphy (1978)
3	PA	FPV(R)[b]	Mouse lung virulence	Scholtissek et al. (1979)
	PA	FPV(D)	Replication in L cells	Israël (1980b)
4	HA	A/NJ/11/76	Replication in swine, chick embryos, MDCK cells	Kilbourne (1978); Kilbourne et al. (1979)
	HA	Avian viruses	Replication in CEF, MDCK and other cells; chicken pathogenicity	Bosch et al. (1979, 1981)
	HA	Avian viruses	Virus spread in chick embryo	Rott et al. (1980)
	HA	A/chick/Jap./24[b]	Lethality for chickens and chick embryos (principal determinant)	Ogawa and Ueda (1981)
	HA	A/PR/8/34(MS)[b]	Mouse lung virulence (principal determinant)	Schulman (1983)

	HA	Avian-human virus reassortant	Replication in ducks	Naeve et al (1984)
	HA	A/chick/Penn/83	Chicken virulence	Webster et al (1986)
6	HA	NWS	Mouse brain virulence (principal determinant)	Hobson et al. (1968)
	NA	WSN	Mouse brain virulence (principal determinant)	Sugiura and Ueda (1980)
6	NA	WSN	Replication in neuroblastoma cells (HA cleaving)	Nakajima and Sugiura (1980)
	NA	WSN	Plaquing in MDBK cells (HA cleaving)	Schulman and Palese (1977)
	NA	H3N2 Rec.[b]	Plaque size in MDCK cells	Markoff et al (1979)
7	?M_1, M_2, M_3	A/reassortants	Replication (yield) in chick embryos	Baez et al (1980)
8	NS_1	CR43-clone 3	Replication in MDCK cells and ferrets	Maassab and De Borde (1983); Buonagurio et al. (1984)
	NS[d]	B/Yamagata/1/73 mutant[c]	Cytolysis in MDCK cells	Tanaka et al. (1984)

[a]FPV(D), Dobson strain, FPV(R), Rostock strain
[b]Reassortant deriving gene from this/these viruses
[c]Derived during influenza A-B virus reassortment attempt
[d]Protein different from both NS_1 and NS_2 of parent

Table 6-3. Determinants of Influenza Viral Virulence

Host cell range (capacity to infect)[a,b]
Productive (nondefective) replication[a,b]
Capacity to damage cells[a,b]
Capacity to damage cells essential to host function and survival[b]

[a]In vitro
[b]In vivo.

in N virus reassortants if, in addition to RNAs 2 and 4, RNA 5 (NP) is also derived from that virus. Extragenic suppressor effects can accompany genetic reassortment (Ghendon et al., 1982) and, indeed, may operate to restore pathogenicity in ts mutants (Murphy et al., 1980; Tolpin et al., 1981; Scholtissek and Spring, 1982). A ts defect can be phenotypically reverted by replacement of a gene other than that bearing the ts defect (Scholtissek and Spring, 1982; reviewed by Rott et al., 1984).

CLONING AND EXPRESSION OF INFLUENZA VIRUS GENES: THE NEW INFLUENZA VIRUS GENETICS

Contributions of Gene Cloning

The study of influenza virus genetics has been revolutionized by the cloning of influenza virus genes in bacterial plasmids or in animal virus vectors. As summarized by Gething and Sambrook (1982a) in their comprehensive review, the cloning in bacterial plasmids of double-stranded DNA copies of RNA segments of the viral genome has greatly expedited the determination of gene nucleotide sequence. This, in turn, has (1) led to elucidation of amino acid sequence of the known virus-coded proteins and to the discovery of polypeptides not previously known to exist, (2) facilitated comparison of primary structures of proteins of different viral subtypes, (3) simplified interpretation of three-dimensional structure of HA and NA proteins, and (4) aided in the mapping of antigenic and glycosylation sites.

Gene Expression

Of equal importance has been the use of vectors to permit expression of influenza viral genes in both prokaryotic and eukaryotic cells. In these experiments, vectors capable of autonomous replication in *E. coli* (plasmids) have been used for expression of viral proteins in that host (in early studies, Emtage et al., 1980; Heiland and Gething, 1981; Davis et al., 1981); mammalian cell vectors have utilized insertion of influenza viral genes into truncated SV40 DNA or into the DNA of vaccinia virus. In mammalian cells, transfection with chimeric SV40 influenza virus DNA results in transient expression of viral protein, or chromosomal integration of DNA enables continuous expression of viral protein at a lower level (Gething and Sambrook, 1982b). Vaccinia virus has also been used as a vector

Table 6-4. Examples of the Importance of Gene Combination in Determining Virulence of Viral Reassortants

		RNA (protein)[a]										
		1	2	3	4	5	6	7	8			
Study	Viruses	(PB2)	(PB1)	(PA)	(HA)	(NP)	(NA)	($M_{1,2}$)	($NS_{1,2}$)	Phenotype		Reference
										Mouse pneumovirulence		Schulman (1983)[f]
										Virus titer[b]	Lesion score[c]	
I	PR8(P)	P	P	P	P	P	P	P	P	7.7	73	
	HK(H)	H	H	H	H	H	H	H	H	6.2	5	
	R1	P	P	P	*H*	P	P	P	P	6.9	33	
	R2	H	H	H	*P*	H	H	H	H	6.8	28	
	R3	P	P	P	P	P	P	P	P	7.9	68	
										Mouse neurovirulence		Sugiura and Ueda (1980)
										Virus titer[d]	d/T[e]	
II	WSN(W)	W	W	W	W	W	W	W	W	5.17	5/5	
	HK(H)	H	H	H	H	H	H	H	H	0	0/5	
	R403	H	H	H	H	H	*W*	*W*	*W*	7.17	3/3	
	R404	H	H	H	H	H	*W*	*W*	H	2.87	0/4	
	R72L	H	H	H	H	H	*W*	H	*W*	3.40	0/4	
	R96	H	H	H	H	H	*W*	H	H	0.65	0/4	
	R52	H	H	H	H	H	H	*W*	*W*	0	0/4	
										Chicken lethality		Rott et al. (1979)
III	FPV(F)	F	F	F	F	F	F	F	F	+		
	Turkey(T)	T	T	T	T	T	T	T	T	+		
	R12	T	T	T	F	T	F	F	T	+		
	R1	T	T	*F*	F	T	T	T	F	0		
	R3	*F*	T	T	F	F	T	F	T	0		

[a]Italicized letters indicate genes critical to virulence
[b]EID_{50} = $\log_{10}$ geometric mean of virus in lungs 3 days after infection
[c]Mean extent of lung lesions 7 days after infection
[d]Virus titer, PFU/brain, $\log_{10}$
[e]Number dead/total animals in group
[f]Tabulated data reproduced with permission

for the transfection and expression of the HA gene in eukaryotic cells (Smith et al., 1983; Paoletti, 1985).

All influenza genes have now been cloned and sequenced. Moreover, expression of cloned viral genes and production of viral proteins occurs with the HA (best studied), NA (Markoff et al., 1984), M (Lamb and Lai, 1982), NS^1 genes (Young et al., 1983), and NP (Andrew et al., 1986). Expression of the three polymerase proteins permitted growth complementation of influenza virus ts mutants (Krystal et al., 1986). Expression of HA has been highly variable in *E. coli*, probably reflecting differences in the efficiency of different promoters and ribosome binding sites as well as differences in the stability of the hybrid mRNA or protein (Gething and Sambrook, 1982b). The HA produced is nonglycosylated, a factor that appears to affect its stability, structure, and antigenicity (Jabbar et al., 1985).

The nonglycosylated protein NS_1 is expressed in *E. coli* at high levels, is immunogenic, and induces the formation of specific antibody capable of reacting with the viral protein in infected cells (Young et al., 1983).

RNA 7 introduced into AGMK cells in an SV40 vector led to expression of M_1 but not M_2 or the putative M_3 protein (Lamb and Lai, 1982).

Expression of HA produced in mammalian or insect cells, on the other hand, has been more successful, resulting in a glycoprotein indistinguishible from that produced during natural infection (Gething and Sambrook, 1981; Hartman et al., 1982; Sveda and Lai, 1981; Kuroda et al., 1986). The recently reported expression of glycosylated HA in yeast cells offers promise for another system for the production of vaccine antigen (Jabbar et al., 1985).

Contribution of the Expression of Isolated Viral Genes to the Understanding of Influenza Virus Infection

The independent expression of cloned influenza virus HA and NA proteins in eukaryotic cells has provided new insight into their synthesis and transport. By deletion of sequences coding for either the amino-terminal signal peptide or the carboxy-terminal anchor region, the putative functions of these sequences have been verified. The anchor-minus HA is glycosylated but is released from the cell as a secretory protein (Gething and Sambrook, 1982a,b; Sveda et al., 1982). In the NA protein, the N-terminal variable hydrophobic region has the anchoring function, and its deletion in cloned NA results in failure of the mutant polypeptide to be translocated into membrane organelles or to be glycosylated (Markoff et al., 1984).

Signal-minus (amino terminal deleted) HA is produced at low levels, is not glycosylated, and remains within the cell (Gething and Sambrook, 1982a,b; Sekikawa and Lai, 1983); HA lacking the carboxy-terminal anchoring sequences is secreted from BPV-transformed cells that constitutively express influenza virus HA (Sambrook et al., 1985). The signal or anchor sequences of VSV are functionally effective in influenza virus–VSV HA chimeras, but chimeric proteins were not transported to the Golgi complex or the plasma membrane, implying that protein tertiary structure may play a role in transport (McQueen et al., 1984). Additional insight into cell-sorting function in viral protein transport has been provided by evidence that the polarized expression of influenza virus HA at the apical surface of certain cells in monolayer culture also occurs with cloned HA and therefore is not dependent on the presence of other viral proteins (Rott et al., 1983).

Most exciting are the prospects for using independently expressed influenza virus genes in the resolution of controversial problems in cell-mediated immune response. Braciale and colleagues (1984) have demonstrated that the cloned HA molecule can serve as a target antigen for both subtype-specific and cross-reactive subpopulations of influenza virus-specific cytotoxic T lymphocytes. Heterosubtypic cross reactivity appeared to be minor in degree when assessed with vaccinia recombinants (Bennink et al., 1986).

Prospects and Potential of Site-Specific Mutagenesis

The prospects exist with present technology for introducing single base changes into nucleotides coding for any of the viral proteins at sites suspected of having structural or functional importance. These techniques should free the investigator from the past tyranny of hit-or-miss genetic manipulations and permit definitive understanding of influenza virus gene structure and function at the molecular and atomic level. Indeed, the technique has already been applied to identify the enzymically active site of the viral NA (Lentz and Air, 1986). A practical outcome of site-specific mutagenesis could be the tailoring of immunogenic antigens free of toxicity.

VIRAL VARIATION AND ATTENUATION IN NATURE

The application of nucleotide sequencing to field isolates of influenza viruses has opened the door to a searching analysis of viral genetic variation in nature not heretofore possible with antigenic analysis alone. Such molecularly defined variation is discussed in relation to viral evolution in the following section and in the discussion of molecular epidemiology in Chapter 10.

INFLUENZA VIRAL ADAPTATION, GENETIC POLYMORPHISM, AND EVOLUTION

Viral Adaptation

Adaptation of influenza viruses by serial propagation in new hosts invariably results in the selection of mutants better equipped to replicate and spread within the new cellular environment in which they find themselves. Thus, influenza viruses, which are inherently cytolytic, "adapt" or become more damaging to the animal host not by changes in capacity to infect (which they already must possess) but by mutational changes that permit attainment of higher titers in the host. A common feature of adaptation of both A and B influenza viruses to the mouse lung has been the emergence of virus characterized by a more rapid growth rate (Wang, 1948; Ledinko, 1956; E. D. Kilbourne, unpublished data) and the capacity to reach higher concentrations in the lung (Schulman, 1970) (Fig. 6-7).

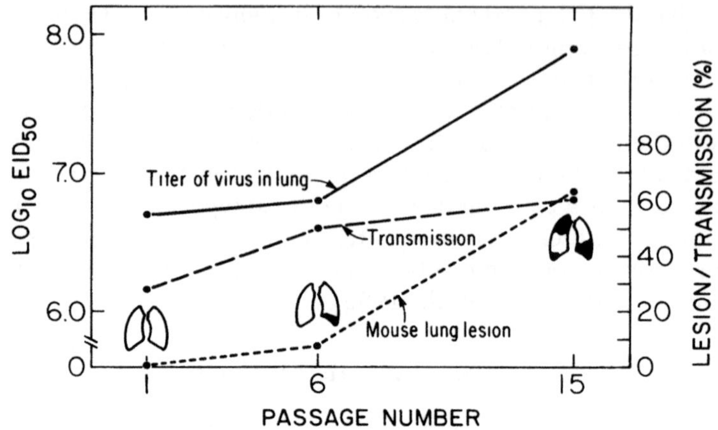

Figure 6-7. Changes in mouse lung virulence and virus transmissibility with adaptation of A/Japan/305/57 virus to mice. Infectivity of virus for mice (not shown) increased fivefold with passage. (From data of Schulman, 1970.)

Adaptation has been studied in the laboratory for many years as the host range and pathogenic effects of influenza viruses have been investigated. Changes in viral phenotype have been noted concomitant with sequential passage of virus. Serial passage of influenza A viruses in the mouse lung has been associated with minor antigenic change (Hirst, 1947; Jensen et al., 1957; Paucker, 1960; Gitelman et al., 1984), increased resistance to mucoprotein inhibitor (Briody et al., 1955; Frösner and Gerth, 1973), and changed sensitivity of viral HA to thermal inactivation (Frösner and Gerth, 1973). Covariation of these phenotypic changes with attainment of mouse lung virulence has not been unequivocally established. PR8 virus may gain mouse lung virulence on passage in monkey kidney cell cultures yet remain unaltered with respect to sensitivity to inhibitor (Heath and Tyrrell, 1958).

Whether adaptation comprises a complex series of mutational steps or primarily entails selection of preexisting viral variants is not clear. Davenport (1954) found that members of a viral population had an "unequal capacity" for adaptation, a finding confirmed by Ledinko (1956). However, adapted virus is characterized by both faster growth rate and the capacity to attain higher titers in the lung (Davenport and Francis, 1951). This suggests that both selection of preexisting mouse lung replication mutants and their subsequent mutation occur. One hundredfold less adapted virus is required to initiate mouse lung infection than unadapted virus (Ginsberg, 1953).

Genetic Dimorphism

As emphasized earlier, influenza viruses, like all viruses, are genetically heterogeneous under ordinary conditions of propagation and passage, and mutants can be demonstrated readily by selective procedures such as passage with monoclonal antibody. However, in certain instances variants of distinctly different phenotype remain stably associated unless cloned and are reisolable from infected individuals. The term "genetic dimorphism" has been applied to such viral populations in which alternative representations of a given gene coexist (Mowshowitz and Kilbourne, 1975; reviewed by Kilbourne, 1979). Apparent dimorphism has been described for the HA and NA genes and must occur with whatever gene programs the spherical or filamentous nature of viral morphology.

L and H (low- and high-yielding) variants of contemporary swine influenza viruses have been recovered from early-passage human isolates (Kilbourne, 1978) and directly from swine (Kilbourne et al., 1979). Segregation of L and H HA genes in reassortants with PR8 virus has identified the HA gene as the determinant of the L and H pleiotropic variation that mediates their antigenic difference (Kilbourne et al., 1983) and influences the capacity of the virus to replicate in chick embryos, MDCK cells, and swine (Kilbourne, 1978; Kilbourne et al., 1979). These dimorphic phenotypes reflect Gly(L)-Glu(H) substitutions at amino acid position 155 in the HA_1 portion of the HA. By analogy with the H3 model (amino acid 158), this residue occupies a site near the tip of the HA monomer adjacent to the receptor binding site (Both et al., 1983b) (Fig. 6-8, arrow). Partial sequences of a number of L and H mutants have identified no other than Gly or Glu substitutions at position 155. In mutants nonreactive with either L-specific or H-specific monoclonal antibody, substitutions at adjacent amino acids have been noted that potentially could shield or alter the L-H epitope (Kilbourne et al., 1984; C. W. Whitaker and E. D. Kilbourne, unpublished data). Without exception, field isolates have been of either L or H phenotype or mixtures of the two.

L-H dimorphism of the swine influenza virus HA appears to be pervasive and has been demonstrated in most early-passage isolates from man and swine since 1957 (Kendal et al., 1977; E. D. Kilbourne, unpublished data) and most recently for turkeys infected with swine influenza virus (E. D. Kilbourne, unpublished data).

The earliest example of influenza viral dimorphism—also involving the HA—are the "plus" and "minus" particles of the RI5 strain of H2N2 influenza virus that differ in sensitivity to mucoprotein inhibitors (Choppin and Tamm, 1960). These variants, which exhibit striking differences in elution from red blood cells, were recovered from the same patient and were ubiquitous in 1957 isolates. Differences in the mutants reside in their different receptor binding specificity: $RI5^+$ binds to Neu Ac $\alpha_{2,6}$ Gal and $RI5^-$ to Neu Ac $\alpha_{2,3}$ Gal linkages (Carroll et al., 1981). Similar mutants of H3N2 virus have been isolated that have amino acid changes at amino acid position 226 of HA_1 in a pocket at the distal tip of the molecule, the proposed receptor binding site (Rogers et al., 1983).

Dimorphism for greater or less neuraminidase activity per virion in relation to the varying amounts of NA protein has been described for several viruses (Kilbourne et al., 1967; Webster et al., 1968; Webster and Campbell, 1972; Palese and Schulman, 1974; Mowshowitz and Kilbourne, 1975). However, although NA phenotype appears to segregate independently of the HA (Mowshowitz and Kil-

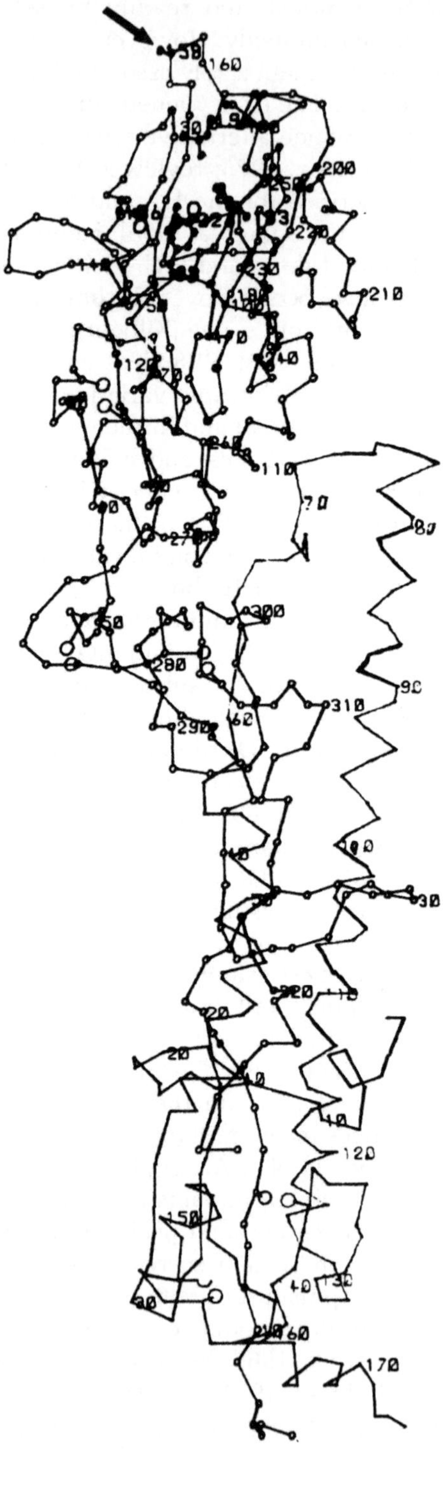

Figure 6-8. Single amino acid substitution at tip of HA monomer (arrow) can pleiotropically affect viral antigenicity and replication (see text). (From data of Both et al., 1983a. Diagram courtesy of D. C. Wiley.)

bourne, 1975), the possible role of suppressor effects of mutations in other genes or mediated by gene combination has not been sufficiently well explored to relate these apparently dimorphic phenotypes to the NA gene alone. Nor has the mutational site yet been defined.

The significance of genetic dimorphism lies in its potential as another mechanism (in addition to reassortment) for the rapid adaptation or interspecific transfer of allelic forms of virus to new or alternative hosts.

Viral Evolution

Because the antigenic evolution of influenza viruses is an everyday reality of great practical concern, consideration of influenza virus evolution is more than an abstraction. Furthermore, the basic nucleotide sequences of all influenza virus genes are known, and prototypes or lineal descendents of viruses from 1918 to 1985 are available for study. The reemergence of a 1946–57 prototype, H1N1 influenza A virus, in 1977 has provided a unique opportunity to follow its adaptation and repeated evolution in a different time frame.

Influenza Virus Evolution is the Study of the Evolution of Genes, Not Viruses

Because influenza viruses have the potential for genetic reassortment and, indeed, do reassort in nature, the definitive ancestry of recent influenza viruses as sequential lineal descendents of single antecedent strains and mere repositories of accumulating point mutations must always be suspect. Rather, in the long view, their identities are ephemeral; they are at best transiently stable packages of genes borrowed from an extended gene pool. Therefore, influenza virus genes must evolve not only in a single combination but in different combinational contexts. In a manner analogous to sexual reproduction, a mutation can be tested in many genotypes. Therefore, it can be selected more nearly according to its average contribution to fitness rather than according to the fitness of the genotype in which it originated (Smith, 1978; Kilbourne, 1981). In a different context, the concept of independent evolution of genes of even distantly related viruses has been advanced previously (Botstein, 1980; Baltimore, 1980).

These assumptions do not deny the probable origin of influenza viruses, including types A, B, and C, from a single primordial ancestor. Rather, they stress that limited divergent evolution of segmented genome viruses does not preclude their subsequent participation in an extended gene pool harbored in a variety of animal hosts and ecological niches (Table 6-5). Thus, the strategy of influenza virus has been to deploy its genes in varied hosts ready for recruitment as ecological pressures (rising population immunity) or chance interspecific contacts dictate. Obviously, divergence into genetically nonhomologous types (A, B, and C) has placed limits on this virological ecumenicism. Even among genetically homologous influenza A virus subtypes, gene sampling probably is infrequent. The nucleotide sequence divergence of RNA segment 3 that has occurred between avian and human viruses exceeds that found among different human strains (Robertson et al., 1984).

Table 6-5. Genetic and Evolutionary Potential of Segmented Genome Viruses[a]

Reassortment of genes is allelic; it occurs in all combinations (2^8)
Reassortant viruses are new viruses instantly created
Reassortment facilitates interspecific infection and adaptation to extend viral host range
Extended gene reservoir can be continuously sampled
Gene reservoir includes "extinct" (stored) genes
Reassortment could salvage "lost" mutations in noninfective particles to enhance mutational pool

[a]Modified from Kilbourne (1981)

Evolutionary Potential of Influenza Viruses as Segmented Genome Viruses

The implications of virual gene reassortment for the evolution of influenza viruses are summarized in Table 6-5. The potential for reassorting genes in any proportion to provide 2^8 combinations exceeds diploidy as a system for testing and selecting genes for evolutionary fitness. A consequence of this selection array is the opportunity it provides for the instantaneous creation of new viruses able to break through species barriers, allowing viral genes to reach new hosts in the context of new genes. Evidence is strong that this occurred some time prior to 1968 with the mixture of seven H2N2 genes with an H3 HA closely related to the HA of an avian virus to create the H3N2 subtype (Fang et al., 1981; Ward and Dopheide, 1981; Nakajima et al., 1982).

The virus laboratory must be recognized as part of the viral ecosystem. This is particularly important with the short-lived influenza viruses, which continually emerge and then disappear. The laboratory thus becomes a repository and cold storage genetic reservoir of the genes of long-extinct influenza viruses (Kilbourne, 1981). Although there are many constraints against the escape of these genes from the laboratory (Kilbourne, 1974), their frozen genes are part of the gene pool potentially available for retrieval and utilization by contemporarily circulating strains.

Although it has not been established that influenza viruses are unusually mutable, reassorting viruses have potentially available a mechanism to provide an increase in available mutations if not a true increase in mutation rate. Endogenous autoreassortment within an infected host might salvage mutations occurring in noninfective virus particles for ultimate incorporation into virions through genetic reassortment during multicycle infection. This event would increase the mutational pool and enhance the opportunity for selection of favorable mutations that might otherwise have been lost (Kilbourne, 1973, 1981).

Evolution of the Genes of Influenza Viruses by Sequential Mutations

Comparison of nucleotide sequences of all eight RNA segments of 14 influenza A viruses isolated at widely separated intervals has revealed an extraordinarily high rate of silent mutation in all genes. All genes appeared to be governed by the same molecular clock, as judged by these non-amino-changing substitutions (Hayashida et al., 1985).

Evolution of the HA Gene

Antigenic drift. Study of antigenic drift in the HA of N3N2 influenza viruses from 1968 to 1980 by nucleotide sequencing has disclosed multiple evolutionary pathways in the strains sampled (Both et al., 1983b) (Fig. 6-9). Amino acid changes in HA_1 were clustered at previously proposed antigenic regions, with other regions containing few changes (Wiley et al., 1981; Both and Sleigh, 1981; Both et al., 1983b). It appears that the sites at which amino acid change can alter an-

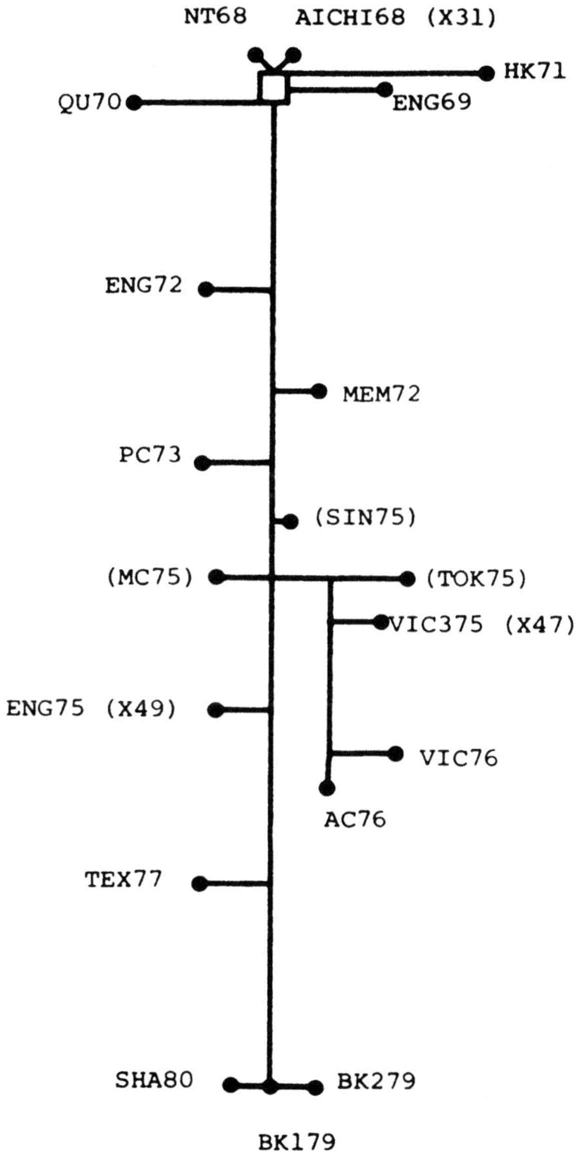

Figure 6-9. Antigenic drift of HA of H3N2 influenza virus (1968–1980) showing multiple evolutionary pathways. (From Both et al., 1983b. Reproduced with permission, from *J. Virol.* **48**:58.)

tigenicity are so limited that in some strains key amino acids at these sites change for a second time as evolution proceeds. These single-site secondary changes did not produce variants having major epidemiologic impact (Both et al., 1983b). Restriction of changes to limited regions of the HA molecule and the limited opportunities for a single base change to provide sequential changes at a single amino acid residue may lead to evolutionary dead ends, reducing the probability of further antigenic drift. In this connection, however, it should be noted that laboratory-induced antigenic variation in apparently terminal field isolates may select mutants showing further changes after their displacement in nature by a new subtype (Haaheim and Schild, 1976).

Less extensive studies of H1N1 viruses are in essential agreement with those of H3N2 strains. The rate of drift for the two subtypes has been estimated respectively at 0.8% and 0.9% per year (Raymond et al., 1983; Both et al., 1983b; Sleigh and Both, 1981). With the H1 HA also, sequential changes at a single defined residue were found as well as reversion (Raymond et al., 1983). The exhibition of alternative evolutionary pathways has already occurred with the H1N1 viruses after their reintroduction in 1977. Oligonucleotide maps of 1977 viruses at the beginning and end of a 6-month period differed more from each other than from maps of 1950 and 1977 viruses isolated 27 years apart (Young et al., 1979). Antigenic characterization of Netherland H1N1 viruses of the past and present suggests that the drift of recent H1N1 strains has abandoned the pattern of the period 1947–1957 (Beyer and Masurel, 1984), a suggestion buttressed by independent studies of HA gene sequence (Raymond et al., 1986).

Variation in the HA antigen of the 1977 H1N1 virus began within weeks of its introduction into a single community (Houston); by the end of the first season few viral isolates were antigenically unchanged (Six et al., 1983).

In the case of influenza C viruses, cumulative HA mutations are not found; rather, nucleotide sequences vary without relation to the year of virus isolation. Thus, epidemiologically dominant variants of influenza C viruses do not emerge successively with time. Furthermore, influenza C virus variants derived from multiple evolutionary pathways cocirculate at any one time (Buonagurio et al., 1985).

Evolutionary changes in the HA affecting other than antigenic sites. Comparison of pathogenic and nonpathogenic strains of FPV (H7 subtype) showed that some of the HA genes of cleavable (i.e., pathogenic) strains of virus were more closely related to those coding for uncleavable HAs than to other cleavable HAs. These data suggested that the overall evolution of the H7 hemagglutinin is different from the evolution of the specific cleavage site (Bosch et al., 1982).

Evolution of Genes for Nonsurface Viral Proteins

NS. Study of NS genes (RNA 8) of influenza A viruses isolated between 1934 and 1972 disclosed nucleotide sequence variation of 2.2–3.4% for 10 years. It appeared that the genes studied had evolved through cumulative base changes along a single common lineage. The evolutionary rate of the H3N2 HA was higher (4.5–6.5% for 10 years). This was also true of silent mutations (Krystal et al., 1983).

Other genes. Nucleotide sequences of other viral genes have been too limited in number to permit valid estimates of evolutionary rates (Krystal et al., 1983). Comparison of the sequence of FPV NP gene with those of two known human influenza viruses revealed 84–85% homology at the nucleotide level and 91–93%

at the amino acid level, suggesting common functional requirements of influenza A virus NPs (Tomley and Roditi, 1984). Comparison of RNAs 5 and 3 in avian influenza A viruses isolated in 1934 and 1968 showed them to be highly homologous and colinear, indicating a common lineage (Fields and Winter, 1982; reviewed by Bishop et al., 1982a,b).

REFERENCES

Allen, H., McCauley, J., Waterfield, M , et al , 1980, Influenza virus RNA segment 7 has the coding capacity for two polypeptides, *Virology* **107**:548-551

Almond, J W., 1977, A single gene determines the host range of influenza virus, *Nature* **270**:617-618

Andrew, M E , Coupar, B. E H , Ada, G L , et al , 1986, Cell-mediated immune responses to influenza virus antigens expressed by vaccinia virus recombinants, *Micro Pathogen* **1**:443-452

Archetti, I , and Horsfall, F. L., 1950, Persistent antigenic variation of influenza A viruses after incomplete neutralization *in ovo* with heterologous immune serum, *J Exp Med.* **92**:441-446.

Baez, M , Taussig, R , Zazra, J. J., et al., 1980, Complete nucleotide sequence of the influenza A/PR/8/34 virus NS gene and comparison with the NS genes of the A/Udorn/72 and A/FPV/Rostock/34 strains, *Nucleic Acids Res.* **8**:5845-5858.

Baltimore, D , 1980, Evolution of RNA viruses, *Ann. N Y. Acad. Sci.* **354**:492-497.

Bennink, J. R., Yewdell, J. W., Smith, G. L , et al., 1986, Recognition of cloned influenza virus hemagglutinin gene products by cytotoxic T lymphocytes, *J. Virol.* **57**:786-791.

Bentley, D. R., and Brownlee, G. G., 1982, Sequence of the N2 neuraminidase from influenza virus A/NT/60/68, *Nucleic Acids Res.* **10**:5033-5042

Beyer, W. E. P., and Masurel, N , 1984, De antigeendrift van het influenza A-H1N1-virus De nieuwe "India-1980"—variant is ook in Nederland actief, *Ned. Tijdschr. Geneeskd* **128**:1369-1372.

Birrer, M. J., Udem, S., Nathenson, S., et al., 1981, Antigenic variants of measles virus, *Nature* **293**:67-69.

Bishop, D. H. L., Huddleston, J. A., and Brownlee, G. G., 1982a, The complete sequence of RNA segment 2 of influenza A/NT/60/68 and its encoded P1 protein, *Nucleic Acids Res* **10**:1335-1343.

Bishop, D. H. L., Jones, K. L., Huddleston, J. A., et al., 1982b, Influenza A virus evolution: Complete sequences of influenza A/NT/60/68 RNA segment 3 and its predicted acidic P polypeptide compared with those of influenza A/PR/8/34, *Virology* **120**:481-489.

Blok, J., and Air, G. M., 1982, Block deletions in the neuraminidase genes from some influenza A viruses of the N1 subtype, *Virology* **118**:229-234.

Bonin, J., and Scholtissek, C., 1983, Mouse neurotropic recombinants of influenza A viruses, *Arch. Virol.* **75**:255-268.

Bosch, F. X., Orlich, M., Klenk, H.-D., et al., 1979, The structure of the hemagglutinin, a determinant for the pathogenicity of influenza viruses, *Virology* **95**:197-207.

Bosch, F. X., Garten, W., Klenk, H.-D., et al., 1981, Proteolytic cleavage of influenza virus hemagglutinins: Primary structure of the connecting peptide between HA_1 and HA_2 determines proteolytic cleavability and pathogenicity of avian influenza viruses, *Virology* **113**:725-735.

Bosch, F. X., von Hoyningen-Huene, V., Scholtissek, C., et al., 1982, The overall evolution of the H7 influenza virus haemagglutinins is different from the evolution of the proteolytic cleavage site, *J. Gen. Virol.* **61**:101-104.

Both, G W., and Sleigh, M. J., 1980, Complete nucleotide sequence of the haemagglutinin gene from a human influenza virus of the Hong Kong subtype, *Nucleic Acids Res.* **8**:2561-2575

Both, G. W., and Sleigh, M. J., 1981, Conservation and variation in the hemagglutinins of Hong Kong subtype influenza viruses during antigenic drift, *J. Virol.* **39**:663-672.

Both, G W , Shi, C. H , and Kilbourne, E D , 1983a, Hemagglutinin of swine influenza virus: A single amino acid change peiotropically affects viral antigenicity and replication, *Proc Nat Acad Sci U.S.A.* **80**:6996-7000.

Both, G. W , Sleigh, M. J , Cox, N. J., et al , 1983b. Antigenic drift in influenza virus H3 hemagglutinin from 1968 to 1980: Multiple evolutionary pathways and sequential amino acid changes at key antigenic sites, *J. Virol.* **48**:52-60.

Botstein, D., 1980, A theory of modular evolution for bacteriophages, *Ann. N.Y. Acad. Sci.* **354**:484–491.

Braciale, T. J., Braciale, V. I., Henkel, T. J., et al., 1984, Cytotoxic T lymphocyte recognition of the influenza hemagglutinin gene product expressed by DNA-mediated gene transfer, *J. Exp. Med.* **159**:341–354.

Brand, C., and Palese, P., 1980, Sequential passage of influenza virus in embryonated eggs or tissue culture: Emergence of mutants, *Virology* **107**:424–433.

Briody, B. A., Cassel, W. A., and Medill, M. A., 1955, Adaptation of influenza virus to mice. III. Development of resistance to β inhibitor, *J. Immunol.* **74**:41–45.

Brown, L. E., Hinshaw, V. S., and Webster, R. G., 1983, Antigenic variation in the influenza A virus nonstructural protein, NS1, *Virology* **130**:134–143.

Buonagurio, D. A., Krystal, M., Palese, P., et al., 1984, Analysis of an influenza A virus mutant with a deletion in the NS segment, *J. Virol.* **49**:418–425.

Buonagurio, D. A., Nakada, S., Desselberger, U., et al., 1985, Noncumulative sequence changes in the hemagglutinin genes of influenza C virus isolates, *Virology* **146**:221–232.

Burnet, F. M., 1956, Structure of influenza virus, *Science* **123**:1101–1104.

Burnet, F. M., and Bull, D. R., 1943, Changes in influenza virus associated with adaptation to passage in chick embryos, *Aust. J. Exp. Biol. Med. Sci.* **21**:55–69.

Carroll, S. M., Higa, H. H., and Paulson, J. C., 1981, Different cell-surface receptor determinants of antigenically similar influenza virus hemagglutinins, *J. Biol. Chem.* **256**:8357-8363.

Caton, A. J., Brownlee, G. G., Yewdell, J. W., et al., 1982, The antigenic structure of the influenza virus A/PR/8/34 hemagglutinin (H1 subtype), *Cell* **31**:417–427.

Choppin, P. W., and Tamm I., 1960, Studies of two kinds of virus particles which comprise influenza A2 virus strains. II. Reactivity with virus inhibitors in normal sera, *J. Exp. Med.* **112**:921–944.

Concannon, P., Kwolek, C. J., and Salser, W. A., 1984, Nucleotide sequence of the influenza virus A/USSR/90/77 neuraminidase gene, *J. Virol.* **50**:654–656.

Daniels, R. S., Douglas, A. R., Skehel, J. J., et al., 1983, Analyses of the antigenicity of influenza haemagglutinin at the pH optimum for virus-mediated membrane fusion, *J. Gen. Virol.* **64**:1657–1662.

Daniels, R. S., Douglas, A. R., Skehel, J. J., et al., 1984, Antigenic analyses of influenza virus haemagglutinins with different receptor-binding specificities, *Virology* **138**:174–177.

Daniels, R. S., Downie, J. C., Hay, A. J., et al., 1985, Fusion mutants of the influenza virus hemagglutinin glycoprotein, *Cell* **40**:431–439.

Davenport, F. M., 1954, The inequality of potential in influenza virus for adaptation to mice, *J. Immunol.* **72**:485–488.

Davenport, F. M., and Francis, T., Jr., 1951, A comparison of the growth curves of adapted and unadapted lines of influenza virus, *J. Exp. Med.* **93**:129–137.

Davis, A. R., Nayak, D. P., Ueda, M., et al., 1981, Expression of antigenic determinants of the hemagglutinin gene of a human influenza virus in *Escherichia coli*, *Proc. Natl. Acad. Sci. U.S.A* **78**:5376–5380.

Deom, C. M., Caton, A. J., and Schulze, I. T., 1986, Host cell-mediated selection of a mutant influenza A virus that has lost a complex oligosaccharide from the tip of the hemagglutinin, *Proc. Natl. Acad. Sci. U.S.A.* **83**:3771–3775.

Desselberger, U., Racaniello, V. R., Zazra, J. T., et al., 1980, The 3′ and 5′-terminal sequences of influenza A, B and C virus RNA segments are highly conserved and show partial inverted complementarity, *Gene* **8**:315–328.

Doms, R. W., Gething, M.-J., Henneberry, J., et al., 1986, Variant influenza virus hemagglutinin that induces fusion at elevated pH, *J. Virol.* **57**:603–613.

Elleman, T. C., Azad, A. A., and Ward, C. W., 1982, Neuraminidase gene from the early Asian strain of human influenza virus, A/RI/5⁻/57 (H2N2), *Nucleic Acids Res.* **10**:7005–7015.

Emtage, J. S., Tacon, W. C. A., Catlin, G. H., et al., 1980, Influenza antigenic determinants are expressed from haemagglutinin genes cloned in *Escherichia coli, Nature,* **283**:171–174.

Erickson, A. H., and Kilbourne, E. D., 1980, Comparative amino acid analysis of influenza A viral proteins, *Virology* **100**:34–42.

Fang, R., Min Jou, W., Huylebroeck, D., et al., 1981, Complete structure of A/duck/Ukraine/63 influenza hemagglutinin gene: Animal virus as progenitor of human H3 Hong Kong 1968 influenza hemagglutinin, *Cell* **25**:315–323.

Fazekas de St. Groth, S., 1978, Antigenic, adaptive and adsorptive variants of the influenza A hemagglutinin, in: *Topics In Infectious Diseases*, Vol. 3 (W. G. Laver, H. Bachmayer, and R. Weil, eds.), Springer Verlag, Berlin, pp. 25–48.

Feng, S.-Y , Tanaka, T , Goto, H., et al. 1983, Genetic recombination between temperature-sensitive and wild-type influenza A virus strains, *Acta Virol.* **27**:21–26.

Fields, S., and, Winter, G , 1982, Nucleotide sequences of influenza virus segments 1 and 3 reveal mosaic structure of a small viral RNA segment, *Cell* **28**:303–313.

Fields, S Winter, G , and Brownlee, G G., 1981, Structure of the neuraminidase gene in human influenza virus A/PR/8/34, *Nature* **290**:213–217.

Flehmig, B , Vallbracht, A , and Gerth, H -J , 1976, Influenza virus Association of mouse-lung virulence with plaque formation in mouse kidney cells, *Intervirology* **7**:201–210

Fraser, K B., 1959, Features of the MEL × NWS recombination systems in influenza A virus. IV Increments of virulence during successive cycles of double infection with the strains of influenza A virus, *Virology* **9**:202–214.

Frosner, G G , and Gerth, H.-J., 1973, Changes in temperature-sensitivity of haemagglutinin and neuraminidase during adaptation of influenza virus to mice, *Med. Microbiol. Immunol.* **158**:161–163.

Gerber, P , Loosli, C G., and Hamre, D., 1955, Antigenic variants of influenza A virus (PR8). I Their development during serial passage in the lungs of partially immune mice, *J Exp Med* **101**:627–638.

Gerhard, W., Yewdell, J W., Lopes, D., et al , 1983, Point mutations in the hemagglutinin molecule of influenza virus· Their frequencies, phenotypic characteristics and biological relevance, in: *Medical Virology II* (L. M. Dela Maza and E. M. Peterson, eds.), Elsevier, New York, pp. 227–249.

Gething, M.-J , and Sambrook, J., 1981, Cell-surface expression of influenza haemagglutinin from a cloned DNA copy of the RNA gene, *Nature* **293**:620–625.

Gething, M.-J , and Sambrook, J., 1982a, The expression of the influenza virus hemagglutinin gene from SV40-HA recombinants, in: *Eukaryotic Viral Vectors* (Y. Gluzman, ed.), Cold Spring Harbor Laboratory, New York, pp. 29–45.

Gething, M -J., and Sambrook, J., 1982b, Construction of influenza haemagglutinin genes that code for intracellular and secreted forms of the protein, *Nature* **300**:598–603.

Gething, M.-J., Bye, J., Skehel, J., et al., 1980, Cloning and DNA sequence of double-stranded copies of haemagglutinin genes from H2 and H3 strains elucidates antigenic shift and drift in human influenza virus, *Nature* **287**:301–306.

Ghendon, Y., Markushin, S , Lisovskaya, K , et al , 1982 Extragenic suppression of a ts phenotype during recombination between ts mutants of two fowl plague virus strains with a ts mutation in gene 1, *J Gen. Virol.* **62**:239–248.

Ginsberg, H. S., 1953, Comparison of quantity of egg and mouse-adapted influenza viruses required to infect each host, *Proc. Soc. Exp. Biol. Med.* **84**:249–252.

Gitelman, A. K , Kaverin, N. V., Kharitonenkov, I. G., et al., 1984, Changes in the antigenic specificity of influenza hemagglutinin in the course of adaptation to mice, *Virology* **134**:230–232.

Gotlieb, T., and Hirst, G K., 1954, The experimental production of combination forms of virus. III. The formation of doubly antigenic particles from influenza A and B virus and a study of the ability of individual particles of X virus to yield two separate strains, *J. Exp Med.* **99**:307–320.

Granoff, A , and Hirst, G. K., 1954, Experimental production of combination forms of virus. IV. Mixed influenza A-Newcastle disease virus infections, *Proc. Soc. Exp. Biol. Med* **86**:84–88.

Haaheim, L. R., and Schild, G C , 1976, Antigenic variants of influenza A virus obtained *in vitro, Bull. WHO* **53**:305–311.

Hartman, J. R , Nayak, D. P., and Fareed, G C., 1982, Human influenza virus hemagglutinin is expressed in monkey cells using simian virus 40 vectors, *Proc. Natl. Acad. Sci. U.S.A.* **79**:233–237.

Hayashida, H., Toh, H., Kikuno, R., et al., 1985, Evolution of influenza virus genes, *Mol. Biol. Evol.* **2**:289–303.

Heath, R B., and Tyrrell, D. A. J , 1958, The behaviour of some influenza viruses in tissue cultures of kidney cells of various species, *Arch. Ges. Virusforsch.* **5**:577–591.

Heiland, I , and Gething, M -J , 1981, Cloned copy of the haemagglutinin gene codes for human influenza antigenic determinants in *E. coli, Nature* **292**:851–852

Henle, W., and Liu, O. C., 1951, Studies on host-virus interactions in the chick embryo-influenza virus system. VI. Evidence for multiplicity reactivation of inactivated virus, *J. Exp Med.*

94:305–322.
Hinshaw, V. S., Webster, R. G., Naeve, C. W., et al., 1983, Altered tissue tropism of human-avian reassortant influenza viruses, *Virology* **128**:260–263.
Hirst, G. K., 1947. Studies on the mechanism of adaptation of influenza virus to mice. *J. Exp. Med.* **86**:357–366.
Hirst, G. K., 1962, Genetic recombination with Newcastle disease virus, polioviruses, and influenza, *Cold Spring Harbor Symp. Quant. Biol.* **27**:303–309.
Hirst, G. K., and Gotlieb, T., 1953, The experimental production of combination forms of virus. 1. The occurrence of combination forms after simultaneous inoculation of the allantoic sac with two distinct strains of influenza virus, *J. Exp. Med.* **98**:41–52.
Hirst, G. K., and Pons, M., 1973, Mechanism of influenza recombination. II. Virus aggregation and its effect on plaque formation by so-called noninfective virus, *Virology* **56**:620–631.
Hiti, A. L., and Nayak, D. P., 1982, Complete nucleotide sequence of the neuraminidase gene of human influenza virus A/WSN/33, *J. Virol.* **41**:730–734.
Hiti, A. L., Davis, A. R., and Nayak, D. P., 1981, Complete sequence analysis shows that hemagglutinins of the H0 and H2 subtypes of human influenza viruses are closely related, *Virology* **111**:113–124.
Hobson, D., Flockton, H. I., and Gould, E. A., 1968, The inhibitor-destroying activity of influenza A virus strains of differing neurovirulence of mice, *J. Gen. Virol.* **3**:445–448.
Holland, J., Spindler, K., Horodyski, F., et al., 1982, Rapid evolution of RNA genomes, *Science* **215**:1577–1585.
Horsfall, F. L., Jr., 1952, Biological selection of antigenic variants, *Fed. Proc.* **11**:804–807.
Horsfall, F. L., Jr., 1954, On the reproduction of influenza virus. Quantitative studies with procedures which enumerate infective and hemagglutinating virus particles, *J. Exp. Med.* **100**:135–161.
Isaacs, A., and Donald, H. B., 1955, Particle counts of hemagglutinating viruses, *J. Gen. Microbiol.* **12**:241–247.
Israël, A., 1980a, A temperature-dependent host range mutant of fowl plague virus (FPV), *Virology* **105**:1–12.
Israël, A., 1980b, Genotypic and phenotypic characterization of a mammalian cell-adapted mutant of fowl plague virus (FPV), *J. Gen. Virol.* **51**:33–44.
Jabbar, M. A., Sivasubramanian, N., and Nayak, D. P., 1985, Influenza viral (A/WSN/33) hemagglutinin is expressed and glycosylated in yeast, *Saccharomyces cerevisiae, Proc. Natl. Acad. Sci. U.S.A.* **82**:2019–2023.
Jackson, D. C., and Webster, R. G., 1982, A topographic map of the enzyme active center and antigenic sites on the neuraminidase of influenza virus A/Tokyo/3/67 (H2N2), *Virology* **123**:69–77.
Jensen, K. E., Minuse, E., and Francis, T., Jr., 1957, Serologic comparisons with lines of influenza virus isolated and serially transferred in different experimental hosts, *J. Immunol.* **78**:356–364.
Jones, K. L., Huddleston, J. A., and Brownlee, G. G., 1983, The sequence of RNA segment 1 of influenza virus A/NT/60/68 and its comparison with the corresponding segment of strains A/PR/8/34 and A/WSN/33, *Nucleic Acids Res.* **11**:1555–1566.
Kaptein, J. S., and Nayak, D. P., 1982, Complete nucleotide sequence of the polymerase 3 gene of human influenza virus A/WSN/33, *J. Virol.* **42**:55–63.
Kendal, A. P., Noble, G. R., and Dowdle, W. R., 1977, Swine influenza viruses isolated in 1976 from man and pigs contain two coexisting subpopulations with antigenically distinguishable hemagglutinins, *Virology* **82**:111–121.
Kilbourne, E. D., 1955, Reactivation of non-infective virus in a cortisone-injected host, *J. Exp. Med.* **101**:437–450.
Kilbourne, E. D., 1957, The influence of cortisone on experimental infection. IV. Negation of interference as the mechanism by which cortisone induces increased virus yields, *J. Exp. Med.* **106**:863–881.
Kilbourne, E. D., 1963, Influenza virus genetics, *Prog. Med. Virol.* **5**:79–126.
Kilbourne, E. D., 1969, Future influenza vaccines and the use of genetic recombinants, *Bull. WHO* **41**:643–645.
Kilbourne, E. D., 1973, Hypothesis—An explanation of the interpandemic antigenic mutability of influenza viruses, *J. Infect. Dis.* **128**:668–669.
Kilbourne, E. D., 1974, Virus research, *Science* **184**:410–411.

Kilbourne, E. D., 1978, Genetic dimorphism in influenza viruses: Characterization of stably associated hemagglutinin mutants differing in antigenicity and biological properties, *Proc Natl. Acad Sci. U S A* **75**:6258-6262.
Kilbourne, E D., 1979, Molecular epidemiology—Influenza as archetype, *Harvey Lect.* **73**:225-258.
Kilbourne, E. D., 1981, Segmented genome viruses and the evolutionary potential of asymmetrical sex, *Perspect Biol. Med.* **25**:66-77.
Kilbourne, E D., 1983, Are new diseases really new? *Nat. Hist.* **92**:28-32.
Kilbourne, E. D , 1984, Molecular determination of the epidemiology of influenza—a reconciliation of approaches, in: *The Molecular Virology and Epidemiology of Influenza* (C. Stuart-Harris and C W Potter, eds.), Academic Press, London, pp. 257-267
Kilbourne, E. D., Lief, F. S., Schulman, J. L., et al , 1967, Antigenic hybrids of influenza viruses and their implications, in *Perspectives in Virology*, Vol V (M Pollard, ed), Academic Press, New York, pp 87-106
Kilbourne, E D , McGregor, S , and Easterday, B. C , 1979, Hemagglutinin mutants of swine influenza virus differing in replication characteristics in their natural host, *Infect. Immun.* **26**: 197-201.
Kilbourne, E D., Gerhard, W , and Whitaker, C. W., 1983, Monoclonal antibodies to the hemagglutinin Sa antigenic site of A/PR/8/23 influenza virus distinguish biologic mutants of swine influenza virus, *Proc Natl Acad Sci U S A* **80**:6300-6402
Kilbourne, E D , Both, G. W., and Gerhard, W., 1984, Pleiotropic effects of a single amino acid change on antigenicity and biologic function of swine influenza virus hemagglutinin mutants, in: *Negative Strand Viruses* (R Compans and D Bishop, eds.), Academic Press, New York, pp 233-237.
Knossow, M., Daniels, R. S., Douglas, A. R., et al., 1984, Three-dimensional structure of an antigenic mutant of the influenza virus haemagglutinin, *Nature* **311**:678-680.
Krystal, M., Buonagurio, D., Young, J. F , et al., 1983, Sequential mutations in the NS genes of influenza virus field strains, *J. Virol* **45**:547-554.
Krystal, M., Li, R , Lyles, D., et al , 1986, Expression of the three influenza virus polymerase proteins in a single cell allows growth complementation of viral mutants, *Proc Natl Acad Sci U S A* **83**:2709-2713.
Kuroda, K , Hauser, C., Rott, R., et al., 1986, Expression of the influenza virus haemogglutinin in insect cells by a baculovirus vector, *Embo J.* **5**:1359-1365
Lamb, R. A , and Lai, C.-J., 1980, Sequence of interrupted and uninterrupted mRNAs and cloned DNA coding for the two overlapping nonstructural proteins of influenza virus, *Cell* **21**:475-485.
Lamb, R. A , and Lai, C.-J., 1981, Conservation of the influenza virus membrane protein (M_1) amino acid sequence and an open reading frame of RNA segment 7 encoding a second protein (M_2) in H1N1 and H3N2 strains. *Virology* **112**:746-751.
Lamb, R. A., and Lai, C.-J., 1982, Spliced and unspliced messenger RNAs synthesized from cloned influenza virus M DNA in an SV40 vector: Expression of the influenza virus membrane protein (M_1), *Virology* **123**:237-256.
Laver, W. G., and Kilbourne, E. D., 1966, Identification in a recombinant influenza virus of structural proteins derived from both parents, *Virology* **30**:493-501.
Laver, W. G., and Webster, R. G., 1968, Selection of antigenic variants of influenza viruses. Isolation and peptide mapping of their hemagglutinating proteins, *Virology* **34**:193-202
Ledinko, N., 1956, An analysis of the process of adaptation of influenza virus B of recent human origin to the mouse lung, *J. Gen. Microbiol.* **15**: 47-60.
Lentz, M. R., and Air, G. M., 1986, Loss of enzyme activity in a site-directed mutant of influenza neuraminidase compared to expressed wildtype protein, *Virology* **148**:74-83.
Lentz, M. R , Air, G. M., Laver, W. G., et al., 1984, Sequence of the neuraminidase gene of influenza virus A/Tokyo/3/67 and previously uncharacterized monoclonal variants, *Virology* **135**:257-265.
Lubeck, M D., Palese, P., and Schulman, J. L., 1979, Nonrandom association of parental genes in influenza A virus recombinants, *Virology* **95**:269-274.
Lubeck, M. D , Schulman, J. L., and Palese, P., 1980, Antigenic variants of influenza viruses: Marked differences in the frequencies of variants selected with different monoclonal antibodies, *Virology* **102**:458-462.
Maassab, H. F., and DeBorde, D. C. 1983, Characterization of an influenza A host range mutant, *Virology* **130**:342-350.
Mackenzie, J. S., 1970, Isolation of temperature-sensitive mutants and the construction of a prelimi-

nary genetic map for influenza virus, *J. Gen. Virol.* **6**:63–75.
Magill, T. P., 1955, Propagation of influenza virus in "immune" environments, *J Exp. Med.* **102**:279–289.
Mahy, B. W. J., 1983, Mutants of influenza virus, in: *Genetics of Influenza Viruses* (P. Palese and D. W. Kingsbury, eds.), Springer-Verlag, New York, pp. 192–254.
Markoff, L., and Lai, C.-J., 1982, Sequence of the influenza A/Udorn/72 (H3N2) virus neuraminidase gene as determined from cloned full-length DNA, *Virology* **119**:288–297.
Markoff, L. J., Murphy, B. R., Kendal, A. J., et al., 1979, Probable association of plaque size with neuraminidase subtype among H3N2 influenza A viruses, *Arch. Virol.* **62**:277–280.
Markoff, L., Lin, B.-C., Sveda, M. M., et al., 1984, Glycosylation and surface expression of the influenza virus neuraminidase requires the N-terminal hydrophobic region, *Mol. Cell. Biol.* **4**:8–16.
Martinez, C., del Rio, L., Portela, A., et al., 1983, Evolution of the influenza virus neuraminidase gene during drift of the N2 subtype, *Virology* **130**:539–545.
Massicot, J. G., Van Wyke, K., Chanock, R. M., et al., 1982, Evidence for intrasegmental complementation between two influenza A viruses having *ts* mutations on their P1 genes, *Virology* **117**:496–500.
Mayer, V., Schulman, J. L., and Kilbourne, E. D., 1973, Non-linkage of neurovirulence exclusively to viral hemagglutinin or neuraminidase in genetic recombinants of A/NWS (H0N1) influenza virus, *J. Virol.* **11**:272–278.
McCauley, J. W., Mahy, B. W. J., and Inglis, S. C., 1982, Nucleotide sequence of fowl plague virus RNA segment 7, *J. Gen. Virol.* **58**: 211–215.
McQueen, N. L., Nayak, D. P., Jones, L. V., et al., 1984, Chimeric influenza virus hemagglutinin containing either the NH_2 terminus or the COOH terminus of G protein of vesicular stomatitis virus is defective in transport to the cell surface, *Proc. Natl. Acad. Sci. U.S.A.* **81**:395–399.
Min Jou, W., Verhoeyen, M., Devos, R., et al., 1980. Complete structure of the hemagglutinin gene from the human influenza A/Victoria/3/75 (H3N2) strain as determined from cloned DNA, *Cell* **19**:683–696.
Mowshowitz, S., and Kilbourne, E. D., 1975, Genetic dimorphism of the neuraminidase in recombinants of H3N2 influenza virus, in: *Negative Strand Viruses*, Vol. 2 (R. D. Barry and B. W. J. Mahy, eds.), Academic Press, London, pp. 765–775.
Murphy, B. R., and Webster, R. G., 1985, Influenza viruses, in: *Virology* (B. N. Fields, D. M. Knide, R. M. Chanock, J. L. Melnick, B. Roizman, and R. E. Shope eds.), Raven Press, New York, pp. 1179–1239.
Murphy, B. R., Tolpin, M. D., Massicot, J. G., et al., 1980, Escape of highly defective influenza A virus mutant from its temperature-sensitive phenotype by extragenic suppression and other types of mutation, *Ann. N.Y. Acad. Sci.* **354**:172–182.
Naeve, C. W., Webster, R. G., and Hinshaw, V. S., 1983, Phenotypic variation in influenza virus reassortants with identical gene constellations, *Virology* **128**:331–340.
Naeve, C. W., Hinshaw, V. S., and Webster, R. G., 1984, Mutations in the hemagglutinin receptor-binding site can change the biological properties of an influenza virus, *J. Virol.* **51**:567–569.
Nakajima, S., and Kendal, A. P., 1981, Antigenic drift in influenza A/USSR/90/77 (H1N1) variants selected *in vitro* with monoclonal antibodies, *Virology* **113**:656–662.
Nakajima, S., and Sugiura, A., 1977, Neurovirulence of influenza virus, *Virology* **81**:486–489.
Nakajima, S., and Sugiura, A., 1980, Neurovirulence as studied in a neuroblastoma cell line, *Virology* **101**:450–457.
Nakajima, K., Nakajima, S., and Sugiura, A., 1982, The possible origin of H3N2 influenza virus, *Virology* **120**:504–509.
Nakajima, S., Nakajima, K., and Kendal, A. P., 1983, Identification of the binding sites to monoclonal antibodies on A/USSR/90/77 (H1N1) hemagglutinin and their involvement in antigenic drift in H1N1 influenza viruses, *Virology* **131**:116–127.
Newton, S. E., Air, G. M., Webster, R. G., et al., 1983, Sequence of the hemagglutinin gene of influenza virus A/Memphis/1/71 and previously uncharacterized monoclonal antibody-derived variants, *Virology* **128**:495–501.
Nohinek, B., Gerhard, W., and Schulze, I. T., 1985, Characterization of host cell binding variants of influenza virus by monoclonal antibodies, *Virology* **143**:651–656.
Ogawa, T., and Ueda, M., 1981, Genes involved in the virulence of an avian influenza virus, *Virology*

113:304–313

Ortin, J , Najera, R., Lopez, C., et al., 1980, Genetic variability of Hong Kong (H3N2) influenza viruses: Spontaneous mutations and their location in the viral genome, *Gene* **11**:319–331.

Palese, P , and Schulman, J., 1974, Isolation and characterization of influenza virus recombinants with high and low neuraminidase activity. Use of 2-(3′-Methoxyphenyl)-*N*-acetylneuraminic acid to identify cloned populations, *Virology* **57**:227–237.

Palese, P., Tobita, K., Ueda, M., et al., 1974, Characterization of temperature sensitive influenza virus mutants defective in neuraminidase, *Virology* **61**:397–410

Paoletti, E , Perkus, M , Piccini, A., et al., 1985, Genetically engineered poxviruses as live recombinant vaccines, in· *Microbiology–1985* (L. Leive, ed), American Society of Microbiology, Washington, pp 251–254

Parvin, J D , Young, J F , and Palese, P , 1983, Nonsense mutations affecting the lengths of the NS1 nonstructural proteins of influenza A virus isolates, *Virology* **128**:512–517

Parvin, J D , Moscona, A , Pan, W T , et al., 1986, Measurement of the mutation rate of animal viruses Influenza A virus and poliovirus-1 *J Virol* **59**:377–383

Paucker, K , 1960, Apparent antigenic change in a laboratory line of influenza A virus during continued propagation in chick embryos, *J Immunol* **85**:148–162.

Porter, A G , Barber, C , Carey, N. H., et al , 1979, Complete nucleotide sequence of an influenza virus haemagglutinin gene from cloned DNA, *Nature* **282**:471–477.

Porter, A G , Smith, J C , and Emtage, J S , 1980, Nucleotide sequence of influenza virus RNA segment 8 indicates that coding regions for NS_1 and NS_2 proteins overlap, *Proc Natl Acad Sci U S A* **77**:5074–5078

Portner, A , Webster, R G , and Bean, W J., 1980, Similar frequencies of antigenic variants in Sendai, vesicular stomatitis, and influenza A viruses, *Virology* **104**:235–238.

Raymond, F. L., Caton, A J., Cox, N. J., et al., 1983, Antigenicity and evolution amongst recent influenza viruses of H1N1 subtype, *Nucleic Acids Res* **11**:7191–7203.

Raymond, F L., Caton, A. J., Cox, N. J., et al., 1986, The antigenicity and evolution of influenza H1 haemagglutinin, from 1950–1957 and 1977–1983 Two pathways from one gene, *Virology* **148**:275–287.

Robertson, J. S., Robertson, M. E. St C , and Roditi, I. J., 1984, Nucleotide sequence of RNA segment 3 of the avian influenza A/FPV/Rostock/34 and its comparison with the corresponding segment of human strains A/PR/8/34 and A/NT/60/68, *Virus Res* **1**:73–79.

Robertson, J. S., Naeve, C. W., Webster, R. G., et al., 1985, Alterations in the hemagglutinin associated with adaptation of influenza B virus to growth in eggs, *Virology* **143**:166–174

Roditi, I. J , and Robertson, J. S., 1984, Nucleotide sequence of the avian influenza virus A/fowl plague/Rostock/34 segment 1 encoding the PB2 polypeptide, *Virus Res.* **1**:65–71

Rogers, G N., Paulson, J. C., Daniels, R. S., et al., 1983, Single amino acid substitutions in influenza haemagglutinin change receptor binding specificity, *Nature* **304**:76–78

Roth, M G., and Compans, R. W., 1981, Delayed appearance of pseudotypes between vesicular stomatitis virus and influenza virus during mixed infection of MDCK cells, *J Virol.* **40**:848–860

Rott, R., Orlich, M., and Scholtissek, C., 1976, Attenuation of pathogenicity of fowl plague virus by recombination with other influenza A viruses, apathogenic for fowl. Non-exclusive dependence of pathogenicity on the hemagglutinin and neuraminidase, *J Virol.* **19**:54–60

Rott, R , Orlich, M., and Scholtissek, C., 1979, Correlation of pathogenicity and gene constellation of influenza A viruses. III. Non-pathogenic recombinants derived from highly pathogenic parent strains, *J. Gen. Virol.* **44**:471–477.

Rott, R , Reinacher, M , Orlich, M , et al , 1980, Cleavability of hemagglutinin determines spread of avian influenza viruses in the chorioallantoic membrane of chicken embryo, *Arch Virol* **65**:123–133

Rott, R., Orlich, M., and Scholtissek, C , 1983, Pathogenicity reactivation of nonpathogenic influenza virus recombinants under von Magnus conditions, *Virology* **126**:459–465.

Rott, R , Orlich, M , Klenk, H -D , et al , 1984, Studies on the adaptation of influenza viruses to MDCK cells, *EMBO J* **3**:3329–3332

Saitou, N., and Nei, M , 1986, Polymorphism and evolution of influenza A virus genes, *Mol Biol Evol* **3**:57–74.

Sambrook, J , Rodgers, L., White, J , et al , 1985, Lines of BPV-transformed murine cells that con-

stitutively express influenza virus hemagglutinin, *EMBO J.* **4**:91–103.
Schild, G. C., Oxford, J. S., and Newman, R. W., 1979, Evidence for antigenic variation in influenza A nucleoprotein, *Virology* **93**:569–573.
Schild, G. C., Oxford, J. S., de Jong, J. C., et al., 1983, Evidence for host-cell selection of influenza virus antigenic variants, *Nature* **303**:706–709.
Scholtissek, C., 1984, Temperature-sensitive (ts) mutants of an influenza A virus for which genetic synergism is required to express their ts phenotypes, *Virus Res.* **1**:677–685.
Scholtissek, C., and Murphy, B. R., 1978, Host range mutants of an influenza A virus, *Arch. Virol.* **58**:323–333.
Scholtissek, C., and Spring, S., 1982, Extragenic suppression of temperature-sensitive mutations in RNA segment 8 by replacement of different RNA segments with those of other influenza A virus prototype strains, *Virology* **118**:28.
Scholtissek, C., Vallbracht, A., Flehmig, B., et al., 1979, Correlation of pathogenicity and gene constellation of influenza A viruses. II. Highly neurovirulent recombinants derived from nonneurovirulent or weakly neurovirulent parent virus strains, *Virology* **95**:492–500.
Schulman, J., 1970, Effects of immunity on transmission of influenza: Experimental studies, *Prog. Med. Virol.* **12**:128–160.
Schulman, J. L., 1983, Virus-determined differences in the pathogenesis of influenza virus infections, in: *Genetics of Influenza Viruses* (P. Palese and D. W. Kingsbury, eds.), Springer-Verlag, New York, pp. 305–320.
Schulman, J. L., and Palese, P., 1977, Virulence factors of influenza A viruses: WSN virus neuraminidase required for plaque production in MDBK cells, *J. Virol.* **24**:170–176.
Schulman, J. L., and Palese, P., 1978, Biological properties of influenza A/Hong Kong and PR8 viruses: Effects of genes for matrix protein and nucleoprotein on virus yield in embryonated eggs, in: *Negative Strand Viruses and the Host Cell* (B. W. J. Mahy and R. D. Barry, eds.), Academic Press, New York, pp. 553–674.
Sekikawa, K., and Lai, C.-J., 1983, Defects in functional expression of an influenza virus hemagglutinin lacking the signal peptide sequences, *Proc. Natl. Acad. Sci. U.S.A.* **80**:3563–3567.
Shaw, M. W., Lamon, E. W., and Compans, R. W., 1982, Immunologic studies on the influenza A virus non-structural protein NS1, *J. Exp. Med.* **156**:243–254.
Shimizu, K., Mullinix, M. G., Chanock, R. M., et al., 1982a, Temperature-sensitive mutants of influenza A/Udorn/72 (H3N2) virus. I. Isolation of temperature-sensitive mutants some of which exhibit host-dependent temperature sensitivity, *Virology* **117**:38–44.
Shimizu, K., Mullinix, M. G., Chanock, R. M., et al., 1982b, Temperature-sensitive mutants of influenza A/Udorn/72 (H3N2) virus. II. Genetic analysis and demonstration of intrasegmental complementation, *Virology* **117**:45–61.
Shimizu, K., Mullinix, M. G., Chanock, R. M., et al., 1983, Temperature-sensitive mutants of influenza A/Udorn/72 (H3N2) virus. III. Genetic analysis of temperature dependent host range mutants, *Virology* **124**:35–44.
Simpson, R. W., and Hirst, G. K., 1968, Temperature-sensitive mutants of influenza A virus: Isolation of mutants and preliminary observations on genetic recombination and complementation, *Virology* **35**:41–49.
Sivasubramanian, N., and Nayak, D. P., 1982, Sequence analysis of the polymerase 1 gene and the secondary structure prediction of polymerase 1 protein of human influenza virus A/WSN/33, *J. Virol.* **44**:321–329.
Six, H. R., Webster, R. G., Kendal, A. P., et al., 1983, Antigenic analysis of H1N1 viruses isolated in the Houston metropolitan area during four successive seasons, *Infect. Immun.* **42**:453–458.
Skehel, J. J., Daniels, R. S., Douglas, A. R., et al., 1983, Antigenic and amino acid sequence variations in the haemagglutinins of type A influenza viruses recently isolated from human subjects, *Bull. WHO* **61**:671–676.
Skehel, J. J., Stevens, D. J., Daniels, R. S., et al., 1984, A carbohydrate side chain on hemagglutinins of Hong Kong influenza viruses inhibits recognition by a monoclonal antibody, *Proc. Natl. Acad. Sci. U.S.A.* **81**:1779–1783.
Sklyanskaya, E. I., Varich, N. L., Ambrosieva, T. V., et al., 1985, Virions and intracellular nucleocapsids produced during mixed heterotypic influenza infection of MDCK cells, *J. Virol.* **53**:679–683.
Sleigh, M. J., and Both, G. W., 1981, The extent of hemagglutinin variation during antigenic drift in

the Hong Kong subtype of influenza from 1968 to 1979, in: *Genetic Variation among Influenza Viruses* (D. P. Nayak, ed.), Academic Press, New York, pp. 341-354.
Sleigh, M. J., Both, G. W., Brownlee, G. G., et al., 1980, The haemagglutinin gene of influenza A virus Nucleotide sequence analysis of cloned DNA copies, in· *Structure and Variation in Influenza Viruses* (G Laver and G. M Air, eds), Elsevier/North Holland, New York, pp. 69-79
Smith, G L , Murphy, B R , and Moss, B., 1983, Construction and characterization of an infectious vaccinia virus recombinant that expresses the influenza hemagglutinin gene and induces resistance to influenza virus infection in hamsters, *Proc. Natl Acad Sci U S.A* **80**:7155-7159.
Smith, J M , 1978, *The Evolution of Sex*, Cambridge University Press, New York
Smorodintsev, A A., Golubev, D. B , and Luzjanina, T. Y., 1982, Principles of rational classification and nomenclature of human influenza A viruses, *Boll. Ist Sieroter Milan* **61**:202-208
Spring, S. B., Nusinoff, S. R , Mills, J V , et al., 1975, Temperature-sensitive mutants of influenza virus VI. Transfer of *ts* lesions from the Asian subtype of influenza A virus (H2N2) to the Hong Kong subtype (H3N2), *Virology* **66**:522-532
Steuler, H , Rohde, W., and Scholtissek, C., 1984, Sequence of the neuraminidase gene of an avian influenza A virus (A/parrot/Ulster/73, H7N1), *Virology* **135**:118-124
Sugg, J. Y , 1949, The variation of antigenic pattern and of mouse virulence in an influenza virus culture, *J Bacteriol.* **58**:399-406
Sugiura, A., 1975, Influenza virus genetics, in: *Influenza Viruses and Influenza* (E D Kilbourne, ed), Academic Press, New York, pp. 171-213
Sugiura, A , and Kilbourne, E D., 1965, Genetic studies of influenza viruses II. Plaque formation by influenza viruses in a clone of a variant human heteroploid cell line, *Virology* **26**:478-488
Sugiura, A , and Kilbourne, E. D., 1966, Genetic studies of influenza viruses· III. Production of plaque type recombinants with A_0 and A_1 strains, *Virology* **29**:84-91.
Sugiura, A., and Ueda, M., 1980, Neurovirulence of influenza virus in mice. I Neurovirulence of recombinants between virulent and avirulent virus strains, *Virology* **101**:440-449.
Sugiura, A., Tobita, K., and Kilbourne, E. D , 1972, Isolation and preliminary characterization of temperature-sensitive mutants of influenza virus, *J. Virol* **10**:639-647
Sveda, M. M., and Lai, C.-J., 1981, Functional expression in primate cells of cloned DNA coding for the hemagglutinin surface glycoprotein of influenza virus, *Proc Natl. Acad Sci U.S.A.* **78**:5488-5492.
Sveda, M. M., Markoff, L. J., and Lai, C.-J., 1982, Cell surface expression of the influenza virus hemagglutinin requires the hydrophobic carboxyterminal sequences, *Cell* **30**:649-656.
Tanaka, T., Urabe, M., Goto, H , et al., 1984, Isolation and preliminary characterization of a highly cytolytic influenza B virus variant with an aberrant NS gene, *Virology* **135**:515-523.
Tolpin, M. D., Massicot, J. G., Mullinix, M. G., et al., 1981, Genetic factors associated with loss of the temperature-sensitive phenotype of the influenza A/Alaska/77-ts-1 A2 recombinant during growth *in vivo*, *Virology* **112**:505-517.
Tomley, F.M., and Roditi, I. J., 1984, Nucleotide sequence of RNA segment 5, encoding the nucleoprotein, of influenza A/FPV/Rostock/34, *Virus Res* **1**:625-630.
Underwood, P. A., 1984, An antigenic map of the haemagglutinin of the influenza Hong Kong subtype (H3N2), constructed using mouse monoclonal antibodies, *Mol. Immunol* **21**:663-671
Van Rompuy, L., Min Jou, W., Huylebroeck, D., et al., 1981, Complete nucleotide sequence of the nucleoprotein gene from the human influenza strain A/PR/8/34 (H0N1), *Eur J Biochem.* **116**:347-353.
Van Rompuy, L., Jou, W M , Huylebroeck, D., et al., 1982, Complete nucleotide sequence of a human influenza neuraminidase gene of subtype N2 (A/Victoria/3/75), *J Mol Biol.* **161**:1-11.
van Wyke, K. L., Hinshaw, V. S., Bean, W. J., Jr , et al., 1980, Antigenic variation of influenza A virus nucleoprotein detected with monoclonal antibodies, *J. Virol.* **35**:24-30.
van Wyke, K. L., Yewdell, J. W., Reck, L. J , et al., 1984, Antigenic characterization of influenza A virus matrix protein with monoclonal antibodies, *J. Virol* **49**:248-252
Verhoeyen, M., Fang, R., Min Jou, W. M., et al., 1980, Antigenic drift between the haemagglutinin of the Hong Kong influenza strains A/Aichi/2/68 and A/Victoria/3/75, *Nature* **286**:771-776
Wang, C -I , 1948, The relation of infectious and hemagglutination titers to the adaptation of influenza virus to mice, *J Exp Med* **88**: 515-519.
Ward, C W , and Dopheide, T. A., 1981, Evolution of the Hong Kong influenza A sub-type, *Biochem.*

J. **195:**337–340.
Webster, R. G., and Bean, W. J., Jr., 1978, Genetics of influenza virus, *Annu. Rev. Genet.* **12:**415–431.
Webster, R. G., and Campbell, C. H., 1972, The *in vivo* production of "new" influenza A viruses, II. *In vivo* isolation of "new" viruses, *Virology* **48:**528–536.
Webster, R. G., Laver, W. G., and Kilbourne, E. D., 1968, Reactions of antibodies with surface antigens of influenza A virus, *J. Gen. Virol.* **3:**315–326.
Webster, R. G., Kendal, A. P., and Gerhard, W., 1979, Analysis of antigenic drift in recently isolated influenza A (H1N1) viruses using monoclonal antibody preparations, *Virology* **96:**258–264.
Webster, R. G., Laver, W. G., Air, G. M., 1980, The mechanism of antigenic drift in influenza viruses, *Ann. N.Y. Acad. Sci.* **354:**142–161.
Webster, R. G., Hinshaw, V. S., Berton, M. T., et al., 1981, Antigenic drift in influenza viruses and association of biological activity with the topography of the hemagglutinin molecule, in: *Genetic Variation among Influenza Viruses* (D. P. Nayak and C. F. Fox, eds.), Academic Press, New York, pp. 309–322.
Webster, R. G., Hinshaw, V. S., and Laver, W. G., 1982, Selection and analysis of antigenic variants of the neuraminidase of N2 influenza viruses with monoclonal antibodies, *Virology* **117:**93–104.
Webster, R. G., Laver, W. G., and Air, G. M., 1983, Antigenic variation among type A influenza viruses, in: *Genetics of Influenza Viruses* (P. Palese and D. W. Kingsbury, eds.), Springer-Verlag, New York, pp. 127–168.
Webster, R. G., Kawaoka, Y., and Bean, W. J., Jr., 1986, Molecular changes in A/chicken/Pennsylvania/83 (H5N2) influenza virus associated with acquisition of virulence, *Virology* **149:**165–173.
Wiktor, T. J., and Koprowski, H., 1980, Antigenic variants of rabies virus, *J. Exp. Med.* **152:**99–112.
Wiley, D. C., Wilson, I. A., and Skehel, J. J., 1981, Structural identification of the antibody binding sites of Hong Kong influenza hemagglutinin and their involvement in antigenic variation, *Nature* **289:**373–378.
Winter, G., and Fields, S., 1980, Cloning of influenza cDNA into M13: The sequence of the RNA segment encoding the A/PR/8/34 matrix protein, *Nucleic Acids Res.* **8:**1965–1974.
Winter, G., and Fields, S., 1982, Nucleotide sequence of human influenza A/PR/8/34 segment 2, *Nucleic Acids Res.* **10:**2135–2143.
Winter, G., Fields, S., and Brownlee, G. G., 1981a, Nucleotide sequence of the haemagglutinin of a human influenza virus H1 subtype, *Nature* **292:**72–75.
Winter, G., Fields, S., Gait, M. J., et al., 1981b, The use of synthetic oligodeoxynucleotide primers in cloning and sequencing segment 8 of influenza virus (A/PR/8/34), *Nucleic Acids Res.* **9:**237–245.
Yamada, A., Brown, L. E., and Webster, R. G., 1984, Characterization of H2 influenza virus hemagglutinin with monoclonal antibodies: Influence of receptor specificity, *Virology* **138:**276–286.
Yewdell, J. W., and Gerhard, W., 1982, Delineation of four antigenic sites on a paramyxovirus glycoprotein via which monoclonal antibodies mediate distinct antiviral activities, *J. Immunol.* **128:**2670–2675.
Yewdell, J. W., Webster, R. G., and Gerhard, W., 1979, Antigenic variation in three distinct determinants of an influenza type A hemagglutinin molecule, *Nature* **279:**246–248.
Yewdell, J. W., Caton, A. J., and Gerhard, W., 1986, Selection of influenza A virus adsorptive mutants by growth in the presence of a mixture of monoclonal antihemagglutinin antibodies, *J. Virol.* **57:**623–628.
Young, J. F., Desselberger, U., and Palese, P., 1979, Evolution of human influenza A viruses in nature: Sequential mutations in the genomes of new H1N1 isolates, *Cell* **18:**73–83.
Young, J. F., Desselberger, U., Palese, P., et al., 1983, Efficient expression of influenza virus NS1 nonstructural proteins in *Escherichia coli*, *Proc. Natl. Acad. Sci. U.S.A.* **80:**6105–6109.
Závada, J., 1976, Viral pseudotypes and phenotypic mixing, *Arch. Virol.* **50:**1–15.
Závada, J., and Rosenbergová, M., 1968, Double neutralization of fowl plague virus reproduced in chick myeloblasts, *Acta Virol.* **12:**282–284.
Závada, J., and Rosenbergová, M., 1972, Phenotypic mixing of vesicular stomatitis virus with fowl plague virus, *Acta Virol.* **16:**103–114.

III

Influenza: Infection and Disease

7

Influenza in Man

CLINICAL RESPONSE TO INFECTION

The clinical expression of infection with influenza virus is variable and is partially influenced by the nature of the infecting virus but to a greater extent is modulated by the age, physiological state, and immunologic experience of the host. Depending on these factors, influenza infection may occur without symptoms, may kill, or, most frequently, may cause an acute, prostrating febrile illness attended by cough and myalgia (Fig. 7-1). Typically, the onset of the disease is rapid, and the degree of prostration and constitutional symptoms are out of proportion to the severity of the respiratory symptoms. Characteristically also, fever is short-lived (3–4 days), and recovery is rapid and uneventful. In its classical form influenza is "an unvarying disease caused by a varying virus" (Kilbourne, 1980) and as such has been recognized through the centuries. However, even the typical case is diagnosed most easily in the context of an epidemic, and, indeed, epidemicity (simultaneous occurrence of other cases) is an important diagnostic clue.

The pattern of symptoms has remained remarkably constant in epidemics caused by all of the human subtype viruses (H1N1, H2N2, and H3N2). A composite rendition of symptom frequency in a number of epidemics since 1957 is presented in Fig. 7-2. Ninety percent or more of patients suffer cough, fever, and moderate prostration. Throbbing, nonlocalized headache is a common complaint, followed in frequency by generalized aching, lumbosacral backache, and myalgia. "Raspy" or "burning" sore throat is remarked by 50% of patients; sharp or stinging substernal pain, although somewhat less frequent, is an important indication of the primary involvement of the trachea in this infection (see Pathogenesis, below). In only one out of three patients is the cough productive, and it should be noted that productive cough does not necessarily indicate secondary bacterial infection.

Other commonly encountered symptoms are extraocular myalgia, chilliness, and nasal and conjunctival burning and discharge. Less common, but diagnostically significant, is epistaxis. Gastrointestinal symptoms are uncommon in adults.

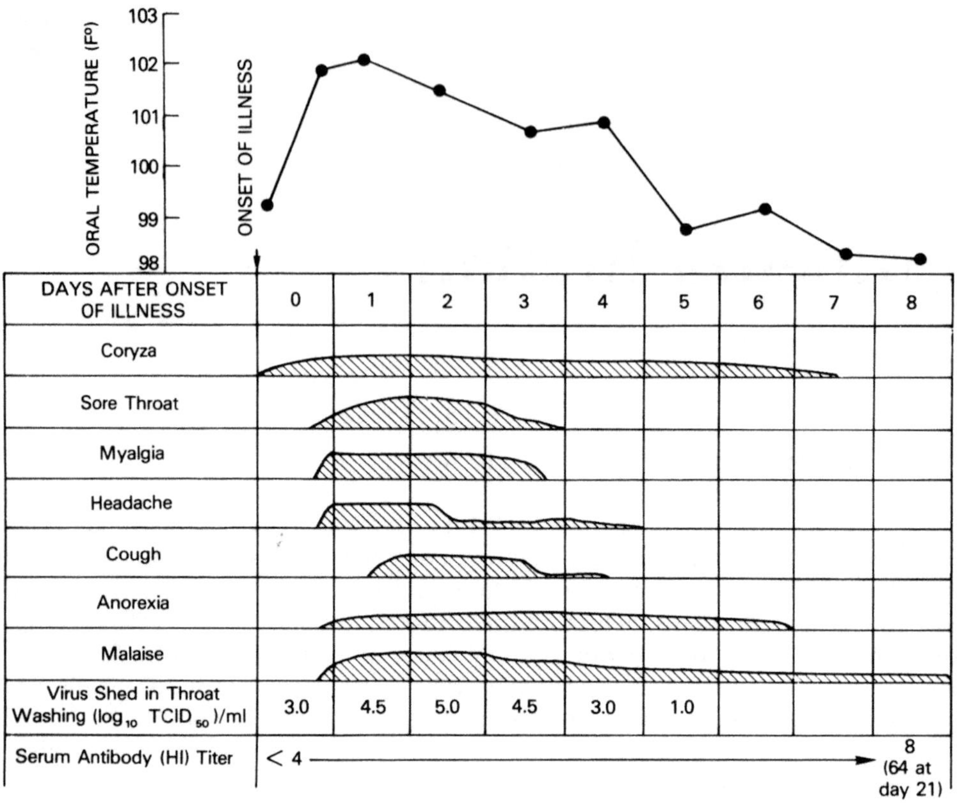

Figure 7-1. Clinical characteristics of a naturally occurring case of influenza caused by infection of a normal 28-year-old man by A/Victoria(H3N2) influenza virus. The height of each shaded curve is proportional to the magnitude of the corresponding symptom. (From Dolin, 1976. Reproduced with permission, from *Am. Fam. Physician* **14**:74.

Findings on Physical Examination

Patients show evidence of malaise and prostration and manifest facial flush and conjunctival injection. Signs referable to the upper respiratory tract include nasal obstruction or discharge and pharygeal injection in one-half to three-quarters of patients. Hoarseness occurs in a few patients. Nontender cervical adenopathy is found in about 25% of patients, and splenomegaly is rare.

Fever is a constant finding. On the average, maximal temperature is 101.3°F (Kilbourne and Loge, 1950), but in some patients it may reach l06°F. A diphasic M-shaped fever curve has been described and, although infrequent, is useful diagnostically when found (Stuart-Harris in Stuart-Harris and Schild, 1976). Pulse rate may be elevated concordant with fever, but relative bradycardia occasionally occurs.

Ordinarily, signs of lower respiratory tract disease are rare, but roughened breath sounds or rales can be found in 1–5% of patients subjected to careful auscultation.

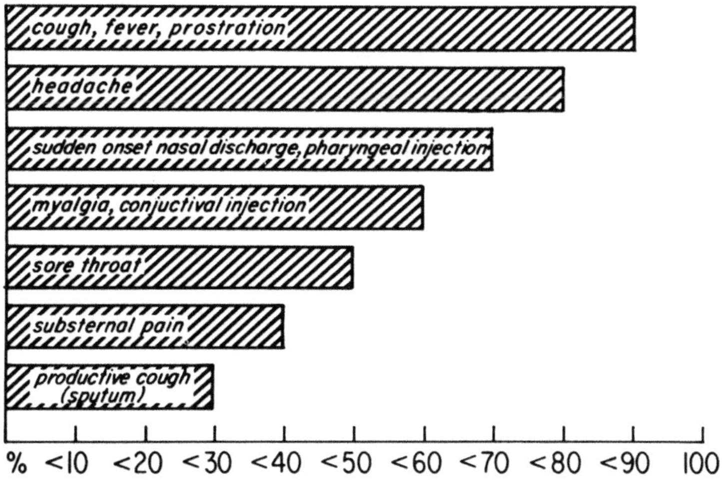

Figure 7-2. Frequency of signs and symptoms of influenza A in adults. (Data of Stuart-Harris, 1961; Kilbourne and Loge, 1950; Jordan et al., 1958; Krylov et al., 1973; Gill, 1973.)

The clinical course of uncomplicated influenza is characterized by prompt defervescence followed by more gradual abatement of respiratory and myalgic symptoms (Fig. 7-1). The febrile period is usually 3 days in adults, and acute symptoms rarely persist beyond 1 week. Protracted postinfluenza asthenia has been described frequently, but its reality and significance are open to question (Imboden et al., 1961). Complete recovery from influenza is the rule.

Host-Determined Variation in Disease Expression and Severity

The same virus that produces a 3-day fever in one patient can kill another in apparent previous good health through generalized involvement of the lungs with primary influenza virus pneumonia. The risk of serious and fatal disease is much higher, however, in the very young or the elderly or in patients with certain well-defined preexisting underlying conditions. These conditions, which are described below, predispose not to the risk of acquiring influenza virus infection but to the risk of suffering severe disease once infection is established. Thus, in general, the severity of influenza is the reciprocal of host susceptibility as influenced by a complex of immunologic, anatomic, and physiological factors (Fig. 7-3).

Variation in Symptoms Related to Age

Influenza in Children

Influenza is a common and potentially serious disease in children. During epidemics, it displaces the usual viral respiratory infections as a leading cause of hospitalization for croup and lower respiratory tract disease (Kim et al., 1979; Glezen et al., 1980).

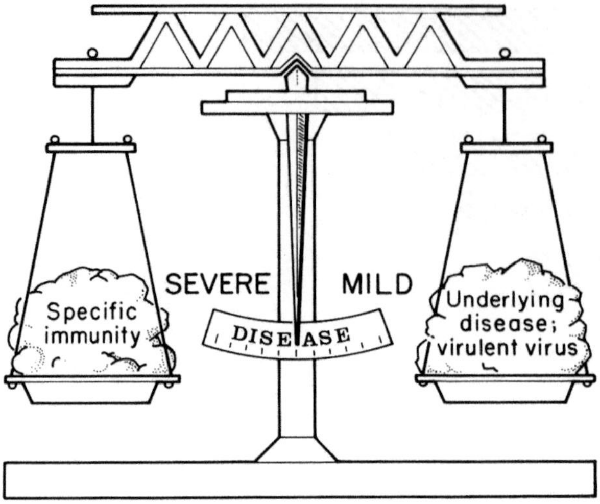

Figure 7-3. The severity of influenza is the resultant of interacting host and viral determinants.

In children less than 5 years of age, the pattern of symptoms differs from that observed in older children and adults (Fig. 7-4). Fever and rhinitis predominate, and vomiting and diarrhea become a significant part of the presenting symptom complex.

Convulsions associated with high fever may lead to hospitalization of children 1–3 years old. Signs of bronchopulmonary disease may occur in one-third of hospitalized patients. As pointed out by Wright and his colleagues (1980), the incidence of systemic symptoms cannot be elicited reliably in infants and young children. Therefore, absence of such characteristic adult symptoms as myalgia cannot be assumed. Delayed recovery from influenza with more protracted excretion of virus has been reported (Hall and Douglas, 1975), but the patients studied had underlying disease predisposing to the development of lower respiratory tract infection.

As in the adult disease, recovery from influenza in children is usually prompt following a brief period of incapacitation. In a Houston study, the average period of absence from school during an influenza B epidemic was 3 days (Glezen, 1980). However, more than 1% of influenza A and B cases require hospitalization, 50% with disease manifestation such as convulsions, encephalopathy, abdominal pain, and other symptoms not directly referable to involvement of the respiratory tract (reviewed by Glezen, 1980).

Influenza in Infants

Because perception of disease in infants depends primarily on objective, visibly manifest signs, and since reliable data have been obtained principally on infants ill enough to require hospitalization, influenza in babies is often unrecognizable in its clinical presentation (Fig. 7-5). Evidence of rhinitis may be the sole clue to respiratory involvement, and fever and prostration simulate neonatal sepsis in early infancy (Meibalane et al., 1977; Dagan and Hall, 1984). In these pa-

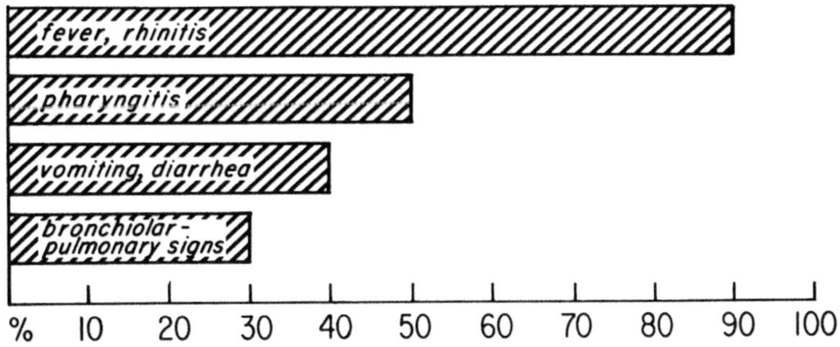

Figure 7-4. Profile of influenza symptomatology in children aged 1-5 yr. (Data of Paisley et al., 1978; Price et al., 1976.)

tients onset of illness is acute, and the infants present with "high fever, lethargy often alternating with irritability, anorexia and signs of upper respiratory tract infection" (Dagan and Hall, 1984). Infection in newborns can occur (Bauer et al., 1973), although maternal antibody lowers the incidence of influenza during the neonatal period (Puck et al., 1980).

Influenza in the Elderly

Although it is well recognized that the risk of dying from influenza and its attendant pneumonic complications increases progressively after middle age, significant differences in uncomplicated disease in older patients have not been observed. Increased mortality in the aged as an epidemiologic observation is discussed in Chapter 10.

Influence of Genetic Factors on the Expression of Disease

Against the background of varying immune experience that characterizes any population, it has not yet been possible to demonstrate genetically determined differences in response to influenza virus infection. Attempts to link antibody re-

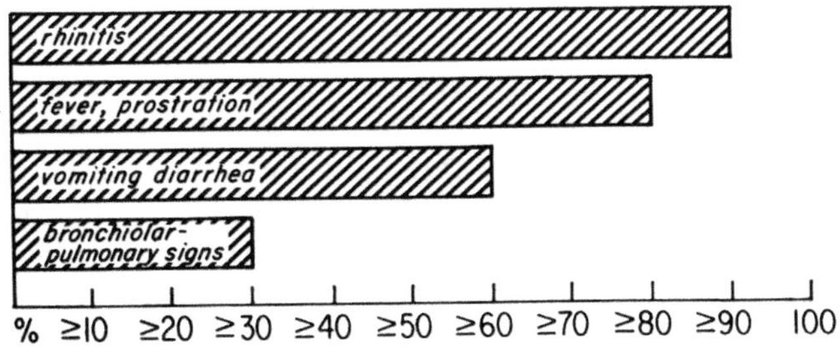

Figure 7-5. Influenza in infants less than 3 months old. Objective criteria of disease. (Data of Paisley et al., 1978.)

sponse or susceptibility to infection with ABO blood groups have been contradictory and are discussed in Chapter 10. However, the demonstration of genetic factors in experimental influenza in animal models suggests that such factors (as yet undefined) probably influence the expression of disease in man.

The Influence of Prior Immunizing Experience on the Severity of Influenza

Specific immunity to influenza not only prevents infection and disease but tempers the severity of disease if infection does occur (see discussion under Immunity to Influenza later in this chapter).

Pregnancy

During interpandemic periods, the blunting effect of partial immunity and the lower total incidence of disease often obscure the perception that pregnant women may be peculiarly vulnerable to influenza. Indeed, the impact of pandemic influenza may be demonstrable only in the first wave. Widelock et al. (1963) noted that although the mortality from influenza in pregnant women 20–29 years of age was ninefold greater than that of nonpregnant women of the same age infected in New York City in October–November 1957, the mortality difference could not be detected in later outbreaks. There is also some evidence that primary influenza virus pneumonia may predominate as the principal type of pneumonia in influenza fatalities in pregnancy (Freeman and Barno, 1959).

In a statistical analysis of 1350 cases of influenza in pregnancy in the 1918 pandemic, Harris (1919) found that 50% developed pneumonia (nature unspecified) and that of these patients with pneumonia, 50% died.

In studies of patients hospitalized with pneumonia in the 1957 pandemic, pregnant women account for about 10% of the cases (Martin et al., 1959a; Petersdorf et al., 1959), findings concordant with the observation that 10% of all deaths from "Asian" (H2N2) influenza in 1957 were in pregnant women. Looked at another way, 50% of influenza-related mortality in women of child-bearing age is in those who are pregnant (Freeman and Barno, 1959; Greenberg et al., 1958; see also Armstrong, 1919).

In the 1968 H3N2 pandemic in Memphis, the approximately 10% contribution of pregnant women to the population hospitalized for influenza with pneumonia was again observed (Bisno et al., 1971). However, in this study, primary influenza virus pneumonia and complicating staphylococcal pneumonia were not found, in contrast to the earlier studies of the 1957 pandemic. One could surmise that earlier acquired antibody to the shared NA antigen of the H2N2 and H3N2 viruses may have been ameliorating but for the fact that typical primary influenza virus pneumonia was seen with Hong Kong subtype virus in previously healthy patients (Feldman et al., 1972; Lindsay et al., 1970; Taylor et al., 1976; Lubkiewicz et al., 1974) and in pregnancy (Lindsay et al., 1970; Taylor et al., 1976).

The severe effects of influenza in pregnancy are more notable because they occur in an age group usually at lowest risk (except in 1918) of fatal consequences of the disease. In view of the profound influence of cardiovascular disease on the outcome of influenza (see next section), it seems most likely that susceptibility in pregnancy is related to hemodynamic changes, particularly because the risk increases as pregnancy proceeds and becomes maximal in the third trimester (Harris, 1919). It has been proposed that increased cardiac stroke volume is an important determinant of influenza mortality; blood volume and stroke volume are increased in pregnancy (Stevens, 1976).

It is not clear whether or not the comparative severity of influenza in pregnancy is shifted up a notch at all levels of disease expression. Only careful prospective studies can answer this question. Mullooly et al. (1986) found excess acute respiratory disease in young pregnant women in 1978 with a reappearance of H1N1 virus.

Underlying Cardiovascular Disease

The association of underlying cardiovascular disease with serious pulmonary manifestations of influenza was first clearly shown in the pandemic of 1957 (Table 7-1). In a study of patients admitted to the New York Hospital with influenza with pulmonary manifestations, 21 of 33 had underlying disease or pregnancy; 14 of the 21 had cardiovascular disease, and of the 14, ten had rheumatic heart disease (Louria et al., 1959). When patients with the diffuse lung involvement of primary influenza virus pneumonia were considered as a separate group, the association with underlying cardiovascular disease was even more striking. In 11 of 15 of such patients, underlying cardiovascular disease was present, which in seven of the 11 was rheumatic heart disease. The frequent association of susceptibility to pneumonic involvement in patients with rheumatic heart disease was reported in a number of other studies of the 1957 epidemic (Rock et al., 1958; Giles and Shuttleworth, 1957; Hers et al., 1957; Martin et al., 1959a) and has also been noted with H3N2 subtype virus (Burk et al., 1971). The prognosis for influenza patients with coincident rheumatic heart disease is grim. Five of ten patients with rheumatic heart disease who developed pneumonia in association with influenza died (Louria et al., 1959). The fact that in patients with cardiovascular

Table 7-1. The Association of Underlying Disease with Pulmonary Complications of Influenza: The Importance of Antecedent Cardiac Disease

		Incidence of underlying disease[a]		
Disease syndrome	Number of patients	Total number of patients	Number with CV disease	Number with RHD
Influenza complicated by bacterial pneumonia	15	8[b]	3	3
Concomitant influenza virus and bacterial pneumonia	9	7[b]	5	3
Primary influenza virus pneumonia	6	6	6	4
Totals	30	21	14	10

[a]CV, cardiovascular disease, RHD rheumatic heart disease
[b]In one of this number, the disease was pregnancy

disease the most common defect was mitral stenosis suggested that hemodynamic rather than general debilitating factors were of cardinal importance and that the pulmonary hypertension secondary to the mitral stenosis might be critical in the pathogenesis of the hemorrhagic edematous pneumonia induced by influenza virus (Louria et al., 1959).

Bronchopulmonary Disease

Underlying chronic bronchopulmonary disease is equal to preexisting cardiac disease as a factor leading to severe or complicated influenza. The nature of such disease is variable; pulmonary insufficiency may be related to chronic bronchitis, bronchiectasis, pulmonary fibrosis, asthma, emphysema, or bronchopulmonary neoplastic disease (Martin et al., 1959a; Petersdorf et al., 1959; Quarles van Ufford and Savelberg, 1959; Lindsay et al., 1970; Bisno et al., 1971; Brocklebank et al., 1972; Lubkiewicz et al., 1974). Asthma is recognized as a high-risk factor leading to increased hospitalization of children for influenza (Mullooly and Barker, 1982).

A study of Boy Scouts in the pandemic of 1957 disclosed that 27% of patients with a history of asthma had abnormal chest x rays in comparison with a background of 8% x-ray abnormalities in nonasthmatics (Podesin and Felton, 1958). Conversely, influenza can provoke asthmatic attacks (Minor et al., 1974, 1976), and the incidence of wheezing in influenza is higher in asthmatics (Rebhan, 1957).

Other Underlying Diseases

The contribution of preexisting diseases or physiological states other than cardiac and pulmonary disease and pregnancy to exacerbating the severity of influenza is less clear than with those notorious conditions. Diabetes is often mentioned as a high-risk condition, but so it is in a variety of infectious diseases.

Although it is not unreasonable to recommend that anyone with any chronic disease should be protected by vaccination from the possible exacerbating effects of influenza and its sequels, there is little evidence that noncardiopulmonary underlying diseases modify the clinical picture of influenza itself. However, in recent years immunosuppression for organ transplantation or incidental therapy of neoplastic diseases has been described as a prelude to severe influenza (Feldman et al., 1972; Karalakulasingam et al., 1977). Two patients receiving corticosteroids for nephrotic syndrome suffered fatal influenza, but both were noted to have concomitant arteriosclerotic cardiovascular disease also (Burk et al., 1971).

Virus-Determined Variation in Disease Expression and Severity

Clinical Responses to Infection with Influenza A, B, and C Viruses Compared

Influenza B. Because influenza B is more common in children than young adults, most published descriptions of the disease involve younger patients. Therefore, it is important in comparing the disease with influenza A to distinguish age-related from virus-related differences in symptomatology. When this is done, influenza A and B are seen to be very similar diseases, both characterized by a predominance of respiratory tract and systemic symptoms (Fig. 7-6). (In making these comparisons the author must say that he yearns for systematized studies with standardized question forms and clinical descriptions to provide assurance that symptoms are truly absent and not simply unreported.)

The published data do suggest that nasal discharge, abdominal pain, and gastrointestinal symptoms are more common in influenza B, but these symptoms also occur in influenza A, and the frequency of their occurrence is also inversely related to age (Fig. 7-2).

Although less severe in their manifestations than influenza A viruses in older subjects, influenza B viruses also have the potential to cause severe and even fatal illness in both young and older patients (Baine et al., 1980) and also have been associated with excess mortality rates (Nolan et al., 1980).

Influenza C. Because isolations of influenza C viruses have been remarkably few, and well-defined epidemics rare, little is known about the clinical expression of infection with this virus. Identification of epidemic disease as influenza C virus related has been further confounded, even in its first description (Francis et al., 1953), by the simultaneous occurrence of other influenza virus infections in the populations under examination (Gerber et al., 1952; Jennings, 1968). Yet serological evidence demonstrates that influenza C virus infection is very common (O'Callaghan et al., 1980; Monto and Troisi, 1981; Tumova et al., 1983).

Individual case reports have suggested that influenza C is milder than either influenza A or B, a suggestion confirmed by a prospective study of a children's residence in Yamagata prefecture, Japan (Katagiri et al., 1983). This study appears

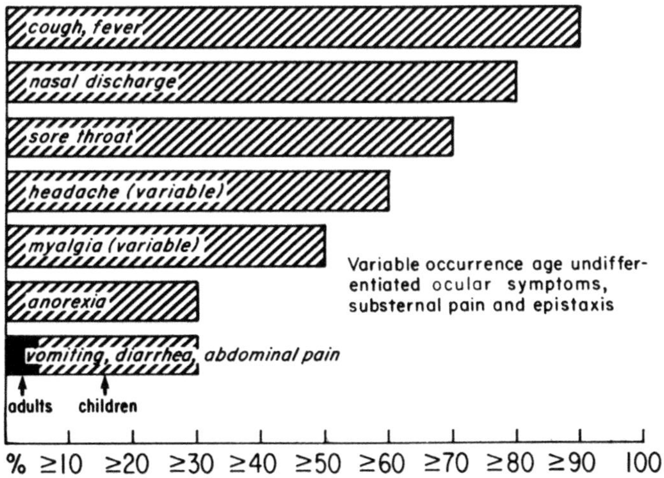

Figure 7-6. Frequency of signs and symptoms of influenza B in children and young adults. (Data of Lennon et al., 1983; DeStefano, 1982; Wright et al., 1980; Clark et al., 1970.)

to be unique in its unambiguous etiology and its detailed definition of clinical manifestations and course of illness in both children and adults (Fig. 7-7).

In these very young children, the disease was characterized by a 2- to 3-day fever of 38–40°C and a protracted (2-week) period of coryza. Cough was present in half the children. The two adult patients described had similar but briefer illnesses accompanied by headache, malaise, and sore throat. In three young adults, other symptoms included headache, myalgia, and conjunctivitis; in these patients also, symptoms were prolonged (Dykes et al., 1980). Influenza C can be the antecedent of bacterial pneumonia (Darke et al., 1957; Chakraverty et al., 1984).

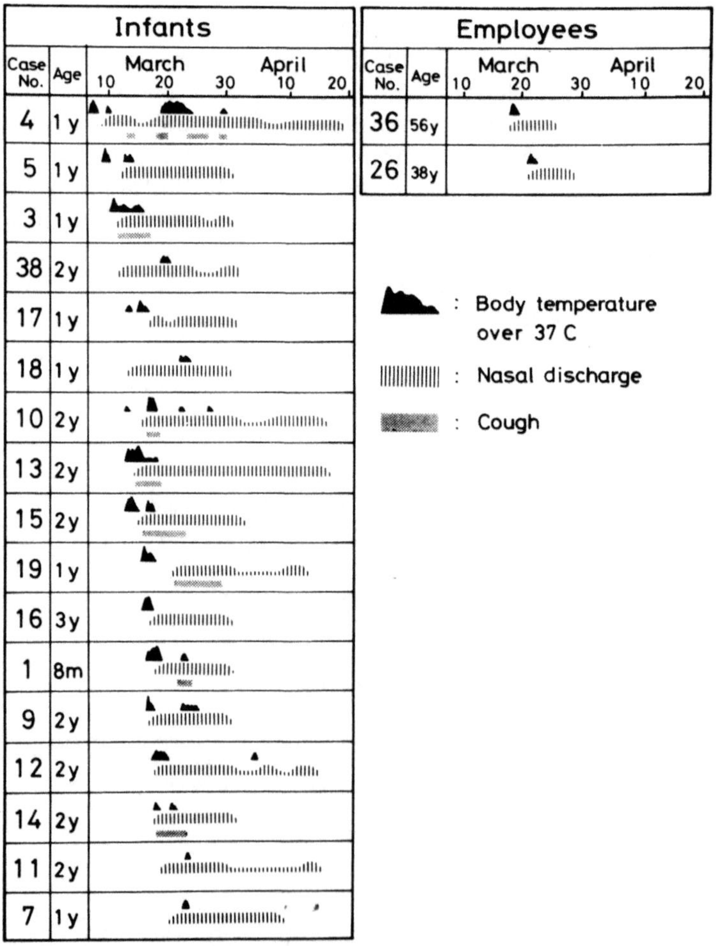

Figure 7-7. Clinical manifestations of influenza C in a childrens' home. Cases are listed in order of onset. (From Katagiri et al., 1983. Reproduced with permission, from *J. Infect. Dis.* **148**:52. Copyright © 1983 by The University of Chicago Press.)

Reinfection with influenza C virus appears to be common despite demonstrable antibody in subjects prior to reinfection (Homma, 1985). Experimental infection of adult volunteers was asymptomatic or productive of mild symptoms (Quilligan et al., 1954) similar to those of the common cold (Joosting et al., 1968).

Variation in Disease Severity Related to Viral Strain or Subtype

All evidence points to the intrinsic capability of all epidemic subtypes or strains of influenza A virus to induce the full panoply of clinical manifestations characteristic of influenza and to provoke the same complications. This is not to say that strains may not vary in virulence. But because disease severity is so obviously influenced by prior immunity, age, and underlying disease, the role of viral strain or subtype variation in determining differences in disease severity can be dissected out only with the greatest difficulty. Only in populations without specific antibody to two or more simultaneously infecting viruses could valid comparisons truly be made.

Because such simultaneous introduction of two distinguishable viruses has not yet occurred in the era of modern virology, no observations in the population at large have been possible. However, the cocirculation of H3N2 and H1N1 viruses since 1977 has afforded an opportunity to compare disease in those members of the community without immunity to either virus, i.e., young children. Indeed, in young children infected with H1N1 virus, Wright et al. (1981) found a striking absence of disease in contrast to a higher clinical attack rate with H3N2 virus in the same season. Other studies have suggested that A/H1N1 influenza is less severe (summarized by Frank et al., 1985). In a suburban population under prospective surveillance, an H1N1-associated illness was less frequently associated with physician visits than was infection with an H3N2 virus (Monto et al., 1985). Peripheral airway function was reduced with H3N2 but not in H1N1 viral infections (Douglas et al., 1981).

Illness during an explosive outbreak of H1N1 infection among Air Force cadets was described as possibly less severe than seen in antecedent H3N2 virus infections and was uncomplicated by pneumonia (Meiklejohn et al., 1980). On the other hand, in an equally abrupt epidemic of H1N1 influenza at a boarding school, Hoskins (1979), although he noted mild or a moderate disease in general, found that 7% of those infected suffered greater prostration, prolonged or recurrent fever, and productive cough; 25% of this group had x-ray evidence of pneumonia, and in one case a bacterial pathogen was identified. The incidence and severity of respiratory complications in this epidemic were stated to be much greater than observed in five previous outbreaks of H3N2 or B influenza virus infection. But these findings demonstrate the importance of distinguishing between severity of primary disease and the possible contributing influence of secondary bacterial infection. Fatalities from H1N1 virus infection of healthy young adults have been reported, but these cases were complicated by staphylococcal pneumonia (Eason et al., 1980). Typical influenza has also been seen in H1N1 virus infection of the elderly despite their earlier experience with H1N1 viruses (Bentley et al., 1983).

On balance, H1N1 virus may cause less severe influenza, but its pathogenic potential appears to be no less than that of other strains.

Unusual Manifestations of Influenza Virus Infection

Primary Viral Pneumonia

In the 1957 pandemic, Louria et al. (1959) distinguished four clinical syndromes in patients who had signs of involvement of the lower respiratory tract:

1. Influenza with physical signs of lung involvement but without roentgenographically detectable pulmonary infiltrates.
2. Focal bacterial pneumonia following influenza.
3. Influenza associated with a diffuse pulmonary process produced by influenza virus infection alone.
4. Influenza associated with both focal and diffuse pulmonary signs resulting from concomitant bacterial and influenza virus infection.

Of these entities, 1 and 3 represent manifestations of primary viral infection (the former not uncommon), whereas 2 and 4, the two syndromes in which bacterial infection occurs, are best considered as complications of the viral infection that will vary in clinical expression dependent on the pathogenic characteristics of the invading bacterium.

In patients with primary influenza virus pneumonia, the characteristic symptoms of influenza are followed within 24 hr by rapidly increasing respiratory distress requiring hospitalization. During this period the initial dry cough becomes productive of bloody sputum. On admission patients appear seriously ill. Extreme dyspnea is attended by cyanosis and respiratory rates of 36 to 54 per min. Focal signs of consolidation of the lungs are absent, and diffuse lung involvement is signaled by fine inspiratory rales, scattered inspiratory wheezes, and prolongation of expiration at the lung bases. Disease progresses rapidly with the development of increasing difficulty in air exchange, which is reflected either by suppression of breath sounds or increase in inspiratory rales and expiratory wheezing. Chest x rays show diffuse bilateral infiltrates fanning out from the hilar regions (Fig. 7-8).

Terminally, coarse rales and rhonchi are prominent, and expirations are extremely prolonged. Respirations are rapid and gasping in these agitated patients, and cyanosis becomes intense (Louria et al., 1959).

Myopathy and Nephropathy

The characteristic generalized myalgia of influenza remains unexplained, particularly because infection appears to be confined principally to the respiratory tract and viremia is rarely demonstrable. It is not clear whether or not this myalgia is pathogenically related to the syndrome of acute myopathy that occurs as an unusual manifestation of either influenza A or B.

Excepting the generalized myalgia characteristic of the disease, myopathy is an unusual manifestation of influenza. It occurs in three distinct clinical presentations.

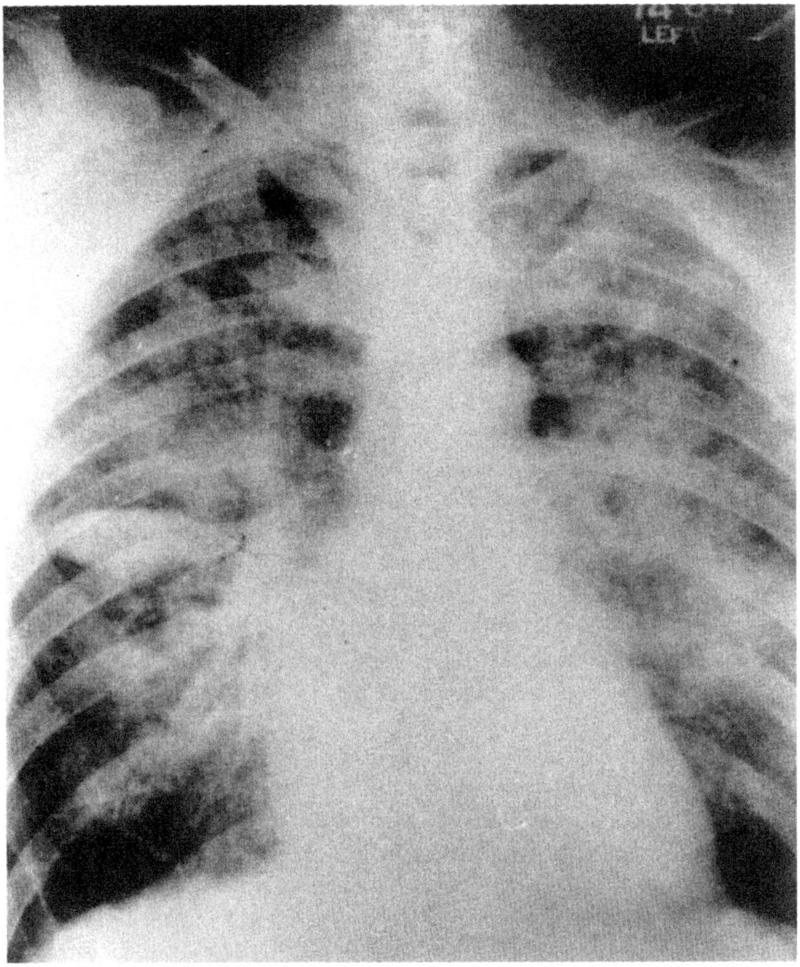

Figure 7-8. Chest x ray of a 49-year-old man with primary influenza virus pneumonia superimposed on rheumatic heart disease with mitral stenosis and insufficiency. (The density in the right midlung field represents a loculated interlobar effusion present before the acute illness.) (From Louria et al., 1959. Reproduced from the *Journal of Clinical Investigation* **38**:242 by copyright permission of The American Society for Clinical Investigation.)

Acute Transient Crural Myopathy in Children. This syndrome appears after the peak of respiratory tract symptoms and is mainly confined to the gastronemius and soleus muscles. Leg pain and tenderness are often sufficiently severe to prevent walking (Middleton et al., 1970; Dietzman et al., 1976; Farrell et al., 1980). Muscle weakness has not been described (Zamkoff and Rosen, 1979). However, biopsies have revealed segmental rhabdomyolysis without inflammation, and influenza B virus was isolated from the muscle of a single patient (Farrell et al., 1980).

Reported cases have been predominantly in influenza B epidemics (Middleton et al., 1970; Mason and Keller, 1975; Dietzman et al., 1976; Farrell et al., 1980), but at least two serologically documented cases have occurred in influenza A (Barton and Chalhub, 1975; Mass, 1985).

In these patients most laboratory tests are normal except for elevation of serum creatine phosphokinase (CPK), characteristic of myopathy or acute carditis; carditis has not been detected in this group of patients. In one study elevation only of CPK_3, the skeletal muscle isoenzyme (Dawson and Fine, 1967), was found (Farrell et al., 1980).

Patients recover promptly from the myopathy in 4–5 days and are left without manifest sequelae.

Acute Myopathy with Rhabdomyolysis and Myoglobinuria with Associated Renal Dysfunction. Although myoglobinuria (Mason and Keller, 1975) and even rhabdomyolysis (DiBona and Morens, 1977) have been detected in children with probable influenza myopathy, severe rhabdomyolysis appears to be more common in the few adult cases that have been reported. Zamkoff and Rosen (1979) described two cases in 23- and 25-year-old brothers with influenza A, one of whom died of oliguric renal failure. As in other similar cases in adults, profound muscle weakness and dark urine appeared during the respiratory illness. Very high serum levels of CPK have been observed (Cunningham et al., 1979). Pathological findings in muscle sections have been variable but include degeneration and lysis of muscle fibers (Cunningham et al., 1979) and absence of inflammatory exudate or viral particles (Zamkoff and Rosen, 1979). Mononuclear cell infiltrates have been observed (Minow et al., 1974). Recovery has occurred in some patients even following myoglobinuric nephropathy.

Myopathy in adults appears most often to be influenza A associated, a fact that may reflect the greater prevalence of influenza A in older persons rather than the pathogenic potential of the virus. The suggestion that myoglobinuria was triggered by other respiratory infections in one case (Simon et al., 1970) and the simultaneous occurrence of influenza-associated myoglobinuria in brothers hint at the possibility of a genetic predisposition to this unusual manifestation of influenza.

Myopathy of the Elderly in the Absence of Classical Symptoms of Influenza. It is notable that the only isolation of influenza viruses from the muscles of patients with influenza-associated myopathy have been from two elderly cardiac patients without respiratory tract symptoms in whom muscle weakness was a primary manifestation. In one case, characterized as myglobinuric polymyositis, influenza B virus infection was demonstrated by virus isolation, specific immunofluorescent staining of muscle tissue, and electron microscopic detection of viruslike particles (Gamboa et al., 1979). From the second, an A/H3N2 influenza virus was isolated from a muscle biopsy specimen obtained on the 27th day of illness. The indolent clinical course in this patient was followed by complete recovery (Kessler et al., 1980). In the first case, 2 months of myalgia preceded hospitalization but may have been unrelated to the influenza-induced myopathy.

Comment. The clinical differences among these three categories of influenza myopathy may represent a spectrum of variation based on similar pathogenetic mechanisms. In patients in all three categories, noninflammatory rhabdomyolysis has been demonstrated, but in the young, myopathy is manifested more by acute myalgia than by the weakness seen in adults, in whom myoglobinuria is more evident. The protracted myopathy described in a few elderly patients may reflect, in addition, the invasion by virus of previously damaged muscles as a consequence of viremia related to impairment of immunologic defenses.

Much remains to be learned about the host and viral factors responsible for the rare myopathies associated with influenza. It is noteworthy that influenza viruses can infect human muscle cells in vitro (Klavinskis et al, 1985).

Carditis

During the course of otherwise uncomplicated influenza, a variable percentage (9–75%) of hospitalized patients have been reported to have transient abnormalities of their electrocardiograms (Walsh et al., 1958; Gibson et al., 1959; Karjalainen, 1980). Even in patients with relatively mild disease treated at home, the ECGs of almost half showed changes, which included ST segment deviation, T wave inversion, sinus bradycardia or tachycardia, and in one case atrial fibrillation (Verel et al., 1976). Often, other evidence of myocarditis is not manifest or sought. In a prospective study of an influenza epidemic in military conscripts, it was found that six of 41 patients with serologically verified influenza had ST segment and T wave changes on serial ECGs that were not responsive to β blockade. In these patients, regional myocardial dysfunction was revealed by multidirectional echocardiography; cardiospecific creatine phosphokinase was increased in the serum of three patients. Taken together, these findings were felt to be diagnostic of acute myocarditis (Karjalainen et al., 1980).

The role of acute myocarditis in influenza fatalities is not readily separated from the contribution of viral pneumonia. However, cases have been described of sudden death in which the degree of cardiac damage alone was consistent with a lethal outcome (Oseasohn et al., 1959; Verel et al., 1976).

Pericarditis has also been reported as an unusual manifestation of influenza (Edelen et al., 1974), and isolation of the virus from pericardial fluid has been reported (Hildebrandt et al., 1962).

The occurrence of myocarditis coincident with the initial respiratory symptoms suggests that the virus is directly involved in its pathogenesis. Indeed, virus has been isolated from the heart in fatal cases (Oseasohn et al., 1959; Woodward et al., 1960; Verel et al., 1976; Engblom et al., 1983). Witzleb et al. (1976) have confirmed that the virus can infect heart muscle by showing that the titer of virus in the heart exceeded that in other tissues and by demonstrating viral antigen in heart muscle and in the endothelium of small blood vessels by immunofluorescent techniques.

Encephalopathy

Although older accounts of influenza epidemics have remarked on the occasional occurrence of disorientation, irritability, stupor, coma, or even psychotic behavior coincident with or immediately following respiratory tract symptoms, very few virologically documented instances of influenza encephalopathy have been published. Credible evidence that encephalopathy is influenza associated must include (1) occurrence of influenzalike illness with respiratory tract symptoms concurrent with or immediately before the onset of encephalopathy, (2) specific diagnosis by virus isolation or a specific serological response to influenza virus infection, (3) absence of known causes of encephalopathy, and (4) brain pathology consistent with encephalopathy in fatal cases.

There is little doubt, however, on the basis of a few well-studied cases, that encephalopathy (as distinguished from encephalitis, in which inflammatory cell response occurs) does occur in some patients with influenza. The relationship of influenza encephalopathy to viral invasion of the brain remains unestablished, although influenza virus isolation from brain tissue of fatal cases of influenza has been reported, and viral antigen has been found in ependymal cells (Franková et al., 1977). Virus has been demonstrated in cerebrospinal fluid in a few cases. Murphy and Hawkes (1970) reported the recovery of A/Hong Kong/68 virus from three postmortem specimens of brain but provided no details; Thraenhart et al. (1975) and Rose and Prabhakar (1982) reported isolation of virus from spinal fluids; in only one of these cases in the former study was a specific antibody rise demonstrated duirng illness.

Although central nervous system involvement during influenza was reported in sporadic cases prior to 1957 (reviewed by Sugiura et al., 1958), the 1957 pandemic was notable for a number of reports of this manifestation of the disease. Among these, those summarized in Table 7-2 fulfill most of the diagnostic criteria set forth above. (In very few instances have other possible causes of encephalopathy been rigidly excluded.) Examination of these cases discloses that stupor or coma was predominant as a symptom initiating hospitalization and that focal signs were usually absent. Influenza-associated encephalopathy occurred in young and middle-aged patients of both sexes. Complete recovery was the rule, but death occurred in one patient from whose lungs virus was recovered. Onset of encephalopathic symptoms occurred soon after onset of respiratory tract disease in most younger patients but after a brief period of remission in the older subjects.

In many case reports from 1957 not included in the Table 7-2 summary, it is probable that encephalopathy was influenza associated as indicated by the presence of antibody to H2N2 influenza virus in a single serum specimen obtained at the epidemic peak, because the population at that time was experiencing H2N2 virus infection for the first time. The Alaskan influenza epidemic in 1973 was attended by both pericarditis and encephalopathy, providing further epidemiologic evidence of encephalopathy as an unusual manifestation of influenza (Edelen et al., 1974).

As in influenza myopathy, there are hints that susceptibility to influenza encephalopathy may be genetically determined. Two siblings were admitted to the hospital on the same day with symptoms of brain involvement during a family epidemic of influenza (McConkey et al., 1958). As with other unusual manifestations of influenza, it is unclear whether or not the sporadic occurrence of encephalopathy reflects a varying pattern of intrinsic host susceptibility, viral variation, or (perhaps most likely) the variable interest and acumen of clinical investigators.

Table 7-2. Encephalopathy in Documented Cases of Influenza 1957–1985

Reference	Case	Age	Sex	Day of onset	Major symptoms	Outcome[a]	Virus isolation[b]	Antibody response	Epidemic date	Probable virus etiology
Sugiura et al (1958)	2	26	F	1	Coma	CR	—	+	6/57	A/H2N2
	3	8	F	1	Coma, convulsions	CR	—	+	11/57	A/H2N2
Bennett and Turk (1957)	1									
Horner (1958)	1	16	M	5	Coma	PR	—	+	10/57	A/H2N2
	2	8	M	15	Confusion, ataxia	CR	—	+	10/57	A/H2N2
	4	14	F	3	Disorientation, stupor	D	+ (L)	–	10/57	A/H2N2
Mellman (1958)	C E	8	F	3	Stupor, dilirium	CR	—	+	10/57	A/H2N2
Dubowitz (1958)	2	10	M	5	Coma, restlessness	CR*	—	+	10/57	A/H2N2
Delorme and Middleton (1979)[c]	1	13	M	2	Confusion, ataxia	PR	+ (NP)	+	2/76	A/H3N2
	2	5	M	3	Agitation, dysarthria	CR	+ (NP)	–	3/74	A/H3N2
Sulkava et al (1981)	1	50	M	7	Stupor, aphasia	CR	—	+	2/80	A/H3N2
	3	43	M	7	Stupor	CR	—	+	2/80	A/H3N2
	4	36	M	7	Stupor	CR	—	+	2/80	A/H3N2

[a] CR, complete recovery, PR, partial recovery, D, death, CR*, recovery from encephalopathy, residual foot drop
[b] L, lung, NP, nasopharyngeal
[c] Four other cases not described in detail

Other Neurological Manifestations

As with many other ubiquitous infections, occasional myelitis and a variety of neuropathies have been temporally associated with influenza, but definitive causal association has not been substantiated.

Miscellaneous

Other rare manifestations of influenza include an unusual interstitial pneumonia leading to reduced pulmonary function (Pinsker et al., 1981). Unusual hemorrhagic manifestations of influenza including hematemesis and hematuria have been described (Thien et al., 1975; Kuberski, 1977; Banks and Tinker, 1977; Khakpour and Nik-Akahtar, 1977; Medical Research Council, 1919). Parotitis is rarely associated with influenza (Brill et al., 1977; Krilov and Swenson, 1985) and has not yet been causally linked to the disease (but see Chapter 1).

COMPLICATIONS OF INFLUENZA

Bacterial Infections

Influenza epidemics are notoriously associated with an increased incidence of bacterial pneumonia (Committee on Acute Respiratory Diseases, 1945; Stuart-Harris et al., 1949; Kilbourne, 1960). In the 1918 pandemic, such pathogenic bacteria as streptococci, staphylococci, *Hemophilus influenzae*, or pneumococci were almost invariably isolated from the lungs of fatal cases of influenza (MacCallum, 1919; Wolbach, 1919; Opie et al., 1921). Even in the subsequent pandemics of 1957 and 1968 in which antibacterial drugs were in wide use, bacteria were major contributors to influenza complications and mortality (Louria et al., 1959; Bisno et al., 1971). The heterogeneity of complicating bacterial pneumonias has been appreciated since 1918 despite the claim at that time that *Hemophilus influenzae* (Pfeiffer's bacillus) was the primary cause of the disease itself. Clearly, the patterns of bacterial infections can vary from one influenza epidemic to another (Fransen and Tunevall, 1969).

It is probable that the antibiotic era has influenced the relative frequency of the various complicating bacterial pneumonias, suppressing streptococci found so frequently in 1918 and favoring the emergence of staphylococcal invaders observed in some studies in 1957 and subsequently (Martin et al., 1959b). Antemortem bacteriological studies confirm that pneumococci remain the most frequent secondary invaders (Louria et al., 1959; Jarstrand and Tunevall, 1974).

Most commonly, bacterial pneumonia develops as an early sequel to ordinary influenza, but it may complicate and coincide with influenza or with acute influenza virus pneumonia (Louria et al., 1959) even in influenza B infections (Connor and Powell, 1985). Reasons for the increased incidence of bacterial pneumonia in influenza patients are described in the following section on Pathogenesis.

In most cases of bacterial pneumonia complicating influenza, influenza symptoms are followed by a 2- to 14-day period of improvement, after which there is an abrupt onset or return of shaking chills, productive cough, pleuritic pain, or other symptoms typical of acute bacterial pneumonia. Localizing signs of pulmonary consolidation are found on physical examination and confirmed by x-ray evidence of focal pneumonia.

In some patients, the acute symptoms of influenza may coincide with signs and symptoms of localized pulmonary disease. From such patients bacterial pathogens can be recovered from blood or sputum, establishing the diagnosis of bacterial pneumonia.

Bacterial pneumonia can also occur simultaneously with influenza viral pneumonia, in which case local consolidation may or may not be superimposed on the diffuse lung involvement caused by the virus (Louria et al., 1959).

Noninfectious Pulmonary Complications

Patients with influenza are vulnerable to more than bacterial infection. Previously healthy adults with influenza developed bronchial hyperreactivity when exposed to an aerosol of sodium nitrate (Utell et al., 1980). In both asthmatic and normal adults, protracted wheezing and dyspnea with eventual resolution may follow influenza (Klein et al., 1978).

In infants and young children, influenza can result in permanent lung damage. In three male patients, chronic pulmonary sequelae followed serologically verified influenza virus infection at 5, 24, and 42 months of age. Varying degrees of interstitial fibrosis, bronchial and bronchiolar damage, obliterative bronchiolitis, and chronic interstitial inflammatory infiltrates were found on lung biopsy. H3N2 influenza virus was isolated from the lung biopsy of one patient 52 days after the onset of illness (Laraya-Cuasay et al., 1977).

Complications of Viral Infection of the Respiratory Tract Not Specific for Influenza

Disseminated intravascular coagulation (Shenouda and Hatch, 1976; Luksza and Jones, 1984) and thrombocytopenic purpura (Wasserstein et al., 1981) are sequels of several virus infections and have been reported to follow influenza.

Asthenia and Depression

A perception that postinfluenza asthenia and depression are frequent complications is common among physicians, but the etiologic linkage of these symptoms to influenza has not been established. No correlation between hospital admissions for depression and influenza epidemics has been found (Flewett, 1976). Delayed recovery following uncomplicated influenza was found in patients who had responded in prior psychological tests in patterns characteristic of depression-prone patients (Imboden et al., 1961).

Effects of Influenza on the Fetus

Viremia is either rare or not easily demonstrable in influenza, but transplacental transmission of virus has been reported without evidence of fetal damage (Yawn et al., 1971; McGregor et al., 1984). However, it is not unreasonable that the outcome of pregnancy might be influenced by influenza virus infection. The contradictory evidence on this question is reviewed briefly in the following paragraphs. Many of the reported studies are retrospective and fail to provide adequate documentation for the diagnosis of influenza.

Abortions and Stillbirths

MacKenzie and Houghton (1974), in a critical review of available data, concluded that there is little evidence that premature termination of pregnancy or the occurrence of stillbirths is causally associated with influenza, citing only the 1918 experience reported by Harris (1919) as an exception.

Teratogenic Effects

No homogeneous pattern of congenital anomalies has been reported in the wake of influenza epidemics such as those found after such known teratogens as rubella and thalidomide (Leck et al., 1969). An increase in anencephaly and neural tube defects was reported in Dublin, Ireland (Coffey and Jessup, 1963) and in Finland (Hakosalo and Saxen, 1971) following the 1957 pandemic. Increased stillbirths related to anencephaly were recorded by the Registrar General of Scotland in 1958 and 1959 but not in a study of pregnancy in northwest London in 1957–58 (Doll et al., 1960). An increased occurrence of anencephaly and spina bifida in 1929–32 has been attributed to prepubertal or pubertal damage in the 1918 pandemic to girls reaching their peak reproductive period in those later years (Janerich, 1971a,b).

Other nonneurological congenital anomalies have been reported to occur inconstantly following influenza (Leck at al., 1969), but a number of negative studies with regard to the occurrence of fetal defects have been reported (Widelock et al., 1963; Wilson and Stein, 1969; Editorial, *The Lancet*, 1971). A "slight excess" of congenital abnormalities associated with influenza in a prospective study was interpreted with caution because the infections occurred during the second and third trimesters of pregnancy, late in gestation for the induction of teratogenesis (Griffiths et al., 1980).

Fetally Induced Neoplastic Disease

A postulated late effect of influenza-induced fetal damage is an increase in leukemia and lymphoma in cohorts of children born during influenza epidemics (Fedrick and Alberman, 1972; Austin et al., 1975), observations not confirmed by others (Randolph and Heath, 1974; reviewed by MacKenzie and Houghton, 1974).

Reye's Syndrome

The occurrence of acute encephalopathy and fatty visceral degeration in children had been noted earlier, but the description of the pathological entity together with its clinical concomitants was first clearly defined by Reye, Morgan, and Baral (1963).

Reye's syndrome is a serious, often fatal illness that occurs as a sequel of acute viral infections of childhood, most frequently after influenza or varicella. Onset within 4–7 days of the start of infection is heralded by altered behavior and episodes of repetitive vomiting. Neither fever nor jaundice is usually present despite clinicopathological evidence of acute impairment of liver function. Lethargy progresses to disorientation and irritability, combativeness, and sometimes coma. In infants, apnea, seizures, and coma may occur without antecedent vomiting or delirium. The primary liver involvement is reflected by laboratory tests showing elevated aminotransferase levels, hyperammonemia, prolongation of prothrombin time, and, in young children, hypoglycemia. Over 90% of cases occur in those less than 18 years of age. Despite recent improvements in therapy, 20–35% of patients die during the acute illness (Barrett et al., 1984).

Because there is no specific diagnostic test for the syndrome, cases have been defined by the Center for Disease Control of the U.S. Public Health Service for epidemiologic purposes by the following three criteria:

1. An acute noninflammatory encephalopathy documented by (a) a clinical picture of an alteration in the level of consciousness and, if available, (b) a cerebrospinal fluid specimen containing fewer than eight leukocytes per mm^3 or (c) histological sections of the brain demonstrating cerebral edema without perivascular or meningeal inflammation.
2. Evidence of hepatic involvement documented by either (a) microvesicular fatty metamorphosis of the liver, diagnosed by biopsy or autopsy, or (b) threefold or greater rise in either the serum glutamic-oxaloacetic transaminase (SGOT), serum glutamic-pyruvic transaminase (SGPT), or serum ammonia, and, if available, (c) a serum total bilirubin of less than 4 mg/dl.
3. No other more reasonable explanation for the cerebral and hepatic abnormalities (Barrett et al., 1984).

A consensus conference sponsored by the National Institutes of Health in 1981 recognized three degress or stages of Reye's syndrome (Consensus Conference, 1981). As somewhat modified and abbreviated by the National Reyes Syndrome Surveillance System, these stages are as follows:

Stage 0—Alert wakefulness
Stage 1—Difficult to arouse, lethargic, sleepy
Stage 2—Delirious, combative, purposeful or semipurposeful motor responses
Stage 3—Unarousable, predominantly flexor motor responses decorticate
Stage 4—Unarousable, predominantly extensor motor responses, decerebrate
Stage 5—Unarousable, flaccid paralysis, areflexia, pupils unresponsive (Barrett et al., 1984)

Although Reye's syndrome has occurred after infection with a large variety of other viruses (Sullivan-Bolyai and Corey, 1981), by far the most common antecedent is influenza, especially that caused by influenza B virus. For influenza B, the incidence per 100,000 infections is 30–60 cases; for influenza A (chiefly H1N1), 2.5–4.3, and for varicella, 0.3–0.4 (Editorial, *The Lancet*, 1982). The temporal and geographic association with influenza B infection was convincingly demonstrated in the United States epidemic of 1973–74 (Corey et al., 1976). There are suggestions that of the influenza A viruses, H1N1 strains are more provocative than H3N2 viruses (Halsey et al., 1980).

Recent studies suggest that the incidence of Reye's syndrome may be higher than originally suspected and that patients with Reye's syndrome who are not comatose outnumber patients in coma three to one. Prospective studies in children presenting with acute onset of vomiting and minimal neurological impairment after upper respiratory tract infections or varicella verified these cases as grade 1 Reye's syndrome by serum enzyme concentration and biopsy (Lichtenstein et al., 1983). The prognosis in such cases is good with early recognition and treatment with intravenous glucose and electrolytes (Heubi et al., 1984). In a previous study, liver function abnormalities had been observed in the course of a type A H1N1 influenza outbreak in Michigan (Monto et al., 1981).

Although occasional isolations of viruses causing the antecedent illness have been demonstrated in patients with Reye's syndrome, the bulk of evidence suggests that direct viral involvement of the liver, brain, or other tissues is not responsible for this syndrome, which has been characterized as "clinically an encephalopathy and chemically a hepatopathy" (DeVivo, 1983). Accordingly, copathogenic factors have been looked for without success. Early studies had suggested a predominance in rural areas (Corey et al., 1977), but in one study a high number of blacks and city dwellers were found (Sullivan-Bolyai et al., 1980). A growing body of evidence implicates salicylates used in the therapy of febrile diseases as a possible cofactor in the causation of the syndrome (reviewed by Alexander, 1982).

The definitive subcellular basis of the syndrome may represent a metabolic response to a universal mitochondrial insult (DeVivo, 1978).

PATHOGENESIS AND PATHOPHYSIOLOGY OF INFLUENZA

Any attempt at a coherent, much less a definitive, statement on the pathogenesis of human influenza immediately discloses large gaps in our knowledge of the process. There is no agreement even on the site of initial deposition of virus in the respiratory tract, and in a disease in which viremia is seldom demonstrable, the genesis of systemic symptoms and the typical myalgia is unknown. However, a synthesis of information derived from studies of cytopathology in natural infections, analysis of viral replication and shedding in experimental human infections, and reasonable extrapolations from animal models permits at least a general picture of the infectious process to be drawn. Imperfect and conjectural though it may be, this construction at least defines present ignorance and provides a framework for future research.

To begin with, there is no doubt that infection occurs via the respiratory route, as was suspected in 1918 (Wolbach, 1919) and even earlier. It is a matter of controversy whether or not in natural infection virus initially infects cells of the upper tract and then descends to its principal replication site, the trachea, or whether it ascends to that site after first being implanted in the alveolae.

Evidence can be found to support either view, and it may be that natural infection can be induced by either method depending on clinical and epidemiologic circumstances.

Sequence of Events

Implantation of Virus in the Respiratory Tract

Clinical observations of the natural disease provide no insight into the sequence of tissue involvement in influenza. The incubation period is short, the onset sudden, and the symptoms reflect, from the first, infection of upper and middle and, less frequently, lower respiratory tract.

Much has been learned from the experimental administration of influenza viruses to human subjects. However, the relevance of such studies to naturally occurring influenza must be interpreted with recognition that they employ cell-culture- or chick-embryo-propagated viruses that potentially are attenuated for man. This is not to say that influenza cannot be induced by artificial challenge. When administered to volunteers who possess little or no specific serum antibody, laboratory-grown viruses can produce the typical signs and symptoms of influenza—a fact established almost a half century ago (Smorodintsev et al., 1937; reviewed by Alford et al., 1966; Douglas, 1975). Experimental infection is most readily induced by virus in small-particle aerosols subject to deposition in the lung (Henle et al., 1946; Alford et al., 1966). The severity of influenza so induced is greater than that following introduction of virus into the nasopharynx (Alford et al., 1966; Douglas, 1975). Only 3 tissue culture infectious doses$_{50}$ ($TCID_{50}$) were required to infect volunteers by aerosol inhalation (Alford et al., 1966), whereas tenfold higher concentrations of the same virus[A/Bethesda/10/63 (H2N2)] were required to infect by the nasopharyngeal route (Jao et al., 1965). Similar observations have been made in influenza virus infection of mice (Larson et al., 1976). The site of initial viral deposition also influences the location and severity of influenza in the ferret (Sweet et al., 1985).

Mulder and Hers (1972) have emphasized the sometimes selective infection and necrosis of alveolar cells in human influenza. After likening the pathology of influenza pneumonia in the mouse to that in man, Hers et al. (1962) suggested that vascular effects in the lung were secondary to alveolar cell destruction and that influenza virus pneumonia in mice can begin in the alveoli. The primacy of alveolar damage had been proposed earlier for human influenza (Winternitz et al., 1920; Hers and Mulder, 1961). Albrecht et al. (1963) also found alveolar cells to be the primary site of influenza virus multiplication in mice, and Schulman and Kilbourne (1963) found higher concentrations of virus in the lungs than in the turbinates and tracheas of mice early in infection. But alveolar infection by influenza virus in the murine model may reflect only the virulence of the mouse-adapted viruses usually employed in experimental infections (Raut et al., 1975; Wyde et al., 1977; Ramphal et al., 1979). Following the pathogenesis of aerosol infection of mice by immunofluorescence, others have concluded that influenza virus infection is a descending process (Yilma et al., 1979).

Progression of Infection

Although the precise progression of influenza virus infection is moot and may indeed be variable (Sweet and Smith, 1980), there is no doubt that a primary target of infection is the ciliated epithelium of the middle respiratory tract (Hers, 1966) and that viral antigen can be found from nasopharynx to alveoli if influenza is studied in the full range of its severity. Point inoculation of virus in chick embryo tracheal explants was followed by upward (laryngeal) spread of virus in the mucociliary stream early in infection before ciliary activity was depressed. However, slower downward (bronchial) spread was also observed (Murakami and Matsuyama, 1985).

The unique studies of Walsh et al. (1961) using tracheobronchial biopsy of acute uncomplicated human influenza have confirmed the rapid onset of epithelial necrosis 24 hr after the onset of illness (Fig. 7-9). Tissue response to the virus was increased progressively down the tracheobronchial tree.

Viral Replication, Shedding, and Persistence

The shedding patterns of infective virus after infection illuminate the progression of the pathogenic process (Fig. 7-10). Sampling of nasopharyngeal or throat

Figure 7-9. Tracheal epithelium 1 day after the onset of symptoms of influenza. Desquamation of epithelium and exposure of the basement membrane have occurred in some areas. There are "hyalinization" and thickening of the basement membrane (arrow), a sparse mononuclear infiltrate, and eosinophilia of the elastic tissue. (From Burch et al., 1961. Reproduced with permission, from *Am. Rev. Respir. Dis.* **83:74.**)

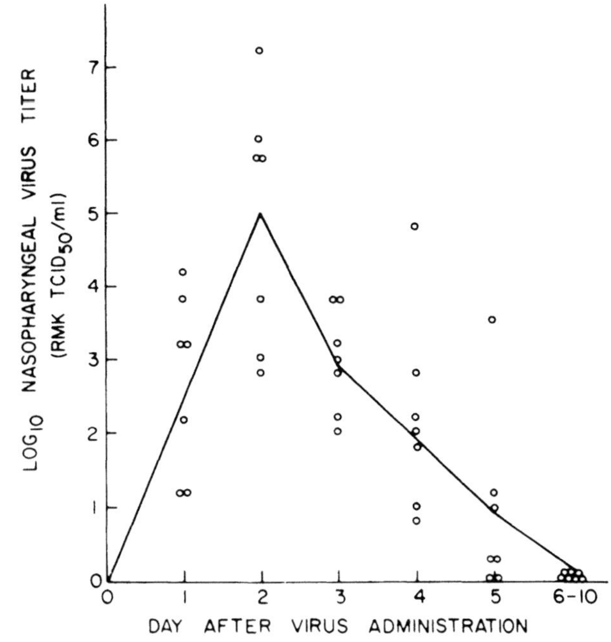

Figure 7-10. Quantities of influenza A (H3N2) recovered from nasal wash specimens obtained daily after inoculation of seven infected volunteers. RMK, rhesus monkey kidney. (From Murphy et al., 1973. Reproduced with permission, from *J. Infec. Dis.* **128**:481. Copyright © 1973 by The University of Chicago Press.)

washings cannot localize the virus to different regions of the respiratory tract as has been done in mice (Schulman and Kilbourne, 1963; Larson et al., 1976) and ferrets (Basarab and Smith, 1969) by tissue extraction. However, in man, severity of illness is directly related to the amount of infective virus sampled from the accessible upper respiratory tract (Couch et al., 1971; Murphy et al., 1973).

Ordinarily, virus is isolable from the human upper respiratory tract no longer than the first week of illness but has been recovered from the lungs of fatal cases as late as 10 days after the appearance of symptoms (Kilbourne, 1959). H3N2 influenza virus was isolated from the lung biopsy of a baby with evidence of persistent pulmonary disease on the 52nd day of infection (Laraya-Cuasay et al., 1977). Persistence of viral antigen in the nasal epithelium has been described in influenza with pulmonary complications (Ebisawa and Kitamoto, 1969). Recent studies in mice have described protracted alveolitis in association with apparent persistence of viral antigen for up to 1 year after infection (Jakab et al., 1983; Astry et al., 1984). These latter studies are provocative but must be confirmed with techniques other than the ELISA system used.

Contribution of Host Response to Disease

There is some evidence that cytotoxic T-cell response to influenza virus may contribute to the pathogenesis of pneumonia in experimental murine infection (Wyde et al., 1977; Wyde and Cate, 1978; Wells et al., 1981a). However, Yap et al. (1979) concluded that influenza immune T cells seem to inhibit rather than contribute to the genesis of virus-mediated pulmonary pathology.

In mice a correlation has been observed between the incidence of metaplasia of the lung alveolar epithelium and serum antibody levels (Krus et al., 1984).

Macrophage release of fibroblast-stimulating factors may play a role in the development of end-stage fibrotic lung disease occasionally seen as a sequel of influenza virus pneumonia (Jennings et al., 1984).

Viremia and Infection of Nonrespiratory Organs

Among the unusual manifestations of influenza that have been discussed earlier in this chapter, the occurrence of myopathy and carditis suggest that peripheral dissemination of virus from the respiratory tract may be more common than the rare recovery of virus in the blood would suggest. If myositis is induced by infective virus, as has been documented in a few cases, then its occurrence during the acute respiratory illness suggests that viremia must occur relatively early in infection. Elevations of serum CPK reflecting skeletal muscle involvement in cases of influenza without recognized myopathy are higher during the acute than the convalescent phase of illness (Friman, 1976; Kessler et al., 1983).

Pathophysiological Effects of Influenza

Effects on Respiratory Function

Evidence is increasing that pulmonary function is impaired during and after uncomplicated influenza. In common with other respiratory virus infections, influenza induces airway hyperreactivity but also results in peripheral airway dysfunction. Johanson et al. (1969) found restrictive ventilatory defects and increased alveolar–arterial oxygen tension gradients unresponsive to bronchodilator aerosols in six of ten patients with normal chest x rays. Increased airway resistance (Leeder et al., 1974) has been found in several studies of serologically confirmed influenza demonstrated as abnormal frequency dependence of total pulmonary resistance (Hall et al., 1976) or diminished density-dependent forced flow rates (Little et al., 1978). In the latter study, airway hyperreactivity to an inhaled cholinergic compound persisted for 7 weeks.

A trend toward bronchial reactivity to cholinergic stimulation was found in infection induced intranasally with an H1N1 virus (Hobbins et al., 1982). Bronchial hyperreactivity has been noted in other normal subjects following naturally occurring influenza (Empey et al., 1976) as well as in asthmatic patients infected with attenuated virus (Laitinen and Kava, 1980). Acute decline in 1-sec forced expiratory volume was found in patients with chronic obstructive pulmonary disease who acquired influenza. This effect was not seen with other respiratory virus infections (Smith et al., 1980).

The significance of these observations is the evidence that they provide of involvement of the lower airways in both normal and chronically diseased patients with influenza. With respect to pathogenesis, it can be inferred that acute and often protracted bronchiolar inflammation occurs, the severity of which may be virus strain dependent (Douglas et al., 1981; Little et al., 1979) and modifiable by viral chemotherapy specific for influenza (Little et al., 1978). Furthermore, airway epithelial damage can result in a vagally mediated reflex causing generalized bronchospasm (Empey et al., 1976), which may explain wheezing and dyspnea as complications of influenza (Klein et al., 1978).

Effects on Ciliary Function

The predilection of influenza virus for ciliated epithelium is reflected by an impairment of ciliary function demonstrable in in vitro systems. Mostow and Tyrrell (1972) related the virulence of several strains of virus for man with the capacity to reduce ciliary activity. In mouse tracheal organ cultures, influenza virus replicates and causes ciliostasis, an effect related to attachment to and clumping of cilia by virus particles (Westerberg et al., 1972). Carson et al. (1985) examined samples of nasal epithelium from children with acute viral infections, including influenza, and found transient configurational change of the normal ciliary microtubular patterns.

In Vivo Effects on Leukocytes

Polymorphonuclear Leukocytes. Slight polymorphonuclear leukocytosis and delayed impairment in chemotaxis were found in experimental influenza (Martin et al., 1981). The oxidative metabolism of granulocytes was increased during naturally occurring influenza (Jarstrand, 1977).

Lymphocytes. Lymphocytopenia is well known in influenza. Its basis is not yet understood, but it is known that T cells are primarily involved, and it is suspected that their normal differentiation may be impaired (Scheinberg et al., 1976). T-cell lymphocytopenia is also induced with noninfective (vaccine) virus, so the effect may not depend on viral replication (Faguet, 1981; Woodruff and Woodruff, 1976).

Depression of skin-test reactivity (Reed et al., 1972) is a possible sequel of T-cell lymphocytopenia (Scheinberg et al., 1976).

Monocytes and Alveolar Macrophages. Depression of chemotaxis in monocytes from influenza patients (Kleinerman et al., 1975) supports the hypothesis that the increased susceptibility to superinfection in influenza is related to virus-induced depression of monocyte function. This depressed function, in turn, may reflect the host's inability to mobilize macrophages to inflammatory sites because of their impaired chemotactic responsiveness (Kleinerman et al., 1976). Direct toxic effects of influenza B virus on human alveolar macrophages have been described (Voisin et al., 1974). Abortive in vitro replication of influenza A viruses in human macrophages without immediate cytopathic effect has been reported (Rodgers and Mims, 1982). Influenza virus-induced immune complexes suppressed alveolar macrophage phagocytosis in mice (Astry and Jakab, 1984), an effect that is obviously relevant to the diminished resistance to bacterial infection discussed below.

Effects on Host Resistance to Bacterial Infections

The effects of influenza virus infection on host resistance to bacterial infection are manifold. Studies of experimental infection of laboratory animals have attempted to evaluate their relative importance, with interesting but not definitive results. Most studies have focused on the mechanism of secondary invasion of the lung infected with influenza virus. Evidence that virus suppresses systemic (general) resistance to intravenous bacterial challenge is contradictory. Jakab et al. (1979) found increased numbers of viable staphylococci in the lung but not in the spleen and liver of influenza virus infected mice; Gardner (1980) found enhanced growth of *Listeria* in the spleen.

Also controversial is the effect of influenza infection of mice on pulmonary macrophage function. Warshauer et al. (1977) found retardation of the ingestion of staphylococci by alveolar macrophages and increased bacterial proliferation within them, whereas Nugent and Pesanti (1982) observed neither effect, although killing of bacteria in the lung was impaired.

Adherence of bacteria to influenza virus-infected human and canine cells in culture is enhanced (George et al., 1983; Davison and Sanford, 1981).

Fever

Fever is a cardinal characteristic of influenza, and its pathogenesis has been discussed previously. In experimental animals, virulent strains are better able to sustain replication during the febrile period (Toms et al., 1977; Rott et al., 1982). The possible protective role of fever in recovery from influenza is suggested by higher levels of nasal virus in fever-suppressed animals (Husseini et al., 1982).

THE IMMUNOLOGIC RESPONSE IN INFLUENZA

Humoral Antibody Response

Nature and Kinetics of the Response

Infection with influenza virus elicits the synthesis of demonstrable serum antibody to the HA, NA, NP, and M viral proteins. The nature of this response has been best defined for the hemagglutination inhibiting (HI) and neutralizing antibodies directed against epitopes of the HA. The antibody formed represents all major classes, IgG, IgM, and IgA (see Tables 7-3 and 7-4).

As in other viral infections, IgM characterizes the primary response, and response in reinfection is restricted to the IgG and IgA classes of antibody. Local and serum HA-specific IgA responses are concordant, suggesting a common mucosal origin for antibody in the two sites (Burlington et al., 1983). After intranasal inoculation of attenuated virus, populations of peripheral blood mononuclear cells that spontaneously secrete IgG and IgA appear, followed later by a second population that secrete IgG antibody when stimulated with mitogen (Yarchoan et al., 1981).

The kinetics of serum and secretory antibody response to intranasally inoculated attenuated influenza A virus has been studied by Murphy et al. (1982) employing an enzyme-linked immunosorbent assay (ELISA) with purified HA antigen (Fig. 7-11). The response was essentially similar to that in other viral infections: IgM response peaked in the second week, and IgG 2 weeks later; IgA, in both serum and nasal washings, also reached maximum titer 2 weeks after in-

Table 7-3. Cross-Reactive Immunity in Influenza: A Glossary

Primary immunity	Virus challenge	Immunity required
A/H1N1	A/H1N1	Homologous (variant-specific)
A/H1N1	A/H1[1]N1	Homosubtypic (heterovariant)
A/H1N1	A/H3N2	Homotypic (heterosubtypic)
A/H1N1	B	Heterotypic

fection. Most of the IgA and IgM but only some of the IgG antibody was actively secreted locally. Serum IgM is found also after natural infection with influenza virus (Buchner et al., 1977) but does not differentiate primary from secondary influenza infection when immunofluorescent antibody against whole virus is measured (Urquhart, 1974). In convalescence from influenza A, 78% of total antibodies were estimated to be of IgG_1 isotype, and other immunoglobin isotypes varied from 3 to 7% of total (Julkunen et al., 1985). In experimental secondary infection, nasal IgA antibodies were secretory and primarily of IgA_1 subclass. In serum virtually all IgA HA antibodies were of the IgA_1 subclass. (Brown et al., 1985).

Table 7-4. Specificity and Cross Reactivity in Antibody-Mediated and Cell-Mediated Immune Response in Influenza from Available Data from Human and Mouse Studies

Viral protein	Antibody class or T cell	Reactions among influenza A viruses: Homologous, variant-specific	Homosubtypic, heterovariant	Homotypic, heterosubtypic	A–B heterotypic
HA	IgG	+[a]	+	+[b]	+[c]
	IgM	+			
	IgA	+	+		
	Th	+	+	+	(?)
	Tc	+	+	+	
NA	IgG	+	+		
	IgA	+			
	Th	+			
NP	IgG			+	+[c]
	IgM			+	
	Tc		+	+	
M	IgG			+	+[d]
	Th	+[e]			
PB_2	Tc	+[f]			
Unknown	Tk				+[g]

[a]For citation, see text.
[b]Cross reactions occur with low frequency.
[c]Influenza A and B virus cross reactivity mediated through HA and NP (Graves et al., 1983).
[d]Bukrinskaya et al. (1985).
[e]Priming with M protein enhances HA antibody response (Russell and Liew, 1980).
[f]Bennink et al. (1982).
[g]A–B cross reactivity probably Th mediated in secondary in vivo reaction (Sitbon et al., 1983).

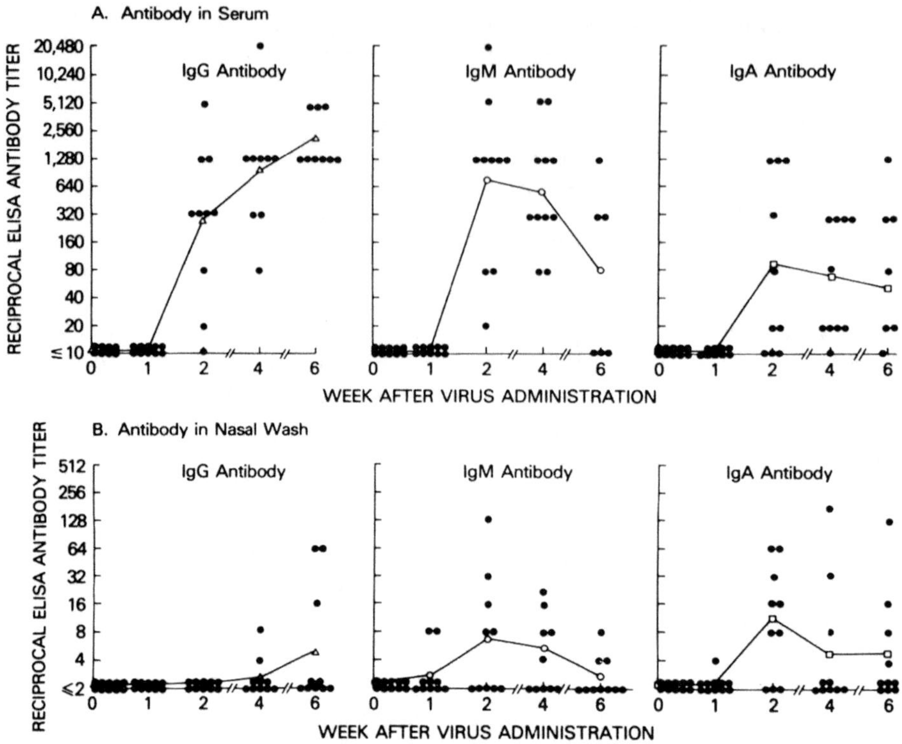

Figure 7-11. Hemagglutinin antibody response of children to the A/Alaska/6/77(H3N2) cold-adapted reassortant virus, showing kinetics of systemic and secretory antibody response to attenuated influenza virus infection. (From Murphy et al., 1982. Reproduced with permission, from Infect. Immun. **36:**1104.)

The dynamics of B lymphocytes in the lungs of influenza virus-infected mice are concordant with the antibody responses in humans described above. Both IgA- and IgM-containing cells increase early, followed by cells staining in immunofluorescence for IgG (Owens et al., 1981). Virus-specific IgA-secreting antibody-forming cells were demonstrated in the mediastinal lymph nodes of mice after influenza viral infection (Reiss and Schulman, 1980). In mice, heterovariant immunity following infection was closely correlated with local secretory IgA antibody (Liew et al., 1984). HA2-specific antibodies may show heterosubtypic (cross-subtype) reactivity in radioimmune precipitation assays (Graves et al., 1983).

The Antigenic Spectrum of Primary Antibody Response in Influenza

In addition to the well-characterized antibody response to the viral HA through which virus neutralization is mediated, antibody response to other viral proteins occurs at the time of primary infection.

Antibody Response to NA. Primary infection in young children is followed by the production of both anti-HA and anti-NA antibodies (Wright et al., 1977). Reports that NA antibody is rarely demonstrable following initial infection (Cough and Kascl, 1983) may reflect the relative insensitivity of the test methods used, neuraminidase enzyme inhibition (NI) and HI, as well as the smaller antigenic mass of the neuraminidase. Primary infection of college students with H1N1 virus resulted in serological response to the NA in only 9% as measured by NI and in 35% as tested by ELISA (E.D. Kilbourne, unpublished data). On the other hand, Schild (1969) found high titers of NI antibody in convalescent human sera following natural infection in a primed population. Chen et al. (1973) reported a different kinetics of NA antibody response from that of HI antibody, with NI antibody rising later and declining more rapidly.

Antibody Response to Internal Antigens of the Virus. Complement fixation tests, which measure antibody of the IgM class, have been used for years to measure antibody response in influenza. These tests, when the antigen comprises extracts of infected tissue, measure principally antibody to the NP. In conventional as well as modified (single radial complement fixation) tests (Haaheim, 1977), antibody response is detectable in about 70% of patients. By the modified test, antibody response to M protein was detectable in 35% of naturally infected subjects (Haaheim, 1977). Heterotypic (A–B) cross reactivity of NP antibodies in radioimmune precipitation assays has been noted (Graves et al., 1983).

In studies of 123 experimentally infected volunteers, Cretescu et al. (1978) demonstrated NP antibody increase in 43 and M antibody increase in 11 subjects with the single radial diffusion test (see section on Laboratory Diagnosis of Influenza). In general, anti-M increases coincided with increases in HI and anti-NP antibody. There was a suggestion that M antibody response was associated with a more serious illness, but further study indicated that antibody response to M protein differed in infections with influenza viruses of different subtypes.

Anamnestic Response in Influenza: "Original Antigenic Sin"

Francis, Davenport, and Hennessey (1953) were the first to recognize that immunologic response in influenza virus infection is oriented towards the initial priming antigens experienced at first infection. Thus, in subsequent infections, cross-reactive epitopes shared by the original and reinfecting virus will stimulate a predominant secondary antibody response while new epitopes in the reinfecting variant induce a primary response. This selective anamnestic response is seen to both HA and NA antigens. Memory also involves the secretory immune system (Wright et al., 1983) as well as antibody-dependent cell-mediated cytotoxicity (ADCC) against influenza virus-infected cells (Hashimoto et al., 1983).

With conventional methods of serological testing (HI tests), significant cross reactions are restricted to strains within a subtype (Morita et al., 1972), although minor cross reactions are encountered with low frequency (Kilbourne, 1976).

Nonspecific Effects of Influenza Virus on Immune Response

Influenza virus HAs are potent lymphocyte mitogens (Butchko et al., 1978;

Armstrong et al., 1981) that specifically stimulate B cells (Anders et al., 1984). Mitogenicity in mice is under control of the major histocompatibility complex (MHC) (Scalzo and Anders, 1985). Conversely, intact influenza virus also inhibits in vitro antibody synthesis (Daniels and Marbrook, 1981). The significance of these contrasting effects for the total immune response in influenza is unknown.

Cell-Mediated Immune Response in Influenza

Macrophage Response

Early in infection macrophages may become activated to increased antiviral activity in response to influenza virus. As part of the inflammatory response, they can gain access to virus and infected cells at mucosal surfaces (reviewed by Couch and Kasel, 1983).

Lymphocyte-Mediated Responses

Helper T Cells. Antibody response to the HA of influenza virus is T-cell dependent (Virelizier et al., 1974; Anders et al., 1979.) The HAs of different viral subtypes share cross-reactive antigenic determinants recognized by T-helper (Th) cells, and within a viral subtype the HAs are broadly cross reactive in Th-cell recognition (Anders et al., 1981). Furthermore, HA-specific T cells recognize determinants different from those seen by anti-HA antibody (Hurwitz et al., 1984; Katz et al., 1985; Sterkers et al., 1985), and an HA region immunodominant for human T-cell recognition was located for an H3 virus (Lamb and Greene, 1983). A similar immunodominant site or the HA of H1 virus is recognized by a murine helper T cell specific for H1 strains. The site was localized to a conserved nine-amino-acid sequence (positions 111–119) in the globular head region that is apparently not a target for neutralizing (and hence selective) antibodies (Hackett et al., 1983).

T-helper cells also play a role in the generation of influenza virus immune cytotoxic T-lymphocyte (Tc) responses (Biddison et al., 1981; Reiss and Burakoff, 1981). T-helper cells primed by internal antigens of influenza virus can cooperate in the antibody response to HA (Russell and Liew, 1979).

Cytotoxic T Lymphocytes. Mice infected with influenza viruses develop cytotoxic T lymphocytes (Tc cells) specific for viral antigens prior to the appearance of virus-specific antibody-forming cells (Reiss and Schulman, 1980). Whereas antibody-forming cells are primarily HA specific (Reiss and Schulman, 1980), Tc cells of lymph nodes and spleen are cross reactive among MHC-compatible cells infected with multiple influenza virus subtypes (Effros et al., 1977; Doherty et al., 1977).

Early studies of cell-mediated immunity (CMI) showed that Tc cells are either narrow or broad in antigenic specificity, one population being specific for the immunizing virus and the other exhibiting extensive cross reactivity among virus subtypes (Braciale, 1977). The cells of restricted specificity were found to recognize the viral HA, and cells of the other population did not (Ennis et al., 1977). Recent studies of Tc lysis of target cells fused with HA- and NA-bearing liposomes suggest that these proteins can function as target antigens for influenza virus-specific T cells (Stitz et al., 1985).

Further research has shown that even more subpopulations of Tc cells exist, including those that react with strain-specific or shared intersubtypic HA determinants (Townsend et al., 1984a,b) (Table 7-4).

The majority of Tc clones with specificity for influenza virus show cross reactivity for different influenza A virus subtypes (Fleischer, 1982). In mice most influenza virus-specific memory cytotoxic T lymphocytes react with antigenic epitopes not contained in viral surface glycoprotein (Kees and Krammer, 1984).

Heterosubtypic Tc reactions appear to be mediated principally through recognition of the NP antigen common to all influenza A viruses (Townsend et al., 1984a,b; Yewdell et al., 1985). These findings are in accord with the ready demonstrability of NP on the surface of infected cells (Yewdell et al., 1981). The NP gene controls the induction of both subtype-specific and cross-reactive Tc cells (Townsend and Skehel, 1984). The Tc cells recognize three distinct epitopes of the NP molecule (Townsend et al., 1985). Influenza virus-specific suppression of immune responsiveness has been reported (Hurwitz and Hackett, 1985).

Antibody-Dependent Cell-Mediated Cytotoxicity against Influenza Virus-Infected Cells

Hashimoto et al. (1983) have shown that human FC IgG receptor-bearing null cells and T lymphocytes can mediate a type of lymphocyte toxicity that is antibody dependent (Greenberg et al., 1977). Antibody mediating ADCC reacts with either NA or HA and is subtype specific on primary response but may exhibit broad heterosubtypic reactivity on secondary stimulation. The ADCC antibody is demonstrable in the sera of young children after natural infection. Its importance in immunity remains to be determined (Hashimoto et al., 1983).

Immunologic Basis of Recovery from Influenza

It is obvious that immunologic response to infection with influenza virus is very complex (Table 7-4). The definitive mechanisms of recovery are not known, but recent studies have identified components of the response that appear to be significant. These studies have been principally in the mouse but can reasonably be extrapolated to human infection.

The Role of Interferon

It has been clear for some time that increase in serum neutralizing antibody occurs relatively late in infection and is poorly correlated with decline in replicating virus. Therefore, factors more closely related to virus decrease have been examined for their importance in recovery.

Iwasaki and Nozima (1977) have defined three phases in influenza virus pneumonia in mice in a simplistic but useful model (Fig. 7-12). The first phase is characterized by an exponential increase in virus titer, the second by a rapid decrease, and the third by a moderate rate of decrease of virus in the lung. Interferon production is temporally associated with the transition from the first to the second phase. The appearance of IgA antibodies may be important in the inhibition of virus replication after the disappearance of interferon; IgM and, subsequently, IgG did not appear until after decline of virus had commenced (not clearly shown in Fig. 7-12). In man also, the appearance of interferon is associated with diminished shedding of virus (Hall et al., 1978).

Macrophages (Ettensohn and Roberts, 1984) and T cells are potential sources of interferon. These cells are present early in infection and therefore, together

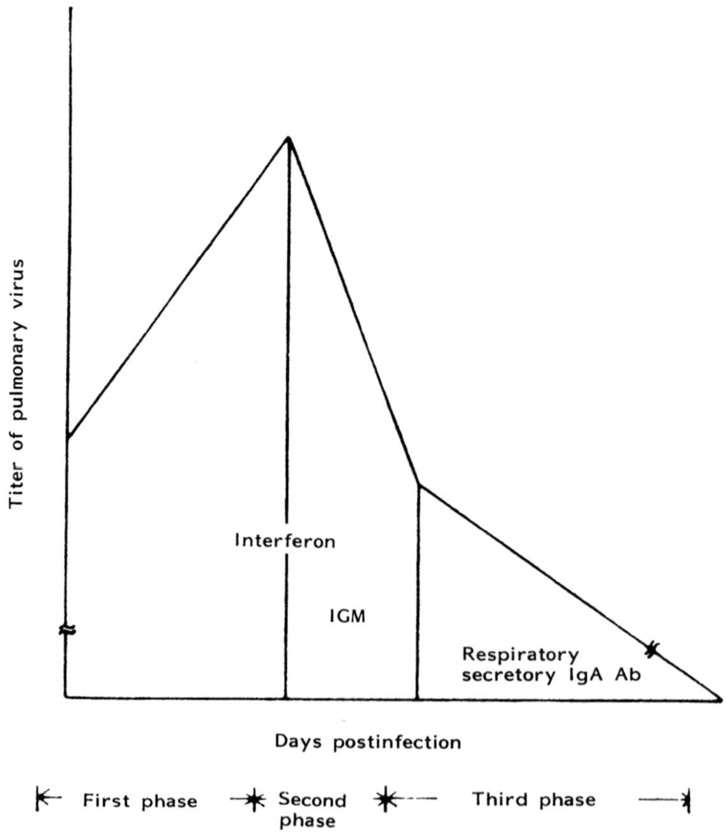

Figure 7-12. Appearance of factors in antiviral responses in relation to pulmonary virus titers (mouse model). (From Iwasaki and Nozima, 1977. Reproduced with permission, from *J. Immunol.* **118**:262.)

with interferon, may have an early role in the termination of influenza virus infection (Wyde et al., 1982).

The Role of Antibody in Recovery from Influenza

Although serum antibody has been shown to be protective in natural and experimental influenza (see below), its role in recovery has not been established. Certainly, it appears relatively late in infection and after virus titers have begun to decline.

Experiments in which antibody response in infected mice was suppressed by antibody to different classes of immunoglobulins have yielded conflicting and inconclusive results. Iwasaki and Nozima (1977) emphasized the contribution of IgA to recovery; Kris et al. (1985), using anti-IgM suppression, found that recovery could occur in the absence of detectable serum and nasal antibody but was delayed. Therefore, the answer may be that any effect of antibody in recovery is contributory but nonessential.

Early Cellular Immune Responses and Recovery

More than any other cellular event, the appearance of Tc cells is most closely correlated with recovery in experimental influenza. Mice infected with influenza virus develop Tc cells specific for viral antigens prior to the appearance of virus-specific antibody-forming cells. The Tc cells were detected coincident with the sharp decline in pulmonary virus titer (as noted also with interferon, above) (Reiss and Schulman, 1980). Adoptive transfer of Tc cells after their in vitro stimulation increased recovery from infection in nude (T-cell-deficient) mice (Wells et al., 1981b). A lesser ability of spleen Tc cells taken directly from mice undergoing primary or secondary infection to clear pulmonary virus correlated with their lower level of influenza virus-specific cytotoxicity (Wells et al., 1983). In human subjects inoculated with wild virus, all who had demonstrable Tc-cell responses cleared virus effectively (McMichael et al., 1983).

Adoptively transferred cloned lines of murine Tc lymphocytes completely protected syngeneic mice with potential lethal infection. Some such clones are subtype specific and others heterosubtypic in their ability to eliminate pulmonary virus and mediate recovery (Lukacher et al., 1984). This specificity of effect favors the concept that Tc cells express their antiviral effects in vivo by direct cytolytic effects rather than by secretion of interferon.

Other Factors That May Influence Recovery

Modulation of the Cellular Immune Response. A transient delayed hypersensitivity (DTH) following influenza virus infection of mice is mediated by Td lymphocytes. These cells increase the mortality rate in infected mice following challenge with homologous virus. This pathogenic effect can be reversed by antigen-specific suppressor Ts cells (Liew and Russell, 1983).

The Importance of an Intact Complement Pathway. Experiments in mice suggest that an intact complement pathway may play an adjunctive role in host recovery from primary influenza virus infection (Hicks et al., 1978).

IMMUNITY TO INFLUENZA

Immunity to influenza is a gradient that, depending on its degree, can suppress infection to the point of undetectability, modulate infection to prevent disease, or reduce the severity of disease.

The recurrent nature of influenza and the capacity of its virus to reinfect might suggest that immunity to the disease is transient or at best imperfect. Therefore, the first point to be made is that solid immunity to influenza can be obtained through natural infection. This immunity is attained as a composite experience derived from sequential infections with homologous and heterovariant or even heterosubtypic viruses. It is clear that individuals respond differently to infection by producing antibodies with different ranges of specificities (Wang et al., 1986). Because the occurrence of infection and the expression of disease are the resultant of intensity of virus dose and the state of multifactorial immunity in each individual, analysis of immunity in influenza has required the use of experimental infection. In artificially induced infection the nature, dose and method of infection can be standardized and defined.

Homologous Immunity

Although a number of such studies could be cited, a study in volunteers conducted by Morris et al. (1966) 6 years after the pandemic introduction of Asian H2N2 virus in 1957, is remarkable for the induction of infection and disease in all seronegative subjects and for experiments with homologous reinfection. An analysis of the results of these studies is presented in Tables 7-5 and 7-6. In summary, (1) all antibody-negative subjects were infected and became ill when challenged with a 1963 variant virus; (2) two-thirds of subjects with low or moderate (10–80) titers of homologous neutralizing antibody were afebrile, although all were infected, and the duration of fever and virus shedding was reduced; (3)

Table 7-5. Relation of Initial Homologous Serum Neutralizing Antibody Titer to Resistance to a Standard Challenge Dose of H3N2 Influenza Virus[a]

Group	N	Mean initial antibody titer	Infected	Number with fever	Mean number of days virus is shed	Mean number of days of fever
I	10	0[b]	100%	100%	2.5	3.1
II	19	15	100%	32%	1.3	0.8
III	4	632	0	0	0	0

[a]Reanalysis of data of Morris et al (1966)
[b]<1 10

Table 7-6. Experimental Testing of Immunity to Reinfection by Homologous Viral Challenge[a]

	Prechallenge I antibody			Postchallenge I antibody			Prechallenge II antibody			Postchallenge II antibody		
	Homologous	Heterovariant		Homologous	Heterovariant		Homologous	Heterovariant		Homologous	Heterovariant	
Subject	1963	1962	1957	1963	1962	1957	1963	1962	1957	1963	1962	1957
1	0	0	0	80	20	0	10	10	10	80	20	20
2	0	0	0	80	20	0	80	10	10	160	20	10
3	0	0	0	80	10	10	80	20	10	320	80	20
4	0	0	0	40	20	10	20	10	10	80	20	10

[a]Data from Morris et al (1966) All subjects were ill on challenge I, three were reinfected, but none was ill, on challenge II (2–10 months after initial infection) The 1963 challenge virus was A/Bethesda/10/63 (H2N2)

subjects with high (320–1280) levels of antibody (all of whom had heterovariant as well as homologous antibody) were solidly immune, i.e., did not shed virus and escaped illness.

Of further note, some subjects with low homologous titers to the challenge virus were protected from illness; others with moderate titers of heterovariant antibody induced by earlier H2N2 viruses were not protected. [The superior protective effect of passively transferred strain-specific antibodies has also been shown in animal experiments (Virelizier, 1975).]

Table 7-6 illustrates that homologous immunity prevents disease but may not prevent reinfection and that antibody response broadens in specificity following second infection, an observation made earlier in experimental animals (Henle and Lief, 1963).

Studies of natural infection confirm the modulating effect of preexisting serum HI antibody on both incidence of influenza and disease severity (Farník and Bruj, 1966). However, it is worth noting that natural reinfection and disease occurred under the close-contact conditions of a boys' boarding school 2 years after primary infection with a closely related virus (Sigel et al., 1950). Prospective studies in children have shown that natural reinfection with homologous or closely related viruses can occur (Fox and Hall, 1980) and that some of these may be symptomatic (Frank et al., 1979). In family contacts, reinfection rates were inversely related to age. In later investigations, reinfection rates were found to be variable, probably reflecting differences in intensity of exposure, age at primary infection, virus antigenicity, and, perhaps, differences in virus virulence (Frank and Taber, 1983).

Heterovariant (Intrasubtypic) Immunity

Cross immunity to infection among strains of a subtype is substantial (Schulman, 1973) (Table 7-3) and, indeed, is probably the significant ecological determinant leading to the progressive decline in the circulation of a viral subtype as it nears extinction at the end of each interpandemic period (Kilbourne, 1978).

Heterosubtypic Immunity

Epidemiologic evidence suggests some cross-protective effects between infections with H1N1 and H2N2 (Slepuskin, 1959), H2N2 and H3N2 (reviewed by Kilbourne, 1975), and H3N2 and H1N1 (Sonoguchi et al., 1985) influenza A subtype viruses. The determinants of such immunity are not know, but shared antigens of several viral proteins recognized either by B or T cells (Table 7-4) could mediate such protection.

Serum Antibody as a Marker for Immunity in Influenza

The widely employed hemagglutination-inhibition (HI) test for serum antibody has established that a titer of 1:32 or 1:40 is associated with protection against homologous virus, whether this level is achieved by infection or parenteral administration of nonreplicating antigen (vaccination). However, this correlation of serum antibody with protection does not necessarily imply that serum antibody itself is a protective factor. More likely, serum antibody concentration simply reflects the total immune response. It is probable that secretory antibody in the respiratory tract makes a major contribution to immunity to influenza and may be responsible for the greater and more durable immunity induced by infection (Kilbourne et al., 1974).

However, Couch et al. (1984) have emphasized that IgG antibody predominates in bronchial lavage fluids of humans and suggest that this finding together with a demonstrable prophylactic effect of passively transferred antibody and the serum derivation of lower respiratory tract antibody point to serum IgG as the primary mediator in the prevention of influenza.

Thus, immunity in influenza is a continuum influenced by the number and intensity of exposures to the same, intratypically related, or even heterotypic viruses. Although protective immunity can be related to levels of serum antibody, it must comprise all of the multiple components of humoral and cellular factors summarized in Table 7-4. The nature and duration of immunity will also reflect how immunity was acquired—whether by natural or attenuated infection or by parenteral injection of antigen (see discussion of vaccines in Chapter 11).

THE PATHOLOGY OF INFLUENZA

A reasonably accurate and comprehensive picture of the pathology of uncomplicated influenza can be put together from (1) autopsy examination of fatal cases in which early lesions are present, (2) biopsy study of uncomplicated cases, and (3) studies of disease in animal models. On this basis it is known that ciliated epithelial cells of the trachea and bronchi are primary targets of the virus in the acute stage of influenza; these cells are extensively involved with necrosis and desquamation. Bronchoscopy reveals acute diffuse inflammation of the larynx, trachea, and bronchi with mucosal injection and edema (Walsh et al., 1961).

An early lesion is swelling of individual cells. Soon thereafter, the normal palisade structure of the epithelium disappears as vacuolization of some cells and shrinkage of others disturbs the normal architecture. Thinning of the epithelium probably is a consequence of both shrinkage and desquamation of diseased cells. At the final stage of desquamation, only a single basal layer of flattened cells may remain, covering the basement membrane (Mulder and Hers, 1972) (see Fig. 7-13). Pathologists have emphasized that all stages of the pathological process may be seen at the same time within the respiratory tract. Deciliated columnar cells may persist, or desquamation of epithelial cells to the basement membrane may occur with only a few cuboidal cells remaining. Inflammatory cells and nuclear debris may be observed in the epithelial layer (Walsh et al., 1961). The process in mice is strikingly similar and has been recorded by scanning electron microscopy (Fig. 7-14).

Cellular regeneration starts early in infection even before desquamation begins (Hers, 1966). The exuberant character of the repair process leading to

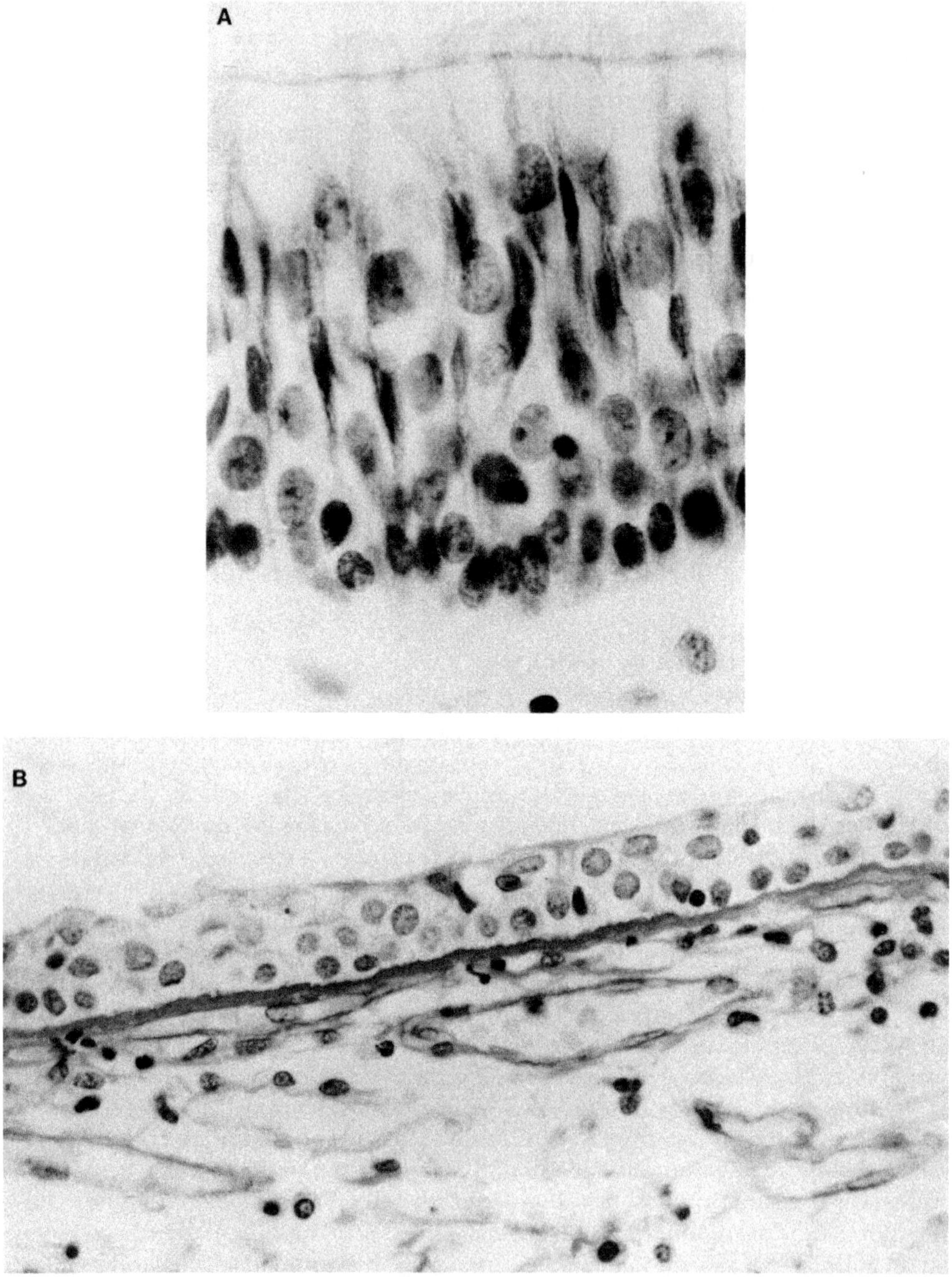

Figure 7-13. (A) Trachea. Normal ciliated epithelium. Hematoxoylin-eosin, ×960. (B) Trachea. In patient with influenza virus pneumonia after 2 days of illness, epithelium is only half normal thickness. New epithelial surface has been formed by desquamation of necrotic cells. Acid hydrolysis-thionine SO_2-PAS, ×960. (C) Trachea. Only basal cells remain as epithelial lining; duration of illness, 2 days. Acid hydrolysis-thionine SO_2-PAS, ×960. (D) Bronchus with cartilage. Regeneration of the epithelial layer of cells. Some su-

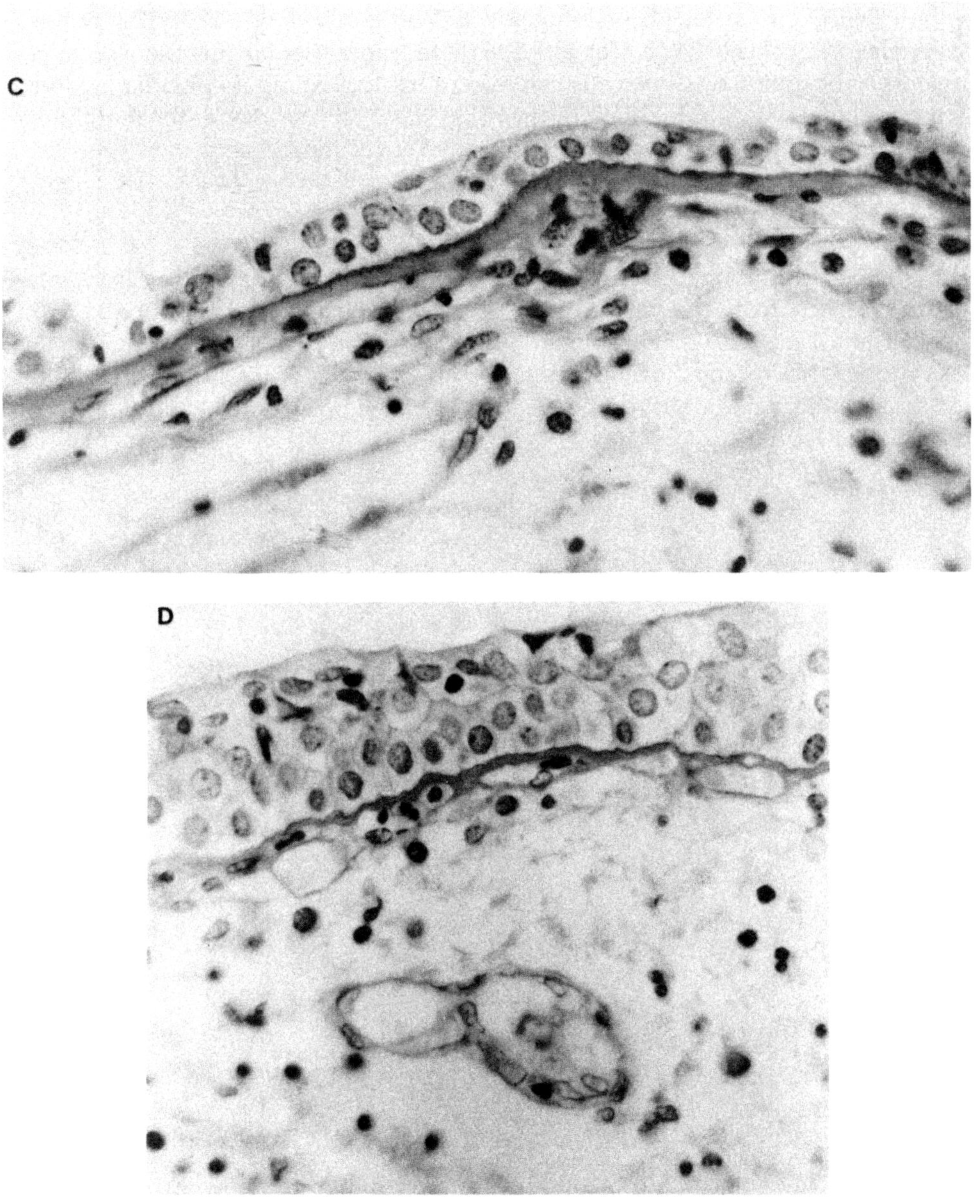

Figure 7-13. *(Continued)* perficial flattened cells show cilia. Degeneration of other cells is visible. Influenza virus pneumonia and staphylococcal bronchopneumonia; duration of illness 16 days. Acid hydrolysis-thionine SO_2-PAS, ×960. (From Mulder and Hers, 1972. Reproduced with permission, from *Influenza—The Air Passages in Fatal Human Influenza*, pp. 12, 15, 16, and 49, Wolters-Noordhoff, Groningen, The Netherlands.)

reepithelialization has long been emphasized (Winternitz et al., 1920). This process resembles metaplasia (Walsh et al., 1961). In experimental infection of mice, metaplasia of squamous-type cells can lead to the formation of epithelial nodules (Stinson et al., 1976). In man, recovery of normal respiratory epithelial architecture apparently is complete in uncomplicated infection.

Influenza Virus Pneumonia

When the lung parenchyma is involved in influenza, the picture is characteristic. Reports from the pandemics of 1918 (Winternitz et al., 1920) and 1957 (Mulder and Hers, 1972; Louria et al., 1959) describe hemorrhagic, edematous, heavy lungs. Microscopic examination reveals evidence of tracheitis, bronchitis, and bronchiolitis as described above as well as submucosal hyperemia, focal

Figure 7-14. (A) Normal uninfected mouse tracheal epithelium visualized by scanning electron microscopy (×1050). (B) Tracheal epithelium 24 hr after infection. The scanning electron micrograph reveals loss of serous and ciliated cells; cell debris is seen lying on the epithelial surface, and the basal cell layer is exposed (×800). (C) Scanning electron micrograph of tracheal epithelium 36 hr after infection. Few desquamating cells remain. The basal cell layer is exposed (×2000). (D) tracheal epithelium 3 days after infection as seen by scanning electron microscopy. Only one ciliated cells and cell debris can be seen on the surface of the otherwise flat and denuded epithelium (×2000). (E) Scanning electron micrograph of tracheal epithelium shows cilia formation 10 days after infection (arrow) (×2000). (From Ramphal et al., 1979. Reproduced with permission, from *Am. Rev. Respir. Dis.* **120:**1313–1324.)

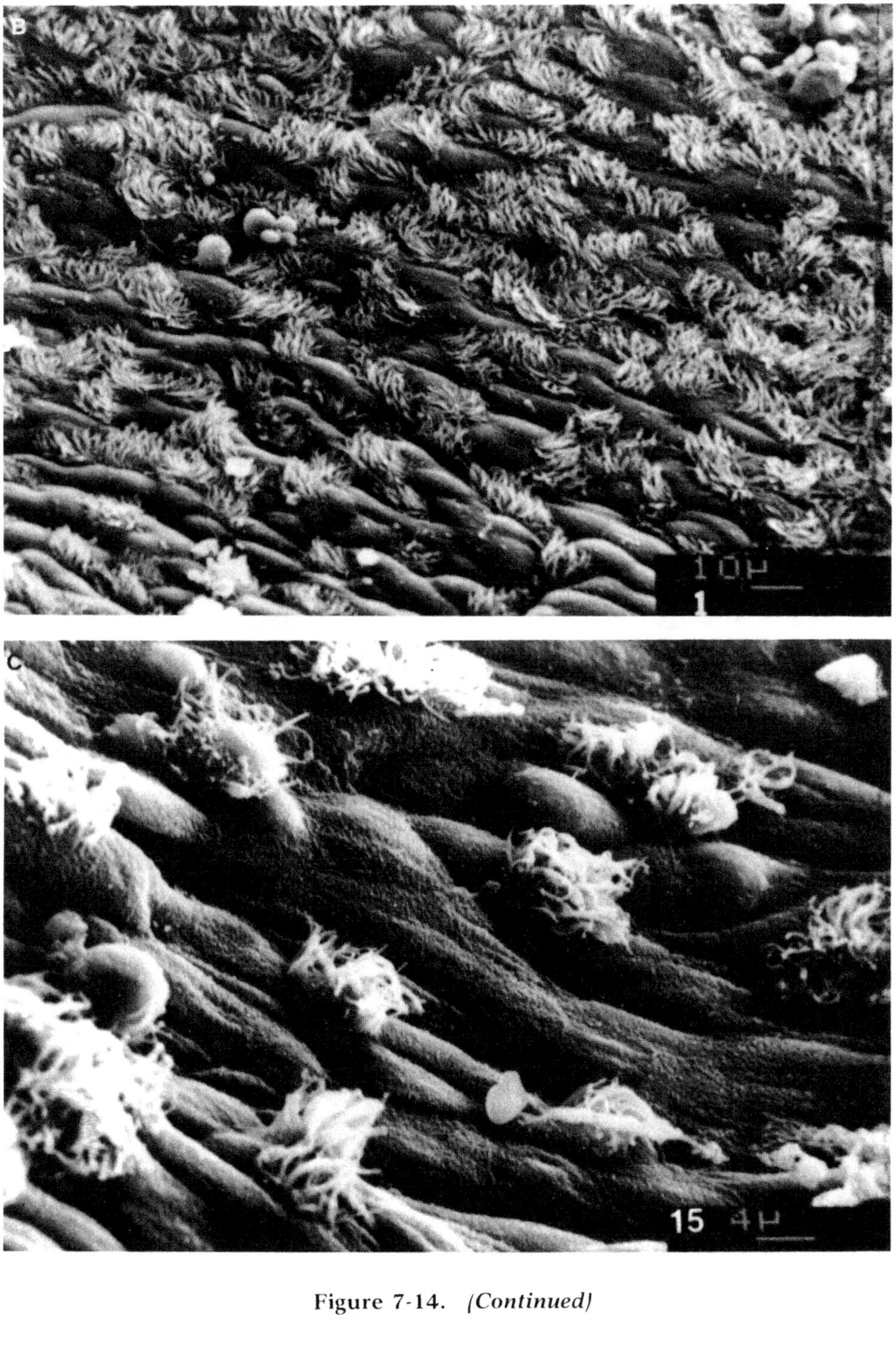

Figure 7-14. *(Continued)*

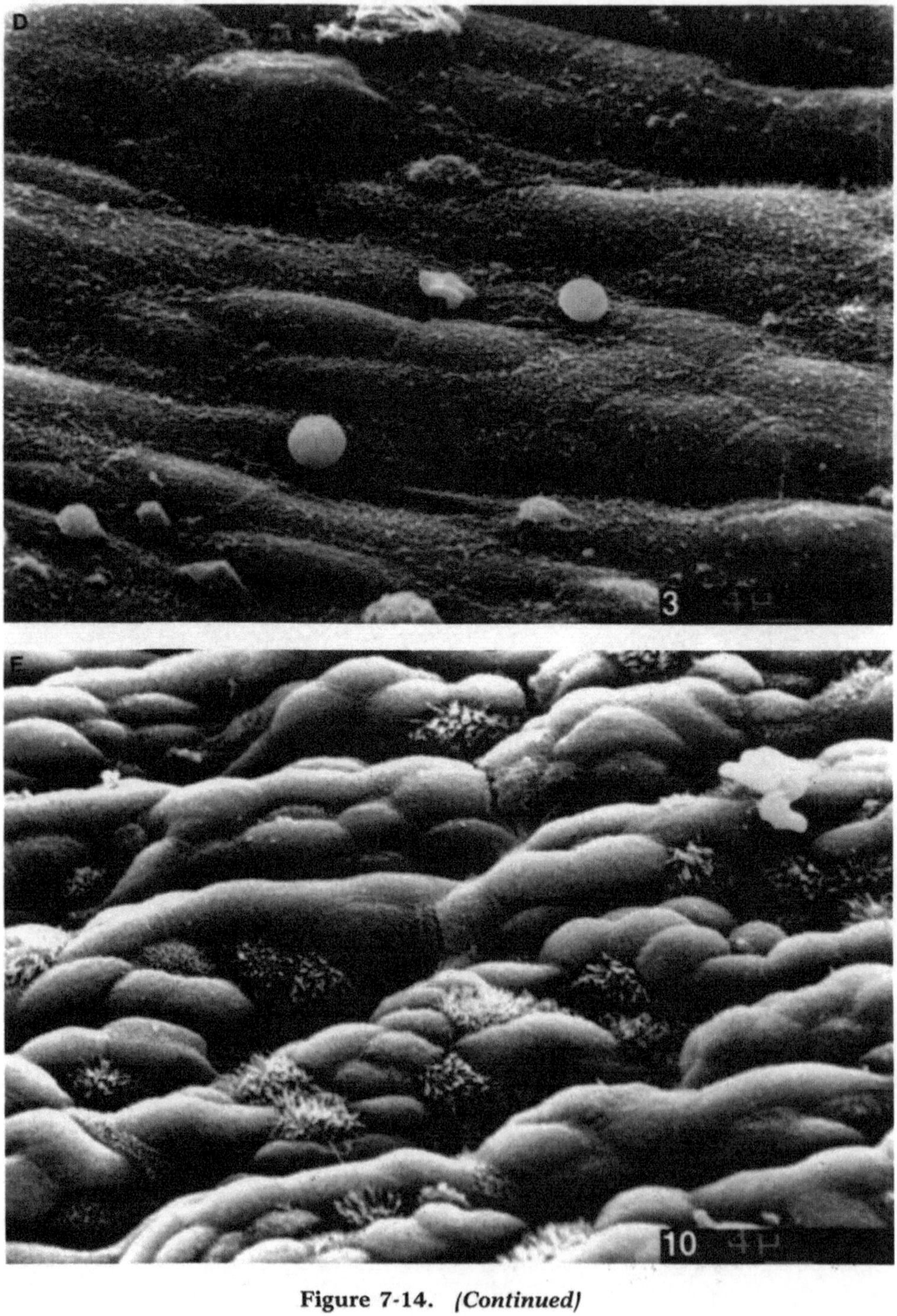

Figure 7-14. *(Continued)*

hemorrhages, edema, and limited cellular infiltration. Alveoli may be denuded of epithelium (Winternitz et al., 1920) and contain both neutrophils and mononuclear cells admixed with fibrin and edema fluid. Alveolar capillaries are hyperemic, and intraalveolar hemorrhage is common (Louria et al., 1959). A characteristic feature of the pneumonia is the presence of acellular hyaline membranes lining alveolar ducts and alveoli (see Fig. 7-15). This finding is unusual

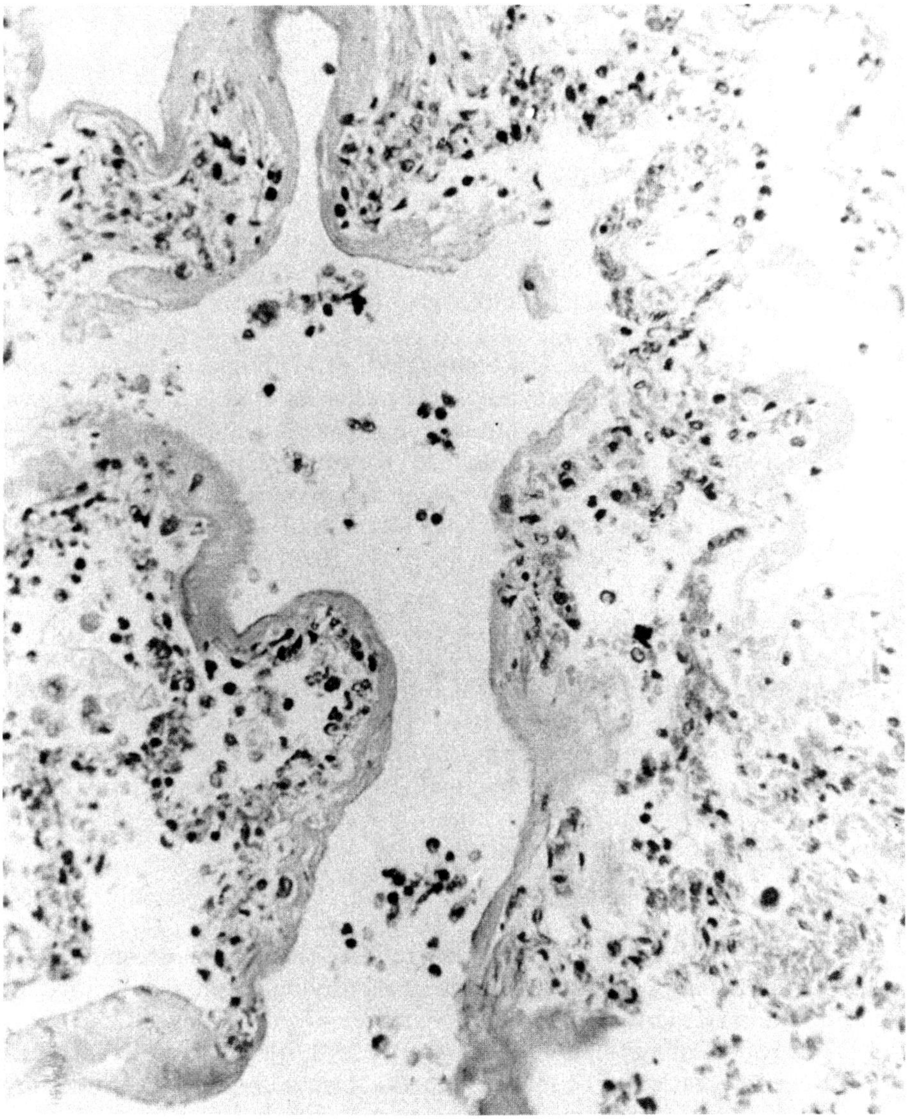

Figure 7-15. Primary influenza virus pneumonia. Hyaline acellular membrane lining alveolar ducts and alveoli of the right upper lobe. Note the relative paucity of inflammatory cells or erythrocytes in the alveoli. Hematoxylin and eosin, ×180. (From Louria et al., 1959. Reproduced with permission, from *J. Clin. Invest.* **38**:236.)

in other viral respiratory infections and is reminiscent of lesions seen following inhalation of poisonous gases (Winternitz et al., 1920. Infection of alveolar cells has been discussed in connection with the initial lesion of influenza. These cells and alveolar macrophages support replication of influenza virus.

The primary inflammatory reaction in the lung involves neutrophils, and only later do lymphocytes and plasma cells appear (Mulder and Hers, 1972). Because most patients with influenza virus pneumonia do not survive into the second and third weeks, the precise process by which recovery occurs in the lung remains obscure. Alveolar epithelial regeneration was observed by Mulder and Hers (1972) in a single case 12 days after onset of influenza.

Clinical Pathology of Influenza

Laboratory tests other than those employed for detection of infective virus, viral antigen, or specific antibody are of little value in the diagnosis of typical influenza. Lymphocytopenia involving both T and B lymphocytes, but primarily T cells, is found in uncomplicated influenza (Criswell et al., 1979), but an increase in total and polymorphonuclear leukocytes occurs in primary influenza virus pneumonia (Louria et al., 1959).

Influenza virus pneumonia is accompanied by a striking reduction in arterial oxyhemoglobin saturation, sometimes to less than 50%, together with elevated partial pressures of CO_2. Serum glutamic-oxaloacetic transaminase levels also are elevated in some patients with influenza virus pneumonia (Louria et al., 1959).

The significance in uncomplicated influenza of reported low serum iron levels (Fernandez, 1980), elevated creatine phosphokinase (Kessler et al., 1983), and increased serum levels of squalene (Gallin et al., 1970) is unknown.

THERAPY OF INFLUENZA

Influenza is a brief disease from which recovery usually is rapid and uneventful. Furthermore, there are few objective criteria other than cough and fever by which to judge therapeutic response; the influence of therapy requires careful evaluation in controlled trials before convincing effects are demonstrable. Thus, it is perhaps not surprising that only recently has a drug developed 20 years ago received limited acceptance for the treatment of influenza.

Specific Chemotherapy: Amantadine

Amantadine hydrochloride, the water-soluble salt of 1-adamantanamine, in oral dosage of 200 mg per day is capable of modifying the clinical course of influenza if administered within 48 hr of onset (reviewed by Jackson and Stanley, 1976; Oxford and Galbraith, 1980; Nicholson, 1984). Although effects on the duration of fever are variable (Younkin et al., 1983), amantadine appears to shorten the course of illness by about one-third in most studies (Nicholson, 1984). Amantadine is effective against all subtypes of human influenza A viruses but is ineffective therapeutically or prophylactically against influenza B. The drug is reported to have activity against influenza C virus in vitro (Davies et al., 1964).

Therapeutic effects of amantadine are presumably related to its inhibitory effects on viral replication that are demonstrable in cell culture and experimental animals. Viral shedding was reduced in some studies in man (Van Voris et al, 1981; Knight et al., 1970). Amantadine does not have antipyretic activity and, indeed, is less effective than aspirin in shortening the febrile period in influenza (Younkin et al., 1983).

The effect of amantadine in influenza virus pneumonia has not been sufficiently studied to allow conclusions on its efficacy. Amantadine is effective when administered as an aerosol by the respiratory route in uncomplicated natural infections (Hayden et al., 1980). Amantadine is reported to accelerate recovery from the peripheral airway dysfunction seen in influenza (Little et al., 1976).

Mechanism of Action

Amantadine and the related compound rimantadine inhibit the uncoating step of influenza virus replication (Kato and Eggers, 1969; Skehel et al., 1977; Koff and Knight, 1979). In the case of rimantadine, it has been suggested that the release of M protein from incoming RNPs is inhibited, thereby preventing their penetration into the nucleus (Bukrinskaya et al., 1982). Recent evidence from study of drug-resistant mutants points to a primary effect of amantadine on the M_2 protein through which either the initiation of infection or virus assembly is affected (Hay et al., 1986).

Pharmacology

Amantadine is rapidly absorbed when administered orally. The drug is bound to cells, and excretion is slow. The half-life is 10–20 hr with complete excretion in 5 days. Therapeutic blood levels are 200–500 g/ml. The half-life of rimantadine is longer, 30 hr, and lower blood levels are achieved at equivalent dosage. However, higher concentrations of drug are found in respiratory secretion (Hayden, 1980).

Toxicity

Amantadine has potential neurotoxicity, and side effects such as headache, confusion, insomnia, and slurred speech occur in 7% or more of patients, with increased incidence in the elderly (Jackson and Stanley, 1976; Editorial, *The Medical Letter*, 1978). Lightheadedness and difficulty in concentration were found in one-third of university students under careful study (Van Voris et al., 1981). However, therapy was discontinued in only 3% of those receiving amantadine, 100–200 mg per day, but in 35% of those who received aspirin in one study (Younkin et al., 1983). In other studies, amantadine was less well tolerated than rimantadine, with 22% of subjects withdrawing from the study compared to 9% who received rimantadine and 9% who received placebo (Dolin, 1976).

Drug Resistance

Influenza A viruses are highly variable in their susceptibility to inhibition by amantadine or rimantadine. Susceptibility is not necessarily linked to viral subtype, although H1N1 viruses and H3N2 strains isolated between 1979 and 1981 tend to greater resistance (Vodeiko et al., 1982). Drug resistance appears to be determined principally by the M protein gene (Lubeck et al., 1978; Hay et al.,

1979; Hamzawi et al., 1981; E.D. Kilbourne, unpublished data), although other viral genes may also be of importance (Lubeck et al., 1978; Scholtissek and Faulkner, 1979). Aman'adine-resistant variants emerge promptly on laboratory passage of virus (Cochran et al., 1965; Oxford and Logan, 1970; Ilyenko, 1975).

Ribavirin

Ribavirin is a triazole nucleotide guanosine analogue that inhibits replication of both DNA and RNA viruses, including influenza viruses. When administered orally, it has proved essentially ineffective in influenza (reviewed by Smith et al., 1980). However, administration of the drug in small-particle aerosols for 2–3 days has been found to shorten the febrile period, reduce systemic symptoms, and decrease virus shedding in influenza A and B (Knight et al., 1981; McClung et al., 1983; Gilbert et al., 1985).

Because treatment requires hospitalization and the administration of aerosols by mask for 12 hr per day, therapy with ribavirin appears to have limited potential. It may prove to be of some value in patients with influenza virus pneumonia.

Supportive and Physiological Therapy

The general measures used in the treatment of other acute febrile viral infections such as bed rest, light diet, and maintenance of adequate hydration should be employed in the therapy of uncomplicated influenza. Symptomatic therapy with analgesics and antipyretics may be given as required. Suppression of cough may be necessary, but antitussives should be used sparingly because of the possibility of secondary bacterial pneumonia as a complication. The routine use of antibiotic prophylaxis is not indicated but may be considered in patients with underlying chronic pulmonary disease. In such cases, full therapeutic doses of penicillin should be used to forestall the development of pneumococcal pneumonia, the most frequent bacterial sequel.

Management of Influenza Virus Pneumonia

The hemorrhagic and exudative pneumonitis of primary influenza virus pneumonia is associated with diffuse pulmonary infiltrates and hypoxemia. The emphasis of therapy, therefore, is on maintenance of adequate oxygenation.

The physiological basis of the hypoxemia is a pulmonary capillary endothelial leak with the development of pulmonary edema. Therapy, therefore, has focused on the reversal of right-to-left pulmonary shunting by delivery of oxygen-rich gas under pressure to previously unventilated areas of the lung (Taylor et al., 1976). Treatment of the oxygenation defect has utilized continuous positive airway pressure during either spontaneous breathing or artificial ventilation. Using ventilatory support and prolonged oxygen therapy, Winterbauer et al. (1977) appeared to influence favorably the course of disease in five of 11 patients with severe influenza viral pneumonia. However, none of four survivors had normal pulmonary function 1 year later, and all showed deficiencies indicative of diffuse pulmonary fibrosis.

Extracorporeal membrane oxygenation has also been tried in this catastrophic illness, in part to reduce the hazard of oxygen toxicity in the lung, but without notable success (Lefrak et al., 1974).

REFERENCES

Albrecht, P., Blaskovic, D., Styk, B., et al , 1963, Course of A2 influenza in intranasally infected mice examined by the fluorescent antibody technique, *Acta Virol.* **7**:405–413.

Alexander, R. E , 1982, Salicylates and Reye's syndrome· Epidemiologic data as a basis of action, *J A M A.* **248**:722–723.

Alford, R H., Kasel, J. A., Gerone, P. J , el al., 1966, Human influenza resulting from aerosol inhalation, *Proc Soc. Exp Biol Med* **122**:800–804.

Anders, M. E., Peppard, P M., Burns, W H , et al., 1979, *In vitro* antibody response to influenza virus. I. T cell dependence of secondary response to hemagglutinin, *J. Immunol.* **123**:1356–1361.

Anders, M. E., Katz, J M., Jackson, D C , et al , 1981, In vitro antibody response to influenza virus II Specificity of helper T cells recognizing hemagglutinin, *J Immunol.* **127**:669–672.

Anders, M. E., Scalzo, A. A., and White, D. O., 1984, Influenza viruses are T cell-independent B cell mitogens, *J. Virol.* **50**:960–963.

Armstrong, R. B , Butchko, G. M., Kiley, S. C , et al , 1981, Mitogenicity of influenza hemagglutinin glycoproteins and influenza viruses bearing H2-hemagglutinin, *Infect. Immun.* **34**:140–143.

Armstrong, W. G., 1919, Report on the influenza epidemic in New South Wales, *Public Health Rep* 144–172

Astry, C L., and Jakab, G. J , 1984, Influenza virus-induced immune complexes suppress alveolar macrophage phagocytosis, *J. Virol.* **50**:287–292.

Astry, C L., Yolken, R. H., and Jakab, G. J., 1984, Dynamics of viral growth, viral enzymatic activity, and antigenicity in murine lungs during the course of influenza pneumonia, *J Med Virol.* **14**:81–90.

Austin, D. F., Karp, S., Dworsky, R , et al., 1975, Excess leukemia in cohorts of children born following influenza epidemics, *Am. J. Epidemiol.* **101**:77–83.

Baine, W. B., Luby, J. P., and Martin, S. M., 1980, Severe illness with influenza B, *Am. J Med.* **68**:181–189.

Banks, R. A., and Tinker, J., 1977, Fulminating haemorrhagic influenza, *Br. Med. J.* **1**:1325.

Barrett, M J., Hurwitz, E. S., Rogers, M. F , et al., 1984, The national Reye syndrome surveillance system, 1983, *Morbid Mortal Weekly Rep.* **33**:9SS–13SS.

Barton, L L., and Chalhub, E. G., 1975, Myositis associated with influenza A infection, *J Pediatr.* **87**:1003–1004.

Basarab, O , and Smith, H., 1969, Quantitative studies on the tissue localization of influenza virus in ferrets after intranasal and intravenous or intracardial inoculation, *Br J. Exp. Pathol.* **50**:612–618.

Bauer, C. R , Elie, K., Spence, L , et al., 1973, Hong Kong influenza in a neonatal unit, *J A.M A.* **223**:1233–1235.

Bennett, A. E., and Turk, R E., 1957, Acute encephalitis and death following Asian influenza, *Calif Med.* **87**:411

Bennink, J. R , Yewdell, J. W., and Gerhard, W., 1982, A viral polymerase involved in recognition of influenza virus-infected cells by a cytotoxic T cell clone, *Nature* **296**:75–76

Bentley, D. M., Watson, S R , and Morris, R. E., 1983, Age-related loss of Lyt-1,2 cells in C58 mice results in susceptibility to lactic dehydrogenase virus-induced polioencephalomyelitis, *Infect Immun.* **41**:1389–1390.

Biddison, W E , Sharrow, S. O., and Shearer, G. M., 1981, T cell subpopulations required for the human cytotoxic T lymphocyte response to influenza virus: Evidence for T cell help, *J Immunol* **127**:487–491

Bisno, A. L., Griffin, J. P., Van Epps, K. A., et al., 1971, Pneumonia and Hong Kong influenza: A prospective study of the 1968–1969 epidemic, *Am. J. Med. Sci* **261**:251–263.

Braciale, T J., 1977, Immunologic recognition of influenza virus-infected cells. II. Expression of influenza a matrix protein on the infected cell surface and its role in recognition by cross-reactive cytotoxic T cells, *J. Exp. Med.* **146**:673–689.

Brill, S. J., and Gilfillan, R. F., 1977, Acute parotitis associated with influenza type A, *N. Engl. J. Med* **296**:1391–1392.

Brocklebank, J. T., Court, S. D. M., McQuillin, J., et al., 1972, Influenza-A infection in children, *Lancet* **2**:497–500.

Brown, T. A., Murphy, B. R., Radl, J., et al., 1985, Subclass distribution and molecular form of immunoglobulin A hemagglutinin antibodies in sera and nasal secretions after experimental secondary infection with influenza A virus in humans, *J. Clin. Microbiol.* **22**:259–264.

Buchner, Y. I., Heath, R. B., Collins, J. V., et al., 1977, Serum IgM antibody and influenza A infection, *J. Clin. Pathol.* **30**:723–727.

Bukrinskaya, A. G., Vorkunova, N. K., Kornilayeva, G. V., et al., 1982, Influenza virus uncoating in infected cells and effect of rimantadine, *J. Gen. Virol.* **60**:49–59.

Bukrinskaya, A. G., Vorkunova, G. K., Liosner, A. L., et al., 1985, Matrix proteins of influenza A and B viruses possess common antigenic determinants, *Lancet* **1**:276–277.

Burch, G. E., Walsh, J. J., and Mogabgab, W., 1961, A study of the response of the cardiovascular system to Asian influenza, *Am. Rev. Respir. Dis.* **83**(2):68–78.

Burk, R. F., Schaffner, W., and Koenig, M. G., 1971, Severe influenza virus pneumonia in the pandemic of 1968–1969, *Arch. Intern. Med.* **127**:1122.

Burlington, B. D., Clements, M. L., Meiklejohn, G., et al., 1983, Hemagglutinin-specific antibody responses in immunoglobulin G, A, and M isotypes as measured by enzyme-linked immunosorbent assay after primary or secondary infection of humans with influenza A virus, *Infect. Immun.* **41**:540–545.

Butchko, G. M., Armstrong, R. B., Martin, W. J., et al., 1978, Influenza A viruses of the H2N2 subtype are lymphocyte mitogens, *Nature* **271**:66–67.

Carson, J. L., Collier, A. M., and Hu, S.-C. S., 1985, Acquired ciliary defects in nasal epithelium of children with acute viral upper respiratory infections, *N. Engl. J. Med.* **312**:463–468.

Chakraverty, P., Adhami, Z., Wise, R., et al., 1984, Influenza C virus in the United Kingdom, *J. Infect.* **8**:177–179.

Chen, S.-H., Huang, C.-H., Yang, Y.-J., et al., 1973, Comparison of influenza virus hemagglutinin, neuraminidase and their antibody responses in infected patients, *J. Formosan Med. Assoc.* **72**:319–324.

Clark, P. S., Feltz, E. T., List-Young, B., et al., 1970, An influenza B epidemic within a remote Alaska community. Serologic, epidemiologic, and clinical observations, *J.A.M.A.* **214**:507–512.

Cochran, K. W., Maassab, H. R., Tsunoda, A., et al., 1965, Studies on the antiviral activity of amantadine hydrochloride, *Ann. N.Y. Acad. Sci.* **130**:432–439.

Coffey, V. P., and Jessop, W. J. E., 1963, Maternal influenza and congenital deformities: A follow-up study, *Lancet* **1**:748–751.

Committee on Acute Respiratory Diseases, 1945, A study of a food-borne epidemic of tonsillitis and pharyngitis due to B-hemolytic streptococcus, type 5, *Bull. Johns Hopkins Hosp.* **77**:143–210.

Connor, E., and Powell, K., 1985, Fulminant pneumonia caused by concomitant infection with influenza B virus and *Staphylococcus aureus*, *J. Pediatr.* **106**:447–450.

Consensus Conference, 1981, Diagnosis and treatment of Reye's syndrome, *J.A.M.A.* **246**:2441–2444.

Corey, L., Rubin, R. J., Hattwick, M. A. W., et al., 1976, A nationwide outbreak of Reye's syndrome. Its epidemiologic relationship to influenza B, *Am. J. Med.* **61**:615–625.

Couch, R. B., and Kasel, J. A., 1983, Immunity to influenza in man, *Annu. Rev. Microbiol.* **37**:529–549.

Couch, R. B., Douglas, R. G., Fedson, D. S., et al., 1971, Correlated studies of a recombinant influenza-virus vaccine. III. Protection against experimental influenza in man, in: *J. Infect. Dis.* **124**:473–480.

Couch, R. B., Kasel, J. A., Six, H. R., et al., 1984, Immunological reactions and resistance to infection with influenza virus, in: *The Molecular Virology and Epidemiology of Influenza* (C. Stuart-Harris and C. W. Potter, eds.), Academic Press, London, pp. 119–152.

Cretescu, L., Beare, A. S., and Schild, G. C., 1978, Formation of antibody to matrix protein in experimental human influenza A virus infections, *Infect. Immun.* **22**:322–327.

Criswell, B. S., Couch, R. B., Greenberg, S. B., et al., 1979, The lymphocyte response to influenza in humans, *Am. Rev. Respir. Dis.* **120**:700–703.

Cunningham, E., Kohli, R., and Venuto, R. C., 1979, Influenza-associated myoglobinuric renal failure, *J.A.M.A.* **242**:2428–2429.

Dagan, R., and Hall, C. B., 1984, Influenza A virus infection imitating bacterial sepsis in early infancy, *Pediat. Infect. Dis.* **3**:218–221.

Daniels, C. A., and Marbrook, J., 1981, Influenza A2 inhibits murine in vitro antibody synthesis, *J. Immunol.* **126:**1737-1741.

Darke, C. S., Watkins, P. H., and Whitehead, J. E. M., 1957, Fulminating staphylococcal pneumonia associated with influenza virus C. Report of a fatal case, *Br. Med. J.* **2:**606-609.

Davies, W L., Grunert, R. R , Haff, R. F., et al , 1964, Antiviral activity of 1-adamantanamine (amantadine), *Science* **144:**862-863.

Davison, V. E., and Sanford, B. A., 1981, Adherence of *Staphylococcus aureus* to influenza A virus-infected Madin-Darby canine kidney cell cultures, *Infect. Immun.* **32:**118-126.

Dawson, M. D., and Fine, I. H., 1967, Creatine kinase in human tissues, *Arch. Neurol.* **16:**175.

Delorme, L , and Middleton, P J., 1979, Influenza A virus associated with acute encephalopathy, *Am J. Dis. Child.* **133:**822-824.

DeStefano, F., 1982, An outbreak of influenza B at an Indiana boarding school: Estimate of vaccine efficacy, *Public Health Rep.* **97:**269-272.

DeVivo, D C , 1978, Reye syndrome: A metabolic response to an acute mitochondrial insult? *Neurology (Minneap.)* **28:**105-108.

DeVivo, D. C., 1983, How common is Reye's syndrome? *N. Engl. J Med* **309:**179-181.

DiBona, F. J., and Morens, D. M., 1977, Rhabdomyolysis associated with influenza A Report of a case with unusual fluid and electrolyte abnormalities, *J. Pediatr.* **91:**943-945.

Dietzman, D. E , Schaller, J. G., Ray, C. G., et al., 1976, Acute myositis associated with influenza B infection, *Pediatrics* **57:**255-258.

Doherty, P. C., Effros, R. B., and Bennink, J., 1977, Heterogeneity of the cytotoxic response of thymus-derived lymphocytes after immunization with influenza viruses, *Proc. Natl. Acad. Sci. U.S A.* **74:**1209.

Dolin, R., 1976, Influenza. Current concepts, *Am. Fam. Physician* **14:**72-77.

Doll, R., Hill, A. B., and Sakula, J., 1960, Asian influenza in pregnancy and congenital defects, *Br. J. Prev. Soc. Med.* **14:**167-172.

Douglas, R. G., Jr., 1975, Influenza in man, *Influenza Viruses and Influenza* (E D Kilbourne, ed), Academic Press, New York, pp. 395-447.

Douglas, R. G., Jr., Hall, W. J., and Betts, R. F., 1981, Changing virulence of influenza A viruses, *Trans. Am. Clin. Climatol. Assoc.* **93:**141-149.

Dubowitz, V., 1958, Influenza encephalitis, *Lancet* **1:**140-141.

Dykes, A. C., Cherry, J. D., and Nolan, C. E., 1980, A clinical, epidemiologic, serologic, and virologic study of influenza C virus infection, *Arch. Intern. Med.* **140:**1295-1298.

Eason, R. J., and Sage, M. D., 1980, Deaths from influenza A, subtype, H_1N_1, during the 1979 Auckland epidemic, *N. Z Med. J.* **91:**129-131.

Ebisawa, I. T., and Kitamoto, O., 1969, Persistence of viral antigen in the nasal epithelium in complicated influenza, *Am. Rev. Respir. Dis.* **99:**275-278.

Edelen, J. S., Bender, T. R., and Chin, T. D. Y., 1974, Encephalopathy and pericarditis during an outbreak of influenza, *Am. J. Epidemiol.* **100:**79-84.

Editorial, 1971, *Br. Med. J.* Influenza and the nervous system **1:**357-358.

Editorial, *Lancet,* 1971, Influenza and the newborn, *Lancet* **2:**911-912.

Editorial, *Lancet*, 1982, Reye's syndrome—epidemiological considerations, *Lancet* **1:**941-943.

Editorial, 1978, Amantadine for high-risk influenza, *Med. Lett.* **20:**25-26.

Effros, R. B., Doherty, P. C., Gerhard, W., et al., 1977, Generation of both cross-reactive and virus-specific T-cell populations after immunization with serologically distinct influenza A viruses, *J. Exp. Med.* **145:**557

Empey, D. W., Laitinen, L. W., Jacobs, L., et al., 1976, Mechanisms of bronchial hyperreactivity in normal subjects after upper respiratory tract infection, *Am. Rev. Respir. Dis.* **113:**131-139.

Engblom, E., Ekfors, T. O., Meurman, O. H., et al., 1983, Fatal influenza A myocarditis with isolation of virus from the myocardium, *Acta Med. Scand.* **213:**75-78.

Ennis, F. A., Martin, W. J., and Verbonitz, M. W., 1977, Hemagglutinin-specific cytotoxic T-cell response during influenza infection, *J. Exp. Med.* **146:**893-898.

Ettensohn, D. B., and Roberts, N. J., Jr., 1984, Influenza virus infection of human alveolar and blood-derived macrophages: Differences in accessory cell function and interferon production, *J Infect. Dis.* **149:**942-949.

Faguet, G. B., 1981, The effect of killed influenza virus vaccine on the kinetics of normal human lymphocytes, *J. Infect. Dis.* **143:**252-258.

Farnik, J., and Bruj, J., 1966, An outbreak of influenza A2 in a population with a known antibody profile, *J. Infect. Dis.* **116**:425–428.
Farrell, M. K., Partin, J. C., Bove, K. E., et al., 1980, Epidemic influenza myopathy in Cincinnati in 1977, *J Pediat.* **96**:545–551.
Fedrick, J., and Alberman, E., 1972, Reported influenza in pregnancy and subsequent cancer in the child, *Br. Med. J.* **2**:485–488.
Feldman, P. S., Cohan, M. A., and Hierholzer, W. J., 1972, Fatal Hong Kong influenza: A clinical microbiological and pathological analysis of nine cases, *Yale J. Biol. Med.* **45**:49.
Fernandez, H., 1980, Low serum iron in influenza, *N. Engl. J. Med.* **302**:865.
Fleischer, B. 1982, Cloned lines of human cytotoxic T lymphocytes with specificity for influenza virus, *J. Immunol.* **129**:1731–1735.
Flewett, T. H., 1976, Postinfluenzal depression, *Br. Med. J.* **2**:815.
Fox, J. P., and Hall, C. E., 1980, *Viruses in Families*, PSG Publishing Co., Littleton, MA.
Francis, T., Jr., Quilligan, J. J., Jr., and Minuse, E., 1950, Identification of another epidemic respiratory disease, *Science* **112**:494–497.
Francis, T., Jr., Davenport, F. M., and Hennessy, A. V., 1953, A serological recapitulation of human infection with different strains of influenza virus, *Trans. Assoc. Am. Physicians* **66**:231–239.
Frank, A. L., and Taber, L. H , 1983, Variation in frequency of natural reinfection with influenza A viruses, *J. Med. Virol.* **12**:17–23.
Frank, A. L., Taber, L. H., Glezen, W. P., et al., 1979, Reinfection with influenza A (H3N2) virus in young children and their families, *J Infect Dis.* **140**:829–836.
Frank, A. L., Taber, L. H., and Wells, J. M., 1985, Comparison of infection rates and severity of illness for influenza A subtypes H1N1 and H3N2, *J. Infect. Dis.* **151**:73–80.
Franková, V., and Rychterová, V., 1975, Inhalatory infection of mice with influenza AO/PR8 virus. II. Detection of the virus in the blood and extrapulmonary organs, *Acta Virol* **19**:35–40.
Franková, V., Jirásek, A., and Tumová, B., 1977, Type A influenza: Postmortem virus isolations from different organs in human lethal cases, *Arch. Virol.* **53**:265–268.
Fransén, H., and Tunevall, G., 1969, Bacteria and serologic reactions against bacteria in patients hospitalized with acute respiratory illness, *Scand. J. Infect. Dis.* **1**:191.
Freeman, D. W., and Barno, A., 1959, Deaths from Asian influenza associated with pregnancy, *Am J. Obstet. Gynecol.* **78**:1172–1175.
Friman, G., 1976, Serum creatine phosphokinase in epidemic influenza, *Scand. J. Infect. Dis.* **8**:13–20.
Gallin, J. I., O'Leary, W. M., and Kaye, D., 1970, Increased serum levels of squalene in patients with acute influenza, *N. Engl. J. Med.* **282**:1225–1227.
Gamboa, E. T., Eastwood, A. B., Hays, A. P., et al., 1979, Isolation of influenza virus from muscle in myoglobinuric polymyositis, *Neurology (Minneap.)* **29**:1323–1335.
Gardner, I. D., 1980, Effect of influenza virus infection on susceptibility to bacteria in mice, *J. Infect. Dis.* **142**:704–707.
George, R. C., Broadbent, D. A., and Drasar, B. S., 1983, The effect of influenza virus on the adherence of *Haemophilus influenzae* to human cells in tissue culture, *Br. J. Exp. Pathol.* **64**:655–659.
Gerber, P., Woolridge, R. L., Seal, J. R., et al., 1952, Epidemic influenza B and C in navy recruits during winter of 1951–52, *Proc. Soc. Exp. Biol. Med.* **81**:624–628.
Gibson, T. C., Arnold, J., Craige, E., et al., 1959, Electrocardiographic studies in Asian influenza, *Am. Heart J.* **57**:661–668.
Gilbert, B. E., Wilson, S. Z., Knight, V., et al., 1985, Ribavirin small-particle aerosol treatment of infections caused by influenza virus strains A/Victoria/7/83 (H1N1) and B/Texas/1/84 *Antimicrob. Agents Chemother.* **27**:309–313.
Giles, C., and Shuttleworth, E. M., 1957, Postmortem findings in 46 influenza deaths, *Lancet* **2**:1224.
Gill, P., Influenza in general practice, 1973, *Med. J. Aust.* [*Suppl.*] **1**:24–28.
Glezen, W. P., 1980, Consideration of the risk of influenza in children and indications for prophylaxis, *Rev. Infect. Dis.* **2**:408–420.
Glezen, W. P., Paredes, A., and Taber, L. H., 1980, Influenza in children: Relationship to other respiratory agents, *J.A.M.A.* **243**:1345–1349.
Graves, P. N., Schulman, J. L., Young, J. F., et al., 1983, Preparation of influenza virus subviral particles lacking the HA1 subunit of hemagglutinin: Unmasking of cross-reactive HA2 determinants, *Virology* **126**:106–116.

Greenberg, M., Jacobziner, H., Pakter, J., et al., 1958, Maternal mortality in the epidemic of Asian influenza, New York City, 1957, *Am J Obstet Gynecol* **76**:897-902
Greenberg, S B., Criswell, B. S., Six, H. R , et al., 1977, Lymphocyte cytotoxicity to influenza virus-infected cells. II Requirement for antibody and non T lymphocytes, *J Immunol* **119**:2100-2106
Griffiths, P. D , Ronalds, C. J., and Heath, R B., 1980, A prospective study of influenza infections during pregnancy, *J. Epidemiol Commun. Health* **34**:124-128.
Haaheim, R , 1977, Single-radial-complement-fixation: A new immunodiffusion technique 2 Assay of the antibody response to the internal antigens (MP and NP) of influenza A virus in human sera after vaccination and infection, *Dev. Biol. Stand.* **39**:481-484.
Hackett, C J , Dietzschold, B , Gerhard, W., et al , 1983, Influenza virus site recognized by a murine helper T cell specific for H1 strains: Localization to a nine amino acid sequence in the hemagglutinin molecule, *J Exp Med* **158**:294-302
Hakosalo, J., and Saxen, L., 1971, Influenza epidemic and congenital defects, *Lancet* **2**:1346-1347
Hall, C B , and Douglas, R B , Jr , 1975, Nosocomial influenza infection as a cause of intercurrent fevers in infants, *Pediatrics* **55**:673-677.
Hall, W J , Douglas, R G , JR , Hyde, R. W , et al., 1976, Pulmonary mechanics after uncomplicated influenza A infection, *Am Rev. Respir. Dis.* **113**:141-147
Hall, C B , Douglas, R. G., Jr , Simons, R. L., et al., 1978, Interferon production in children with respiratory syncytical, influenza, and parainfluenza virus infections, *J Pediatr* **93**:28-32
Halsey, N A , Hurwitz, E. S , Meiklejohn, G., et al , 1980, An epidemic of Reye syndrome associated with influenza A (H1N1) in Colorado, *J. Pediatr* **97**:535-539
Hamzawi, M , Jennings, R , and Potter, C W , 1981, The amantadine-sensitivity of recombinant and parental influenza virus strains, *Med Microbiol Immunol* **169**:259-268.
Harris, J W , 1919, Influenza occurring in pregnant women A statistical study of thirteen hundred and fifty cases, *J.A.M.A* **72**:978-980.
Hashimoto, G , Wright, P. F., and Karzon, D. T., 1983, Antibody-dependent cell-mediated cytotoxicity against influenza virus-infected cells, *J Infect Dis* **148**:785-794
Hay, A J , Kennedy, N C. T., Skehel, J. J., et al, 1979, The matrix protein gene determines amantadine-sensitivity of influenza viruses, *J Gen. Virol.* **42**:189-191.
Hay, A. J., Wolstenholme, A. J , Skehel, J J , et al , 1985, The molecular basis of the specific anti-influenza action of amantadine, *EMBO J* **4**:3021-3024
Hayden, F G , Hall, W J , and Douglas, R. G , Jr , 1980, Therapeutic effects of aerosolized amantadine in naturally acquired infection due to influenza A virus, *J Infect Dis.* **141**:535-542
Henle, W , and Lief, F S , 1963, The broadening of antibody spectra following multiple exposures to influenza viruses, *Am. Rev. Respir Dis.* **88**:379-386.
Henle, W., Henle, G., Stokes, J., Jr., et al., 1946, Experimental exposure of human subjects to viruses of influenza, *J Immunol.* **52**:145-165.
Hers, J F P , 1966, Disturbances of the ciliated epithelium due to influenza virus, *Am. Rev Respir Dis* **93**:162
Hers, J F. P , and Mulder, J., 1961, Broad aspects of the pathology and pathogenesis of human influenza, *Am Rev. Respir. Dis.* **83**:84-94.
Hers, J. F. P., Goshings, W. R. O., Masurel, N., et al, 1957, Deaths from Asiatic influenza in Netherlands, *Lancet* **2**:1164-1165.
Hers, J. F P., Mulder, J., Masurel, N., et al, 1962, Studies on the pathogenesis of influenza virus pneumonia in mice, *J. Pathol. Bacteriol* **83**:207-217
Hers, J. F P , van der Kuip, L., and Masurel, N., 1971, Rapid diagnosis of respiratory virus infection in infected tissue cultures, *Ann. N.Y Acad. Sci* **177**:70-77
Heubi, J. E., Daugherty, C. C., Partin, J. S., et al., 1984, Grade I Reye's syndrome—outcome and predictors of progression to deeper coma grades, *N. Engl J. Med.* **311**:1539-1542
Hicks, J T., Ennis, R. A , Kim, E , et al., 1978, The importance of an intact complement pathway in recovery from a primary viral infection: Influenza in decomplemented and in C5-deficient mice, *J. Immunol* **121**:1437-1445
Hildebrandt, H. M , Maassab, H. F , and Willis, P W., 1962, Influenza virus pericarditis: Report of a case with isolation of Asian influenza virus from the pericardial fluid, *Am J Dis Child* **104**:579-582
Hobbins, T E., Hughes, T. P , Rennels, M. B , et al , 1968, Bronchial reactivity in experimental in-

fections with influenza virus, *J. Infect Dis.* **146**:468-471.

Homma, M., 1985, Epidemiological characteristics of type C influenza, *J. Cell. Biochem.* [*Suppl.*] **9C**:278.

Horner, F. A., 1958, Neurologic disorders after Asian influenza, *N. Engl. J. Med.* **258**:983–985.

Hoskins, T. W., 1979, Severity of influenza, *Lancet* **1**:381–382.

Hurwitz, J. L., and Hackett, C. J., 1985, Influenza-specific suppression: Contribution of major viral proteins to the generation and function of T suppressor cells, *J. Immunol.* **135**:2134–2139.

Hurwitz, J. L., Heber-Katz, E., Hackett, C. J., et al., 1984, Characterization of the murine T_H response to influenza virus hemagglutinin: Evidence for three major specificities, *J. Immunol.* **133**:1–7.

Husseini, R. H., Sweet, C., Collie, M. H., et al., 1982, Elevation of nasal viral levels by suppression of fever in ferrets infected with influenza viruses of differing virulence, *J. Infect. Dis.* **145**:520–524.

Ilyenko, V. I., 1975, On the study of formation of amantadine and rimantadine-resistant variants of influenza A2 subtype virus, *Vopr. Virusol.* **2**:199–202.

Imboden, J. B., Canter, A., and Cluff, L. E., 1961, Convalescence from influenza, *Arch. Intern. Med.* **108**:393–399.

Iwasaki, T., and Nozima, T., 1977, Defense mechanisms against primary influenza virus infection in mice. I. The roles of interferon and neutralizing antibodies and thymus dependence of interferon and antibody production, *J. Immunol.* **118**:256–263.

Jackson, G. G., and Stanley, E. D., 1976, Prevention and control of influenza by chemoprophylaxis and chemotherapy: Prospects from examination of recent experience, *J.A.M.A.* **235**:2739–2742.

Jakab, G. J., Warr, G. A., and Knight, M. E., 1979, Pulmonary and systemic defenses against challenge with staphylococcus aureus in mice with pneumonia due to influenza A virus, *J. Infect. Dis.* **140**:105–108.

Jakab, G. J., Astry, C. L., and Warr, G. A., 1983, Alveolitis induced by influenza virus, *Am. Rev. Respir Dis.* **128**:730–739.

Janerich, D. T., 1971a, Relationship between the influenza pandemic and the epidemic of neurological malformations, *Lancet* **1**:1165.

Janerich, D. T., 1971b, Influenza and neural-tube defects, *Lancet* **2**:551–552.

Jao, R. L., Wheelock, E., and Jackson, G. G., 1965, *J. Clin. Invest.* **44**:1062.

Jao, R. L., Wheelock, E. F., and Jackson, G. G., 1970, Production of interferon in volunteers infected with Asian influenza, *J. Infect. Dis.* **121**:419–426.

Jarstrand, C., 1977, The NBT (nitroblue tetrazolium) activity of neutrophil granulocytes in patients with influenza A infection, *Scand. J. Infect. Dis.* **9**:5–7.

Jarstrand, C., and Tunevall, G., 1974, The significance of bacterial superinfection in influenza, *Scand. J. Infect. Dis.* **6**:137–144.

Jennings, R., 1968. Respiratory viruses in Jamaica: A virologic and serologic study. III. Hemagglutination-inhibiting antibodies to type B and C influenza viruses in the sera of Jamaicans, *Am. J. Epidemiol.* **87**:440–446.

Jennings, S. T., Ettensohn, D. B. and Roberts, N. J., Jr., 1984, Influenza virus infection of human alveolar and peripheral blood-derived macrophages: Production of factors that alter fibroblast proliferation, *Am. Rev. Respir. Dis.* **130**:98–102.

Johanson, W. G., Jr., Pierce, A. K., and Sanford, J. P., 1969, Pulmonary function in uncomplicated influenza, *Am. Rev. Respir. Dis.* **100**:141.

Joosting, A. C. C., Head, B., Bynoe, M. L., et al., 1968, Production of common colds in human volunteers by influenza C virus, *Br. Med. J.* **4**:153–154.

Jordan, W. S., Jr., Denny, F. W., Jr., Badger, G. F., et al., 1958, The occurrence of Asian influenza, *Am. J. Hyg.* **68**:190–212.

Julkunen, I., Hovi, T., Seppala, I., et al., 1985, Immunoglobulin G subclass antibody responses in influenza A and parainfluenza type 1 virus infections, *Clin. Exp. Immunol.* **60**:130–138.

Karalakulasingam, R., Schacht, R. A., Lansing, A. M., et al., 1977, Influenza virus pneumonia after renal transplant, *Postgrad. Med.* **62**:164–167.

Karjalainen, J., Nieminen, M. S., and Heikkila, J., 1980, Influenza A1 myocarditis in conscripts, *Acta Med. Scand.* **207**:27–30.

Katagiri, S., Ohizumi, A., and Homma, M., 1983, An outbreak of type C influenza in a children's home, *J. Infect. Dis.* **148**:51–56.

Kato, N., and Eggers, H. J., 1969, Inhibition of uncoating of fowl plague virus by 1-adamantanamine hydrochloride, *Virology* **37**:632–641.

Katz, J. M., Laver, W. G., White, D O , et al., 1985, Recognition of influenza virus hemagglutinin by subtype-specific and cross-reactive proliferative T cells: Contribution of HA1 and HA2 polypeptide chains, *J. Immunol.* **134**:616–622.

Kees, U., and Krammer, P H., 1984, Most influenza A virus-specific memory cytotoxic T lymphocytes react with antigenic epitopes associated with internal virus determinants, *J. Exp Med* **159**:365–377.

Kessler, H. A., Trenholme, G. M , Harris, A. A., et al., 1980, Acute myopathy associated with influenza A/Texas/1/77 infection Isolation of virus from a muscle biopsy specimen, *J A M A* **243**:461

Kessler, H A , Trenholme, G. M , Vogelzang, N J , et al., 1983, Elevated creatine phosphokinase levels associated with influenza A/Texas/1/77 infection, *Scand J. Infect Dis* **15**:7–10

Khakpour, M , and Nik-Akhtar, B., 1977, Epidemics of haemorrhagic cystitis due to influenza A virus, *Postgrad. Med J.* **53**:251–253

Kilbourne, E D , 1959, Studies on influenza in the pandemic of 1957–58 III Isolation of influenza A (Asian strain) viruses from influenza patients with pulmonary complications Details of virus isolation and characterization of isolates, with quantitative comparison of isolation methods, *J Clin. Invest* **38**:266–274

Kilbourne, E D , 1960, The inaccessibility of genetically active but noninfective influenza virus to the intracellular action of ribonuclease, *Virology* **11**:291–294

Kilbourne, E D., 1975, Epidemiology of influenza, in *The Influenza Viruses and Influenza* (E D. Kilbourne, ed), Academic Press, New York, pp 483–538

Kilbourne, E D , 1976, Comparative efficacy of neuraminidase-specific and conventional influenza virus vaccines in the induction of anti-neuraminidase antibody in man *J Infect Dis* **134**:384–394

Kilbourne, E D., 1978, Influenza as a problem in immunology, *J Immunol.* **120**:1447–1452

Kilbourne, E. D , 1980, Influenza Viral determinants of the pathogenicity and epidemicity of an invariant disease of variable occurrence, *Phil Trans. R Soc. Lond.* [*Biol*] **288**:291–297.

Kilbourne, E D , and Loge, J P , 1950, Influenza A prime A clinical study of an epidemic caused by a new strain of virus, *Ann Intern Med* **33**:371–379

Kilbourne, E D , Chanock, R. M , Choppin, P W., et al., 1974, Influenza vaccines—Summary of Influenza workshop V, *J. Infect Dis.* **129**:750–771

Kim, H. W., Brandt, C D., Arrobio, J. O., et al., 1979, Influenza A and B virus infection in infants and young children during the years 1957–1976, *Am. J. Epidemiol.* **109**:464–479.

Klavinskis, L S , Patterson, S , Whiting, P. J., et al , 1985, Infection of cultured human muscle cells by influenza virus, *J. Gen. Virol* **66**:2335–2345.

Klein, R C., Bobear, J B., and Gohd, R , 1978, Influenza, respiratory distress and clear chest films: Report of four cases, *South. Med. J.* **71**:5–6.

Kleinerman, E. S , Synderman, R., and Daniels, C. A , 1975, Depressed monocyte chemotaxis during acute influenza infection, *Lancet* **2**:1063–1066

Kleinerman, E S , Daniels, C A., Polisson, R P , et al , 1976, Effect of virus infection on the inflammatory response, *Am. J Pathol.* **85**:373–382.

Knight, V., Fedson, D., Baldini J., et al., 1970, Amantadine therapy of epidemic influenza A2 (Hong Kong), *Infect Immun.* **1**:200–204

Knight, V., Wilson, S. Z , Quarles, J M , et al., 1981, Ribavirin small-particle aerosol treatment of influenza, *Lancet* **2**:945–949.

Koff, W. C., and Knight, V., 1979, Inhibition of influenza virus uncoating by rimantadine hydrochloride, *J. Virol* **31**:261–263

Krilov, L. R., and Swenson, P., 1985, Acute parotitis associated with influenza A infection, *J Infect Dis.* **152**:853.

Kris, R. M., Asofsky, R , Evans, C B., et al., 1985, Protection and recovery in influenza virus-infected mice immunosuppressed with anti-IgM, *J Immunol* **134**:1230–1235.

Krus, S., Wilczynski, J , and Semkow, R., 1984, Metaplasia of the lung alveolar epithelium in mice repeatedly challenged with influenza virus, *Acta Microbiol. Pol* **33**:41–43

Krylov, V F , Obrosova-Serova, N P , Zilina, N. N., et al , 1973, Clinical and epidemiological features of an outbreak of influenza in Ulan Bator in 1971, *Bull. Org. Mond. Sante* **49**:567–570

Kuberski, T T , 1977, Hemorrhage associated with influenza, *Lancet* **1**:709.

Laitinen, L. A., and Kava, T., 1980, Bronchial reactivity following uncomplicated influenza A infection in healthy subjects and in asthmatic patients, *Eur J Respir. Dis* **61**:51–58

Lamb, J. R., and Green, N., 1983, Analysis of the antigen specificity of influenza haemagglutinin-immune human T lymphocyte clones: identification of an immunodominant region for T cells, *Immunology* **50:**659–666.

Laraya-Cuasay, L. R., DeForest, A., Huff, D., et al., 1977, Chronic pulmonary complications of early influenza virus infection in children, *Am. Rev. Respir. Dis.* **116:**617–625.

Larson, E. W., Dominik, J. W., Rowberg, A. H., et al., 1976, Influenza virus population dynamics in the respiratory tract of experimentally infected mice, *Infect. Immun.* **13:**438–447.

Leck, I., Hay, S., Witte, J. J., et al., 1969, Malformations recorded on birth certificates following A2 influenza epidemics, *Public Health Rep.* **84:**971–979.

Leeder, S. R., Gell, P. W., and Peat, J. K., 1974, Short and long term effects of influenza A on lung function, *Med. J. Aust.* **2:**812–814.

Lefrak, E. A., Stevens, P. M., Pitha, J., et al., 1974, Extracorporeal membrane oxygenation for fulminant influenza pneumonia, *Chest* **66:**384–388.

Lennon, D. R., Cherry, J. D., Morgenstein, A., et al., 1983, Longitudinal study of influenza B symptomatology and interferon production in children and college students, *Pediatr. Infect. Dis.* **2:**212–215.

Lichtenstein, P. K., Heubi, J. E., Daugherty, C. C., et al., 1983, Grade I Reye's syndrome: A frequent cause of vomiting and liver dysfunction after varicella and upper-respiratory-tract infection, *N Engl. J. Med.* **309:**133–139.

Liew, F. Y., and Russell, S. M., 1983, Inhibition of pathogenic effect of effector T cells by specific suppressor T cells during influenza virus infection in mice, *Nature* **304:**541–543.

Liew, F. Y., Russell, S. M., Appleyard, G., et al., 1984, Cross-protection in mice infected with influenza A virus by the respiratory route is correlated with local IgA antibody rather than serum antibody or cytotoxic T cell reactivity, *Eur. J. Immunol.* **14:**350–356.

Lindsay, M. I., Jr., Herrmann, E. C., Jr., Morrow, G W., Jr., et al., 1970, Hong Kong influenza. Clinical, microbiologic, and pathologic features in 127 cases, *J.A.M.A.* **214:**1825–1832.

Little, J. W., Hall, W. J., Douglas, R. G. Jr., et al., 1976, Amantadine effect on peripheral airways abnormalities in influenza: A study in 15 students with natural influenza A infection, *J. Intern. Med.* **85:**177–182.

Little, J. W., Hall, W. J., Douglas, R. G. Jr., et al., 1978, Airway hyperreactivity and peripheral airway dysfunction in influenza A infection, *Am. Rev. Respir. Dis.* **118:**295–303.

Little, J. W., Douglas, R. G., Jr., Hall, W. J., et al., 1979, Attenuated influenza produced by experimental intranasal inoculation, *J. Med. Virol.* **3:**177–188.

Louria, D. B., Blumenfeld, H. L., Ellis, J. T., et al., 1959, Studies of influenza in the pandemic of 1957–1958. II. Pulmonary complications of influenza, *J. Clin. Invest.* **38:**213–265.

Lubeck, M. D., Schulman, J. L., and Palese, P., 1978, Susceptibility of influenza A viruses to amantadine is influenced by the gene coding for M protein, *J. Virol.* **28:**710–716.

Lubkiewicz, K., Elie, K., Spence, L., et al., 1974, Laboratory and clinical features of influenza A_2 1971–72 in Montreal, *Chest* **66:**671–674.

Lukacher, A. E., Braciale, V. L., and Braciale, T. J., 1984, In vivo effector function of influenza virus-specific cytotoxic T lymphocyte clones is highly specific, *J. Exp. Med.* **160:**814–826.

Luksza, A. R., and Jones, D. K., 1984, Influenza B virus infection complicated by pneumonia, acute renal failure and disseminated intravascular coagulation, *J. Infect.* **9:**174–176.

MacCallum, W. G., 1919, Pathology of the pneumonia following influenza, *J.A.M.A.* **72:**720–723.

MacKenzie, J. S., and Houghton, M., 1974, Influenza infections during pregnancy. Association with congenital malformations and with subsequent neoplasms in children, and potential hazards of live virus vaccines, *Bacteriol. Rev.* **38:**356–370.

Martin, C. M., Kunin, C. M., Gottlieb, L. S., et al., 1959a, Asian influenza in Boston, 1957–1958. I. Observations in thirty-two influenza-associated fatal cases, *Arch. Intern. Med.* **103:**515–531.

Martin, C. M., Kunin, C. M., Gottlieb, L. S., et al., 1959b, Asian influenza A in Boston, 1957–1958. II. Severe staphylococcal pneumonia complicating influenza, *Arch. Intern. Med.* **103:**532–542

Martin, R. R., Couch, R. B., Greenberg, S. B., et al., 1981, Effects of infection with influenza virus on the function of polymorphonuclear leukocytes, *J. Infect. Dis.* **144:**279.

Mason, W., and Keller, E., 1975, Acute transient myositis with influenza-like illness, *J. Pediatr.* **86:**813–815.

Mass, A., 1985, Severe influenza myositis, *Med. J. Austral.* **142:**330–331.

McClung, H. W , Knight, V., Gilbert, B E., et al., 1983, Ribavirin aerosol treatment of influenza B virus infection, *J.A M.A.* **249**:2671-2674.
McConkey, B., Oxon, B M., and Daws, R. A., 1958, Neurological disorders associated with Asian influenza, *Lancet* **2**:15-17.
McGregor, J. A., Burns, J. C , Levin, M J., et al., 1984, Transplacental passage of influenza A/Bangkok (H3N2) mimicking amniotic fluid infection syndrome, *Am. J Obstet Gynecol.* **149**:856-859.
McMichael, A J , Gotch, F. M , Noble, G. R , et al , 1983, Cytotoxic T-cell immunity to influenza, *N Engl J Med* **309**:13-17
Medical Research Council, 1919, *Special Report Series No 36*, MRC, London, p 11.
Meibalane, R., Sedmak, G V , Sasidharan, P., et al., 1977, Outbreak of influenza in a neonatal intensive care unit, *J Pediatr* **91**:974-976
Meiklejohn, G., Graves, P , Eickhoff, T. C , et al , 1980, An influenza A epidemic caused by H1N1, *Mil Med* **145**:129-131
Mellman, W J , 1958, Influenza encephalitis, *J Pediatr* **53**:292-297
Middleton, P J , Alexander, R M , and Szymanski, M T , 1970, Severe myositis during recovery from influenza, *Lancet* **2**:533-535
Minor, T E , Dick, E. C , DeMeo, A N., et al., 1974, Viruses as precipitants of asthmatic attacks in children, *J A M A* **227**:292-298
Minor, T E., Dick, E C , Baker, J W , et al , 1976, Rhinovirus and influenza type A infections as precipitants of asthma, *Am Rev Respir Dis* **113**:149-153
Minow, R , Gorbach, S , Johnson, L , et al , 1974, Myoglobinuria associated with influenza A infection, *Ann Intern Med* **80**:359-361
Monto, A S , and Troisi, C L , 1981, Comparison of enzyme-linked immunosorbent assay and hemagglutinin inhibition in a seroepidemiological study of influenza type C infection, *J Clin Microbiol* **14**:516-521
Monto, A. S , Ceglarek, J P , and Hayner, N S , 1981, Liver function abnormalities in the course of a type A (H1N1) influenza outbreak Relation to Reye's syndrome *Am J Epidemiol* **114**:750-759.
Monto, A S , Koopman, J S , and Longini, I. M , Jr , 1985, Tecumseh study of illness XIII Influenza infection and disease, 1976-1981, *Am. J. Epidemiol* **121**:811-822
Morita, M , Suto, T., and Ishida, N , 1972, Antigenic memory in man in response to sequential infections with influenza A viruses, *J Infect Dis* **126**:61-68.
Morris, J. A , Kasel, J. A , Saglam, M., et al., 1966, Immunity to influenza as related to antibody levels, *N Engl J Med.* **274**:527-535.
Mostow, S. R , and Tyrrell, D. A J , 1972, Detection of attenuation of recombinant influenza viruses in vitro, *Lancet* **2**:116-117.
Mulder, J., and Hers, J F. P., 1972, *Influenza*, Wolters-Noordhoff, Groningen, The Netherlands
Mullooly, J P., and Barker, W. H , 1982, Impact of type A influenza on children A retrospective study, *Am J Public Health* **72**:1008-1016
Mullooly, J P , Barker, W H , and Nolan, T F , Jr., 1986, Risk of acute respiratory disease among pregnant women during influenza A epidemics, *Public Health Rep* **101**:205-211
Murakami, T , and Matsuyama, T , 1985, Behaviour of influenza virus point-inoculated onto the mucous membrane of the chick tracheal explant, *Br J Exp. Pathol.* **66**:233-241
Murphy, A. M , and Hawkes, R A., 1970, Neurological complications of influenza A2/Hong Kong/68 virus, *Med J. Aust.* **2**:511.
Murphy, B R , Chalhub, E G , Nusinoff, S. R , et al., 1973, Temperature-sensitive mutants of influenza virus III. Further characterization of the *ts*-1[E] influenza A recombinant (H3N2) virus in man, *J Infect. Dis* **128**:479-487
Murphy, B R., Nelson, D. L., Wright, P. F., et al, 1982, Secretory and systemic immunological response in children infected with live attenuated influenza A virus vaccines, *Infect Immun* **36**:1102-1108
Nicholson, K. G , 1984, Antiviral therapy Respiratory infections, genital herpes, and herpetic keratitis, *Lancet* **2**:617-621
Nolan, T F , Goodman, R A., Hinman, A R , et al , 1980, Morbidity and mortality associated with influenza B in the United States, 1979-1980 A report from the Center for Disease Control, *J Infect Dis.* **142**:360-362
Nugent, K M , and Pesanti, E , 1982, Staphylococcal clearance and pulmonary macrophage function

during influenza infection, *Infect. Immun.* **38**:1256–1262.
O'Callaghan, R. J., Gohd, R. S., and Labat, D. D., 1980, Human antibody to influenza C virus: Its age-related distribution and distinction from receptor analogs, *Infect. Immun.* **30**:500–505.
Opie, E. L., Blake, F. G., Small, J. C., et al., 1921, *Epidemic Respiratory Disease*, C. V. Mosby, St. Louis.
Oseasohn, R., Adelson, L., and Kaji, M., 1959, Clinicopathologic study of thirty-three fatal cases of Asian influenza, *N. Engl. J. Med.* **260**:509–518.
Owens, S. L., Osebold, J. W., and Zee, Y. C., 1981, Dynamics of B-lymphocytes in the lungs of mice exposed to aerosolized influenza virus, *Infect. Immun.* **33**:231–238.
Oxford, J. S., and Galbraith, A., 1980, Antiviral activity of amantadine: A review of laboratory and clinical data, *Pharmacol. Ther.* **11**:181–262.
Oxford, J. S., and Logan, I. S., 1970, In vivo selection of an influenza A A2 strain resistant to amantadine, *Nature* **226**:82–83.
Paisley, J. W., Bruhn, F. W., Lauer, B. A., et al., 1978, Type A_2 influenza viral infections in children, *Am. J. Dis. Child.* **132**:34–36.
Petersdorf, R. G., Fusco, J. J., Harter, D. H., et al., 1959, Pulmonary infections complicating Asian influenza, *Arch. Intern. Med.* **103**:262–272.
Pinsker K. L., Schneyer, B., Becker, N., et al., 1981, Usual interstitial pneumonia following Texas A2 influenza infection, *Chest* **80**:123–126.
Podesin, R. L., and Felton, W. L., 1958, The clinical picture of Far East influenza occurring at the Fourth National Boy Scout Jamboree; report of 616 cases, *N. Engl. J. Med.* **258**:778–782.
Price, D. A., Postlethwaite, R. J., and Longson, M., 1976, Influenzavirus A_2 infections presenting with febrile convulsions and gastrointestinal symptoms in young children, *Clin. Pediatr.* **15**:361–367.
Puck, J. M., Glezen, W. P., Frank, A. L., et al., 1980, Immunoprophylaxis. Protection of infants from infection with influenza A virus by transplacentally acquired antibody, *J. Infect. Dis.* **142**:844–849.
Quarles van Ufford, W. J., and Savelberg, P. J., 1959, Asiatic influenza in allergic patients with bronchial asthma, *Int. Arch. Allergy Appl. Immunol.* **15**:189.
Quilligan, J. J., Jr., Minuse, E., Francis, T., Jr., et al., 1954, Type C influenza virus. II. Intranasal inoculation of human individuals, *J. Lab. Clin. Med.* **43**:43–47.
Ramphal, R., Fischlschweiger, W., Shands, J. W., Jr., et al., 1979, Murine influenzal tracheitis: A model for the study of influenza and tracheal epithelial repair, *Am. Rev. Respir. Dis.* **120**:1313–1324.
Randolph, V. L., and Heath, C. W., Jr., 1974, Influenza during pregnancy in relation to subsequent childhood leukemia and lymphoma, *Am. J. Epidemiol.* **100**:399–409.
Raut, S., Hurd, J., Cureton, R.-J., et al., 1975, The pathogenesis of infections of the mouse caused by virulent and avirulent variants of an influenza virus, *J. Med. Microbiol.* **8**:127–136.
Rebhan, A., 1957, An outbreak of Asian influenza in a girl's camp, *Can. Med. Assoc. J.* **77**:797.
Reed, W. P., Olds, J. W., and Kisch, A. L., 1972, Decreased skin-test reactivity associated with influenza, *J. Infect. Dis.* **125**:398–402.
Reiss, C. S., and Burakoff, S. J., 1981, Specificity of the helper T cell for the cytotoxic T lymphocyte response to influenza viruses, *J. Exp. Med.* **154**:541.
Reiss, C. S., and Schulman, J. L., 1980, Cellular immune responses of mice to influenza virus infection, *Cell. Immunol.* **56**:502–509.
Reye, R. D., Morgan, G., and Baral, J., 1963, Encephalopathy and fatty degeneration of the viscera, a disease entity in childhood, *Lancet* **2**:749–752.
Rock, J. A., Braude, A. I., and Moran, T. J., 1958, Asian influenza and mitral stenosis. Report of a case with autopsy, *J.A.M.A.* **166**:1467.
Rodgers, B. C., and Mims, C. A., 1982, Influenza virus replication in human alveolar macrophages, *J. Med. Virol.* **98**:177–184.
Rose, E., and Prabhakar, P., 1982, Influenza A virus associated neurological disorders in Jamaica, *W.I. Med. J.* **31**:29–33.
Rott, R., Orlich, M., and Scholtissek, C., 1982, Differences in the multiplication at elevated temperature of influenza virus recombinants pathogenic and nonpathogenic for chicken, *Virology* **120**:215–224.
Russell, S. M., and Liew, F. Y., 1979, T cells primed by influenza virion internal components can co-operate in the antibody response to haemagglutinin, *Nature* **280**:147.
Russell, S. M., and Liew, F. Y., 1980, Cell cooperation in antibody responses to influenza virus. I. Priming of helper T cells by internal components of the virion, *Eur. J. Immunol.* **10**:791–796.

Scalzo, A. A., and Anders, E. M., 1985, Influenza viruses as lymphocyte mitogens: I. B cell mitogenesis by influenza A viruses of the H2 and H6 subtypes is controlled by the I-E/C subregion of the major histocompatibility complex, *J. Immunol.* **134**:757–760.

Scheinberg, M. A., Blacklow, N. R., Goldstein, A. L., et al., 1976, Influenza: Response of T-cell lymphopenia to thymosin, *N. Engl. J. Med.* **294**:1208–1211.

Schild, G. C., 1969, Antibody against influenza A2 virus neuraminidase in human sera, *J. Hyg. (Camb.)* **67**:353–365.

Scholtissek, C., and Faulkner, G. P., 1979, Amantadine-resistant and sensitive influenza A strains and recombinants, *J. Gen. Virol.* **44**:807–815.

Schulman, J. L., 1973, Effect of 'antigenic drift' on resistance of mice to influenza A virus challenge. International Symposium on Influenza Vaccines for Men and Horses, London 1972, *Symp. Ser. Immunobiol. Stand.* **20**:106–111.

Schulman, J. L., and Kilbourne, E. D., 1963, Experimental transmission of influenza virus infection in mice. I. The period of transmissibility, *J. Exp. Med.* **118**:257–266.

Schenouda, A., and Hatch, F. E., 1976, Influenza A viral infection associated with acute renal failure, *Am. J. Med.* **61**:697–702.

Sigel, M. M., Kitts, A. W., Light, A. B., et al., 1950, The recurrence of influenza A prime in a boarding school after 2 years, *J. Immunol.* **64**:33–38.

Simon, N. M., Rovner, R. N., and Berlin, B. S., 1970, Acute myoglobinuria associated with type A2 (Hong Kong) influenza, *J.A.M.A.* **212**:1704–1705.

Sitbon, M., Gomard, E., Hannoun, C., et al., 1983, Anti-influenza human T killer cells present an intertypic activity anti-A and -B type viruses in a secondary reaction in vitro, *Clin. Exp. Immunol.* **54**:49–58.

Skehel, J. J., Hay, A. J., and Armstrong, J. A., 1977, On the mechanism of inhibition of influenza virus replication by amantadine hydrochloride, *J. Gen. Virol.* **38**:97–110.

Slepushkin, A. N., 1959, The effect of a previous attack of A1 influenza on susceptibility to A2 virus during the 1957 outbreak, *Bull. WHO* **20**:297–301.

Smith, C. B., Kanner, R. E., Golden, C. A., et al., 1980, Effect of viral infections on pulmonary function in patients with chronic obstructive pulmonary diseases, *J. Infect. Dis.* **141**:271–279.

Smorodintsev, A. A., Tushinsky, M. D., Drobyshevskaya, A. I., et al., 1937, Investigation on volunteers infected with the influenza virus, *Am. J. Med. Sci.* **194**:159–170.

Sonoguchi, T., Naito, H., Hara, M., et al., 1985, Cross-subtype protection in humans during sequential, overlapping, and/or concurrent epidemics caused by H3N2 and H1N1 influenza viruses, *J. Infect. Dis.* **151**:81–88.

Sterkers, G., Michon, J., Henin, Y., et al., 1985, Fine specificity analysis of human influenza-specific cloned cell lines, *Cell. Immunol.* **94**:394–405.

Stevens, K. M., 1976, Cardiac stroke volume as a determinant of influenzal fatality, *N. Engl. J. Med.* **295**:1363–1366.

Stinson, S. F., Ryan, D. P., Hertweck, M. S., et al., 1976, Epithelial and surfactant changes in influenzal pulmonary lesions, *Arch. Pathol. Lab. Med.* **100**:147–153.

Stitz, L., Huang, R. T. C., Hengartner, H., et al., 1985, Cytotoxic T cell lysis of target cells fused with liposomes containing influenza virus haemagglutinin and neuraminidase, *J. Gen. Virol.* **66**:1333–1339.

Stuart-Harris, C. H., 1961, Twenty years of influenza epidemics, *Am. Rev. Rispir. Dis.* **83**:54–67.

Stuart-Harris, C. H., and Schild, G. C., 1976, *Influenza: The Viruses and the Disease*, Edward Arnold, London.

Stuart-Harris, C. H., Laird, J., Tyrrell, D. H., et al., 1949, The relationship between influenza and pneumonia, *J. Hyg. (Camb.)* **7**:434.

Sugiura, A., Tateno, I., and Kitamoto, O., 1958, Cerebral involvement by influenza: Report of three indicative cases, *Jpn. J. Exp. Med.* **28**:337–344.

Sulkava, R., Rissanen, A., and Pyhala, R., 1981, Post-influenzal encephalitis during the influenza A outbreak in 1979/1980, *J. Neurol. Psychol.* **44**:161–163.

Sullivan-Bolyai, J. Z., and Corey, L., 1981, Epidemiology of Reye syndrome, *Epidemiol. Rev.* **3**:1–26.

Sullivan-Bolyai, J. S., Marks, J. S., Johnson, D., et al., 1980, Reye syndrome in Ohio 1973–1977, *Am. J. Epidemiol.* **112**:629–638.

Sweet, C., and Smith, H., 1980, Pathogenicity of influenza virus, *Microbiol. Rev.* **44**:303–330.

Sweet, C., Bird, R. A., Howie, A. J., et al., 1985, Further studies of the reasons for the lack of alveolar infection during influenza in ferrets, *Br. J. Exp. Pathol.* **66**:217–231.

Taylor, G. J., Brenner, W., and Summer, W. R., 1976, Severe viral pneumonia in young adults: Therapy with continuous positive airway pressure, *Chest* **69**:722–728.

Thein, S., Ming, C. K., Thaung, U., et al., 1975, Haemorrhagic manifestations of influenza A infection in children, *J. Trop. Med. Hyg.* **78**:78–80.

Thraenhart, O., Schley, G., and Kuwert, E., 1975, Isolation of influenza virus A/Hong Kong/1/68 (H3N2) from liquor cerebrospinalis of patients with CNS involvement, *Med. Klin.* **70**:1910–1914.

Toms, G. L., Davies, J. A., Woodward, C. G., et al., 1977, The relation of pyrexia and nasal inflammatory response to virus levels in nasal washings of ferrets infected with influenza viruses of differing virulence, *Br. J. Exp. Pathol.* **58**:444–458.

Townsend, A. R. M., and Skehel, J. J., 1984, The influenza A virus nucleoprotein gene controls the induction of both subtype specific and cross-reactive cytotoxic T cells, *J. Exp. Med.* **160**:552–563.

Townsend, A. R. M., McMichael, A. J., Carter, N. P., et al., 1984a, Cytotoxic T cell recognition of the influenza nucleoprotein and hemagglutinin expressed in transfected mouse L cells, *Cell* **39**:13–25.

Townsend, A. R. M., Skehel, J. J., Taylor, P. M., et al., 1984b, Recognition of influenza A virus nucleoprotein by an H-2-restricted cytotoxic T-cell clone, *Virology* **133**:456–459.

Townsend, A. R. M., Gotch, F. M., and Davey, J., 1985, Cytotoxic T cells recognize fragments of the influenza nucleoprotein, *Cell* **42**:457–467.

Tumova, B., Scharfenorth, H., and Adamczyk, G., 1983, Incidence of influenza C virus in Czechoslovakia and German Democratic Republic, *Acta Virol.* **27**:502–510.

Urquhart, G. E. D., 1974, Serum IgM and IgA responses in influenza A infections, *J. Clin. Pathol.* **27**:198–201.

Utell, M. J., Aquilina, A. T., Hall, W.J., et al., 1980, Development of airway reactivity to nitrates in subjects with influenza, *Am. Rev. Respir. Dis.* **121**:233–241.

Van Voris, L. P., Betts, R. F., Hayden, F. G., et al., 1981, Successful treatment of naturally occurring influenza A/USSR/77 H1N1, *J.A.M.A.* **245**:1128–1131.

Vella, E. E., 1985, Pandemic influenza and the viral haemorrhagic fevers, *J. R Soc Med* **78**:175.

Verel, D., Warrack, A. J. N., Potter, C. W., et al., 1976, Observations on the A2 England influenza epidemic: A clinicopathological study, *Am. Heart J.* **92**:290–296.

Virelizier, J.-L., 1975, Host defenses against influenza virus: The role of anti-hemagglutinin antibody, *J. Immunol* **115**:434–439.

Virelizier, J. L., Allison, A. C., and Schild, G. C., 1974, Antibody responses to antigenic determinants of influenza virus hemagglutinin. II. Original antigenic sin: A bone marrow-derived lymphocyte memory phenomenon modulated by thymus-derived lymphocytes, *J. Exp. Med.* **140**:1571–1578.

Vodeiko, G. M., Zazimko, L. A., Farashyan, V. R., et al., 1982, Study of parent and recombinant influenza A virus strains differing in sensitivity to remantadine, *Vopr. Virusol.* **2**:86–90.

Voisin, C., Aerts, C., and Tonnel, A. B., 1974, Mechanism of post-influenza respiratory tract microbial super-infection: In vitro study on the effect of influenza virus on human alveolar macrophages, *Lille Med.* **19**:440–444.

Walsh, J., Burch, G. E., White, A., et al., 1958, A study of the effects of type A (Asian strain) influenza on the cardiovascular system of man, *Ann. Intern. Med.* **49**:502–528.

Walsh, J. J., Dietlein, L. F., Low, F. N., et al., 1961, Bronchotracheal response in human influenza *Arch. Intern. Med.* **108**:376–388.

Wang, M. -L., Skehel, J. J., and Wiley, D. C., 1986, Comparative analyses of the specificities of anti-influenza hemagglutinin antibodies in human sera, *J. Virol.* **57**:124–128.

Warshauer, D., Goldstein, E., Akers, T., et al., 1977, Effect of influenza viral infection on the ingestion and killing of bacteria by alveolar macrophages, *Am. Rev. Respir. Dis.* **115**:269–277.

Wasserstein, A., Hill, G., Goldfarb, S., et al., 1981, Recurrent thrombocytopenic purpura after viral infection: Clinical and histologic simulation of chronic glomerulonephritis, *Arch. Intern. Med.* **141**:685–687.

Wells, M. A., Albrecht, P., and Ennis, F. A., 1981a, Recovery from a viral respiratory infection: I.

Influenza pneumonia in normal and T-deficient mice, *J Immunol* **126:**1036–1041.
Wells, M A , Ennis, F. A , and Albrecht, P , 1981b, Recovery from a viral respiratory infection. II. Passive transfer of immune spleen cells to mice with influenza pneumonia, *J Immunol.* **126:**1042–1046.
Wells, M A , Daniel, D , Djeu, J. Y., et al., 1983, Recovery from a viral respiratory tract infection: IV. Specificity of protection by cytotoxic T lymphocytes, *J. Immunol* **130:**2908–2914.
Westerberg, S C., Smith, C. B , Wiley, B B , et al, 1972, Pathogenesis of influenza virus infection in mouse tracheal organ cultures (36565), *Proc Soc Exp Biol. Med* **140:**846–851.
Widelock, D , Csizmas, L , and Klein, S , 1963, Influenza, pregnancy, and fetal outcome, *Public Health, Rep* **78:**1–12
Wilson, M G , and Stein, A. M , 1969, Teratogenic effects of Asian influenza, *J A M.A* **210:**336–337
Winterbauer, R H , Ludwig, W R , and Hammar, S P , 1977, Clinical course, management, and long-term sequelae of respiratory failure due to influenza viral pneumonia, *Johns Hopkins Med J* **141:**148–155
Winternitz, M C , Wason, I M , and McNamara, F P , 1920, *The Pathology of Influenza,* Yale University Press, New Haven
Witzleb, W , Witzleb, H , Mehihorn, J , et al , 1976, Demonstration of influenza virus A in human heart by semiquantitative virus assay and immunofluorescence, *Acta Virol* **20:**168
Wolbach, S B., 1919, Comments on the pathology and bacteriology of fatal influenza cases as observed at Camp Devens, Mass , *Bull Johns Hopkins Hosp* **30:**104
Woodruff, J J , and Woodruff J F , 1976, Influenza A virus interaction with murine lymphocytes II Changes in lymphocyte surface properties induced by influenza virus A/Japan 305 (H2N2), *J Immunol* **117:**859–864.
Woodward, T. E., Mc Crumb, F. R , Carey, T. N , et al., 1960, Viral and rickettsial causes of cardiac disease, including the coxsackie virus etiology of pericarditis and myocarditis, *Ann Intern Med* **53:**1130–1150
Wright, P F , Ross, K. B., Thompson, J., et al , 1977, Influenza A infections in young children Primary natural infection and protective efficacy of live-vaccine-induced or naturally acquired immunity, *N Engl. J Med.* **296:**829–834
Wright, P. F , Bryant, J D , and Karzon, D T , 1980, Comparison of influenza B/Hong Kong virus infections among infants, children, and young adults, *J Infect Dis.* **141:**430–435
Wright, P. F., Thompson, J., and Karzon, D T , 1981, Differing virulence of H1N1 and H3N2 influenza strains, *Am J Epidemiol.* **112:**814–819.
Wright, P F., Murphy, B. R., Kervina, M., et al , 1983, Secretory immunological response after intranasal inactivated influenza A virus vaccinations. Evidence for immunoglobulin A memory, *Infect Immun* **40:**1092–1095.
Wyde, P. R , and Cate, T. R., 1978, Cellular changes in lungs of mice infected with influenza virus Characterization of the cytotoxic responses, *Infect. Immun* **22:**423–429
Wyde, P. R , Couch, R. B., Mackler, B. F., et al., 1977, Effects of low and high-passage influenza virus infection in normal and nude mice, *Infect Immun.* **15:**221–229
Wyde, P. R., Wilson, M. R., and Cate, T. R., 1982, Interferon production by leukocytes infiltrating the lungs of mice during primary influenza virus infection, *Infect. Immun* **38:**1249–1255.
Yap, K. L., Braciale, T J , and Ada, G. L., 1979, Role of T-cell function in recovery from murine influenza infection, *Cell Immunol* **43:**341–351.
Yarchoan, R., Murphy, B R , Strober, W , et al , 1981, In vitro production of anti-influenza virus antibody after intranasal inoculation with cold-adapted influenza virus, *J. Immunol* **127:**1958–1963
Yawn, D H , Pyeatte, J C , Joseph, J M , et al , 1971, Transplacental transfer of influenza virus, *J A M A.* **216:**1022–1023.
Yewdell, J W , Frank, E., and Gerhard, W., 1981, Expression of influenza A virus internal antigens on the surface of infected P815 cells, *J Immunol.* **126:**1814–1819
Yewdell J W., Bennink, J. R , Smith, G L , et al., 1985, Influenza A virus nucleoprotein is a major target antigen for cross-reactive anti-influenza A virus cytotoxic T lymphoctyes, *Proc Natl Acad Sci U S.A* **82:**1785–1789.

Yilma, T., Zee, Y. C., and Osebold, J. W., 1979, Immunofluorescence determination of the pathogenesis of infection with influenza virus in mice following exposure to aerosolized virus, *J. Infect. Dis.* **139:**458–464.

Younkin, S. W., Betts, R. F., Roth, F. K., et al., 1983, Reduction in fever and symptoms in young adults with influenza A/Brazil/78 H1N1 infection after treatment with aspirin or amantadine, *Antimicrob. Agents Chemother.* **23:**577–582.

Zamkoff, K., and Rosen, N., 1979, Influenza and myoglobinuria in brothers, *Neurology (Minneap.)* **29:**340–345.

8

The Laboratory Diagnosis of Influenza

INTRODUCTION

The specific diagnosis of human influenza depends on the recovery of virus from human tissue or secretions, the demonstration of viral antigens in infected cells or secretions, or the measurement of specific immunologic response. The present chapter is not intended as a laboratory manual but reviews and summarizes currently available diagnostic methods and approaches. Procedural details can be found in *Concepts and Procedures for Laboratory Based Influenza Surveillance* (Centers for Disease Control, 1982) and in *Laboratory Diagnosis of Viral Infections* (Lennette, 1985) (also see Dowdle and Coleman, 1974).

VIRUS ISOLATION AND IDENTIFICATION

As emphasized in Chapter 6, recovery of virus from its natural host involves selection from a universe of genetically heterogeneous virus particles of virus best equipped to replicate in the laboratory host. Virus so selected may not be representative of the virus population in the human respiratory tract, but the problem is academic with respect to virus detection. To the extent that selection of antigenic variants is host determined (Chapter 6), then serological diagnosis using such virus might be influenced. A further consideration in the isolation of virus is its possible use in vaccines. Virus isolated in aneuploid cell cultures cannot be used in vaccines.

Animal Inoculation

The original recovery of human influenza virus was in the ferret, and that animal still has a place in influenza research in providing strain-specific antisera and in testing of antiviral agents. It is also possible to isolate virus directly in mice (Francis and McGill, 1937) or hamsters (Taylor, 1940) by intranasal inoculation, but use of intact animals is costly and cumbersome and has no place in ordinary laboratory diagnosis.

Chick Embryos

The chick embryo was first used for the propagation of influenza virus in 1935 (Smith, 1935) and for its isolation in 1940 (Burnet, 1940). Since then, it has been the standard host for the isolation and growth of virus. Amniotic and chorio-allantoic membrane endodermal cells of the developing embryo are maximally susceptible to virus, and yields of infective influenza A virus up to 10^{10} $EID_{50}/1.0$ ml of allantoic fluid are not uncommon; 500–1400 virus particles per cell are produced during infection with laboratory-adapted strains of virus (Horsfall, 1955). Inoculation of the allantoic sac is more easily performed and is the procedure most frequently used, but amniotic inoculation is the more sensitive method (Hirst, 1945). Ten- to 11-day-old embryos are usually employed, although 13- to 15-day embryos are reported to be most susceptible to amniotic inoculation (Beveridge and Burnet, 1946; French and Dineen, 1958). With H2N2 and H3N2 viruses, the allantoic and amniotic routes of inoculation appear to be equally effective in isolation of viruses from man (Kilbourne, 1959; E.D. Kilbourne, unpublished data). For isolation of influenza C virus, the amniotic route of inoculation is required.

Cell Culture

Primary Cell Culture

As indicated in Chapter 5 (Table 5-2), a variety of primary cell cultures will support replication and plaquing of influenza viruses. However, only primary rhesus monkey kidney cell cultures have been widely employed. Such cultures prove to be more efficient than chick embryos in the isolation of a wide variety of influenza viruses (Smith and Reichrath, 1974), especially with respect to influenza B virus (Monto et al., 1981). However, the use of primary cultures has been superseded by the use of a continuously propagable cell line (MDCK, see following section), which is equally efficient in detecting virus (Davies et al. 1978).

Continuous Cell Lines

The Madin–Darby line of canine kidney (MDCK) cells was first shown by Green (1962) to support influenza virus replication. When used with trypsin to facilitate viral HA cleavage, this cell line is the present system of choice for the isolation of influenza A and B viruses (Tobita, 1975; Tobita et al., 1975; Meguro et al., 1979; Dumitrescu et al., 1980; Reizin et al., 1981).

Influenza C viruses are reported to produce plaques in MDCK cells when relatively high concentrations of trypsin are used (Nerome and Ishida, 1978). Plasminogen is less damaging to cells and facilitates the production of plaques in MDCK cell monolayers (Racaniello, 1979).

The human conjunctival cell variant clone 1-5C-4 has also been used for influenza virus isolation (Rytel and Sedmack, 1973; E.D. Kilbourne, unpublished data), but its efficiency with added protease compared to the MDCK line has not been evaluated.

In cell cultures, virus replication is recognized by CPE, by hemadsorption, by the release of hemagglutinating virus, or by the detection of antigens by ELISA (Watanabe and MacKenzie, 1981) or by immunofluorescence (Canil et al., 1985). The presence of virus may also be recognized by the activity of the NA enzyme in allantoic fluid (Quash et al., 1982) or MDCK cell culture supernatants (Kiyotani, 1984), but little practical use has been made of this approach.

Virus Identification

Hemagglutinating or hemadsorbing viruses in chick embryo allantoic fluid or cell cultures are identified by type-specific antibody in hemagglutination-inhibition (HI), hemadsorption-inhibition, or, less commonly, in immunodiffusion tests. Typing by HI is facilitated by the use of HA-specific antisera made to antigenically hybrid virus that precludes cross reactivity mediated through the viral NA by steric inhibition (Schulman and Kilbourne, 1969; Styk et al., 1971).

Recently, monoclonal antibodies broadly reactive with NP or M antigens of influenza A or B viruses have proved useful in enzyme immunoassay procedures, identifying a wide variety of viruses of various subtypes (Walls et al., 1986).

In situ typing of cell culture isolates by fluorescent antibody has been reported (McQuillan et al., 1970; Hockney et al., 1974; Fishaut et al., 1979) but has not been widely used.

DIRECT DEMONSTRATION OF INFLUENZA VIRUSES OR VIRAL PROTEINS IN PATIENTS' SPECIMENS

Because the amplification of virus through cultivation requires at least 24–48 hr, methods for the rapid diagnosis of infection by direct demonstration of virus or viral antigens have received continuing attention but have not yet been perfected.

Electron Microscopy

Direct visualization of influenza in nasopharyngeal secretions or washings has only rarely been accomplished, but two successes reported (Joncas et al., 1969; McLean and Wong, 1984) offer hope for future development of the approach, perhaps by direct ultracentrifugation of specimens onto grids (Hammond et al., 1981). Immunoelectron microscopy has not been effective (Evans and Olson, 1982).

Immunofluorescence

Fluorescent antibody staining of exfoliated respiratory epithelial cells was first used by Liu (1956, 1961) for the diagnosis of influenza. In Japan, Tateno-(Ebisawa) and his colleagues (1962, 1966a) extended Liu's studies to include observations on pathogenesis (Tateno et al., 1965) and showed that cells exhibiting specific fluorescence were present in 89% of serologically proven cases of H2N2 influenza (Tateno, 1966b). False positive tests in seronegative patients have been a problem (Ebisawa et al., 1968; Thomson et al., 1972) , and skilled microscopy is required. General application of influenza diagnosis by immunofluoresence microscopy has been frustrated by a shortage of antisera of high quality. However, promising results have been obtained with selected mouse monoclonal antibodies for the NP antigen (McQuillan et al., 1985), and in the diagnosis of influenza B (Pothier et al., 1986).

Enzyme Immunoassay

Influenza virus has been demonstrated in nasopharyngeal washes using enzyme immunoassay systems (Berg et al., 1980; Harmon and Pawlik, 1982). In one study it was comparable to immunofluorescence in specificity and sensitivity (Sarkkinen et al., 1981). Comparison of several rapid diagnostic methods by Evans and Olson (1982) showed immunofluoresence to be superior to immunoassay in sensitivity. Maximal sensitivity, comparable to conventional virus isolation, was provided by a tissue culture amplification method in which short-term cultures were stained by immunofluorescent antibodies.

MEASUREMENT OF SPECIFIC ANTIBODY RESPONSE

Increase in serum antibody response during influenza is usually demonstrable no earlier than the end of the first week of illness. Therefore, specific response to infection diagnostic of influenza can be documented by comparative measurement of antibody in serum obtained during the acute phase and in convalescent-phase serum obtained 10 days or more following the onset of illness. (For details on the kinetics of the immune response, see Chapter 7.) Specific antibody to the HA, NA, NP or M protein can be measured by the tests described in the following sections. A single serum specimen is inadequate for diagnosis because the mere presence of antibody permits no valid inference about when antibody to these ubiquitous viruses may have been acquired (Table 8-1).

Hemagglutination Inhibition

Neuraminic-acid-containing receptors for influenza viruses are present on the erythrocytes of many species. In the presence of virus, cells to which virus is bound are agglutinated in the hemagglutination reaction. Just as this reaction serves as an in vitro measure for the quantitation of serially diluted viruses, its inhibition can be used to measure specific antibody in serum. Because serum may contain nonspecific inhibitors to which virus may bind, prior treatment of serum with a receptor-destroying enzyme (RDE) is necessary before it is used in HI tests.

The HI test is the simplest and most commonly used technique for the measurement of antibody to influenza virus and, because it detects principally antibody to the hemagglutinin that mediates viral attachment, usually is concordant

Table 8-1. Principal Serological Diagnostic Tests for Influenza

Test	Viral antigen	Principle	Specificity
HI (hemagglutination inhibition)	HA	Ab[a] inhibits agglutination of RBC by virus	Strain and subtype
Neutralization	HA	Ab inhibits infection of host cells by virus	Strain and subtype
CF (complement fixation)	NP	Ab fixes complement and inhibits RBC lysis	Type
SRH (single radial hemolysis)	HA NA	Hemolysis of virus-coated RBC by complement (coat RBC with antigenically hybrid reassortant)	Strain and subtype Strain and subtype
ELISA (enzyme-linked immunoassay)	HA NA NP M	Ab bound to virus or antigen-coated plates is detected by assay for enzyme conjugated with anti-FC ab	Dependent on antigen used
NI (neuraminidase inhibition)	NA	Ab inhibits release by NA of sialic acid from high-molecular-weight substrate	Strain and subtype

[a]Antibody

with neutralization tests. However, the test can be influenced by the binding avidity of the virus and by the species of RBC employed (Kilbourne et al., 1951). Antibody to viral NA can sterically block the access of the HA of the test virus to RBC receptors and thereby inhibit hemagglutination (Kilbourne et al., 1968a). Indeed, certain reassortant viruses can be used in HI for measurement of neuraminidase antibody (Kilbourne et al., 1968b). Therefore, antigenically hybrid reassortants carrying NA antigens alien to human experience are required for the specific measurement of anti-HA response. This point was especially critical in 1968, when the Hong Kong H3N2 virus appeared in a population primed to the N2 antigen. There is increasing evidence that the detection of HI antibody response is influenced by host origin of virus used as reagent in the reaction. Especially in the case of influenza B, cell culture growth antigens appear to be superior to egg-grown virus in the demonstration of antibody response (Lathey et al., 1986; Schild et al., 1983; Robertson et al., 1985).

As with any twofold dilution series, titer increases of fourfold or more are considered significant.

Neutralization

Measurement of antibody by neutralization is inconvenient and requires the ability of the laboratory to propagate virus. Because such titrations require the replication of virus to produce some significant endpoint, such as CPE or plaque formation, several days are required for performance of the test. Usually dilutions of serum are mixed with a standard dose of virus before inoculation, but serum can also be incorporated into agar overlays and added after virus inoculation. With few exceptions, neutralization tests are more specific and usually more sensitive than HI (Gross and Davis, 1979; Zakay-Rones et al., 1980).

Complement Fixation

Complement fixation (CF) tests for influenza can be carried out by any laboratory familiar with the procedure, employing antigen preparations that include the viral NP. Because the NP antigens of all influenza A subtypes are cross reactive, the test is type specific. However, subtype- or strain-specific CF tests have been employed using intact virus (Lief et al., 1958). Complement fixation tests are relatively insensitive in detection of antibody response.

Single Radial Hemolysis

The single radial hemolysis (SRH) test is carried out in agarose gels employing RBC to which virus has been adsorbed and complement has been added. In this system, test serum specimens are added to punched-out wells. Incubation of these "immunoplates" leads to complement-mediated lysis of cells to which antibody is attached; thus, the radius of the resultant hemolyzed zone is proportional to the concentration of antibody (Schild et al., 1975).

The test compares favorably to HI in sensitivity (Farrohi, 1977; Al-Khayatt et al., 1984) and exceeds it in specificity (Pettersson, 1980). The SRH test was more sensitive than HI in detecting small amounts of antibody to influenza B and H1N1 virus (Mancini et al., 1983).

Enzyme Immunoassay

Enzyme-linked immunosorbant assays (ELISA) have recently been introduced for the diagnosis of influenza. Depending on the virus preparation or test antigen used, they can be either type or subtype specific. When ELISAs are carried out with purified NP (Masihi and Lange, 1980) or with a mixture of viral proteins released from infected cells (Bishai and Galli, 1978; Hammond et al., 1980), the test, like conventional CF tests, measures type-specific antibody directed chiefly against the NP antigen. Sensitivity is superior to CF, however. With purified antigens, ELISA can be HA or NA (subtype) specific (Murphy et al., 1981; Kahn et al., 1982). These ELISA assays were more sensitive than HI or NI tests, respectively. An ELISA using whole-virus vaccine as antigen was superior in sensitivity to CF, HI, and neutralization (Madore et al., 1983; Van Voris et al., 1985). Antibody response to M protein in ELISA was dissociated from response to HA or NA antigens (Kahn et al., 1982). When whole viruses were used as ELISA antigens, subtype specificity was not observed, and cross reactions between influenza A and B antibody were found (Julkunen et al., 1985).

Broad cross reactivity is to be expected with ELISA because the test measures antibodies to all antigenic sites on virion proteins, not just those sites that are functionally or sterically associated with biological activity or availability on the virion surface. For this reason, ELISA antibody tests may be less relevant as a measure of protection. However, ELISA does provide a common system for the comparative measurement of antibody responses to proteins with different functions such as HA and NA.

Neuraminidase Inhibition

Responses to HA and NA surface proteins are not concordant (Kilbourne et al., 1968a). Detection of antibodies to neuraminidase has relied principally on the neuraminidase-inhibition (NI) test, which requires that they sterically block access of neuraminidase to substrate (Fig. 3-12). This test does not measure other NA antibodies that might function in the aggregation of virus particles, etc. As indicated above, measurement of NA antibody by ELISA is more sensitive (Kahn et al., 1982). Antibody to NA can also be measured by HI tests employing certain reassortant viruses susceptible to inhibition by neuraminidase antibody (Kilbourne et al., 1968a,b; Davenport et al., 1970). Single radial hemolysis has also been used to detect NA antibody response (Callow and Beare, 1976; Grilli and Smith, 1983). In general, measurement of neuraminidase antibody response is too costly and imprecise to have a significant place in diagnostic virology. Its importance lies chiefly in epidemiologic and clinical research.

REFERENCES

Al-Khayatt, R , Jennings, R , and Potter, C W., 1984, Interpretation of responses and protective levels of antibody against attenuated influenza A viruses using single radial haemolysis, *J. Hyg (Camb)* **92**:301–312.

Berg, R. A., Yolken, R. H., Rennard, S. I. et al., 1980, New enzyme immunoassays for measurement of influenza A/Victoria/3/75 virus in nasal washes, *Lancet* **1**:851–853.

Beveridge, W. I. B., and Burnet, F. M., 1946, *The Cultivation of Viruses and Rickettsiae in the Chick Embryo. Medical Research Council Special Report Series No. 256*, His Majesty's Stationery Office, London, pp. 1–92.

Bishai, F. R., and Galli, R., 1978, Enzyme-linked immunosorbent assay for detection of antibodies to influenza A and B and parainfluenza type 1 in sera of patients, *J. Clin, Microbiol.* **8**:648–656.

Burnet, F. M., 1940, Influenza virus infections of the chick embryo by the amniotic route. I. General character of the infections, *Aust. J. Exp. Biol. Med.* **18**:353–360.

Callow, K. A., and Beare, A. S., 1976, Measurement of antibody to influenza virus neuraminidase by single radial hemolysis in agarose gels, *Infect. Immun.* **13**:1–8.

Canil, K. A., Pratt, D., Sungu. M. S., et al., 1985, Influenza surveillance: Alternative laboratory techniques for a developing country, *Bull. WHO* **63**:79–82.

Centers for Disease Control, 1982, *Concepts and Procedures for Laboratory Based Influenza Surveillance*, Centers for Disease Control, U.S. Department of Health and Human Services, Washington.

Davenport F. M., Minuse, E., and Hennessy, A. V., 1970, Antibody response to influenza virus enzyme in man, *Arch. Environ. Health* **21**:307–311.

Davies, H. W., Appleyard, G., Cunningham, P., et al., 1978, The use of a continuous cell line for the isolation of influenza viruses, *Bull. WHO* **56**:991–993.

Dowdle, W. R., and Coleman M. T., 1974, Influenza virus, in: *Manual of Clinical Microbiology* E. H. Lennette, E. H. Spaulding, and J. P. Truant, eds., American Society for Microbiology, Washington, pp. 678–685.

Dumitrescu, M. -R., Grobnicu, M., and Zilisteanu, E., 1980, Use of MDCK cell line for isolation of influenza virus strains, *Arch. Roum. Pathol. Exp. Microbiol.* **39**:245–248

Ebisawa, I. T., Takeuchi, Y., and Makino, M., 1968, Immunocytologic and serologic diagnosis of influenza A2, *Jpn. J. Exp. Med.* **38**:225–234.

Evans, A. S., and Olson, B., 1982, Rapid diagnostic methods for influenza virus in clinical specimens: A comparative study, *Yale J Biol. Med.* **55**:391–403.

Farrohi, K., Farrohi, F. K., Noble, G. R., et al., 1977, Evaluation of the single radial hemolysis test for measuring hemagglutinin-and neuraminidase-specific antibodies to H3N2 influenza strains and antibodies to influenza B, *J. Clin. Microbiol.* **5**:353–360.

Fishaut, M., McIntosh, K., and Meiklejohn, G., 1979, Rapid subtyping of influenza A virus isolates by membrane fluorescence, *J. Clin. Microbiol.* **9**:269–273.

Francis, T., Jr., and Magill, T. P., 1937, Direct transmission of human influenza virus to mice, *Proc. Soc. Exp. Biol. Med.* **36**:132–133.

French, E. L., and Dineen, J. K., 1958, Primary isolation of influenza A virus: The optimal conditions for amniotic infection, *Aust. J. Exp. Biol.* **36**:257–266.

Green, I. J., 1962, Serial propagation of influenza B (Lee) virus in a transmissible line of canine kidney cells, *Science* **138**:42–43.

Grilli, E. A., and Smith, A.J., 1983, The use of a radial haemolysis test for neuraminidase antibodies in the diagnosis of influenza A infection, *J. Hyg. (Camb.)* **91**:147–156.

Gross, P. A., and Davis, A. E., 1979, Neutralization test in influenza: Use in individuals without hemagglutination inhibition antibody, *J. Clin. Microbiol.* **10**:382–384.

Hammond, G. W., Smith, S. J., and Noble, G. R., 1980, Sensitivity and specificity of enzyme immunoassay for serodiagnosis of influenza A virus infections, *J. Infect. Dis.* **141**:644–651.

Hammond, G. W., Hazelton, P. R. , Chuang, I., et al., 1981, Improved detection of viruses by electron microscopy after direct ultracentrifuge preparation of specimens, *J. Clin. Microbiol.* **14**:210–221.

Harmon, M. W., and Pawlik, K. M., 1982, Enzyme immunoassay for direct detection of influenza type A and adenovirus antigens in clinical specimens, *J. Clin. Microbiol.* **15**:5–11.

Hirst, G. K., 1945, Direct isolation of influenza virus in chick embryos, *Proc. Soc. Exp. Biol. Med.* **58**:155–157.

Hockney, J., Bellenger, L., and Harper, I. A., 1974, A simple scheme for the early detection of influenza virus type A, *Med. Lab. Technol.* **31**:3–6.

Horsfall, F. L., Jr., 1955, Reproduction of influenza viruses. Quantitative investigations with particle enumeration procedures on the dynamics of influenza A and B virus reproduction, *J Exp. Med* **102**:441-473.

Ignatieva, T V., Demidova, S. A., Polyakova, T. G., et al , 1985, Detection and identification of influenza virus in nasopharyngeal washings by solid-phase radioimmunossay, *Vopr Virusol.* **3**:287-290.

Joncas, J. H., Berthiaume, L , Williams, R , et al , 1969, Diagnosis of viral respiratory infections by electron microscopy, *Lancet* **1**:956-959

Julkunen, I , Pyhala, R., and Hovi, T , 1985, Enzyme immunoassay, complement fixation and hemagglutination inhibition tests in the diagnosis of influenza A and B virus infections. Purified hemagglutinin in subtype-specific diagnosis, *J Virol. Methods* **10**:75-84.

Khan, M W , Bucher, D J Koul, A K., et al., 1982, Detection of antibodies to influenza virus M protein by an enzyme-linked immunosorbent assay, *J Clin Microbiol* **16**:813-820

Kilbourne, E D., 1959, Studies on influenza in the pandemic of 1957-58 III. Isolation of influenza A (Asian strain) viruses from influenza patients with pulmonary complications. Details of virus isolation and characterization of isolates, with quantitative comparison of isolation methods, *J Clin Invest* **38**:266-274

Kilbourne, E D , Anderson H C., and Horsfall, F L., Jr., 1951, Concurrent infection with influenza A and B viruses in a single epidemic, *J. Immunol* **67**:547-558.

Kilbourne, E D., Laver, W G., Schulman, J. L , et al., 1968a, Antiviral activity of antiserum specific for an influenza virus neuraminidase, *J Virol.* **2**:281-288

Kilbourne, E D., Christenson, W. N , and Sande, M., 1968b, Antibody response in man to influenza virus neuraminidase following influenza, *J. Virol* **2**:761-762

Kiyotani, K , Takei, N., and Matsuo, Y., 1984, Fluorometric measurement of neuraminidase activity of influenza viruses, *Hiroshima J. Med Sci.* **33**:287-292.

Lathey, J L , Van Voris, L. P , and Belshe, R B , 1986, Superiority of tissue-culture-grown antigens over egg-grown antigens for serologic diagnosis of influenza B virus infections, *J Med Virol.* **19**:155-159

Lennette, E. H , 1985, *Laboratory Diagnosis of Virus Infections,* Marcel Dekker, New York.

Lief, F S., Ostapiak, M., Fabiyi, A., et al., 1958, Antigenic analysis of influenza viruses by complement fixation. III Rapid identification of new isolates, *J. Immunol.* **81**:478.

Liu, C , 1956, Rapid diagnosis of human influenza infection from nasal smears by means of fluorescein-labelled antibody, *Proc. Soc Exp. Biol Med.* **92**:883-887.

Liu, C , 1961, Diagnosis of influenzal infection by means of fluorescent antibody staining, *Am Rev. Respir Dis.* **83**:130-133.

Madore, H. P., Reichman, R. C., and Dolin, R., 1983, Serum antibody responses in naturally occurring influenza A virus infection determined by enzyme-linked immunosorbent assay, hemagglutination inhibition, and complement fixation, *J. Clin. Microbiol.* **18**:1345-1350

Mancini, G., Donatelli, I., Arangio-Ruiz, G., et al., 1983, Comparison of haemagglutination-inhibition and single radial haemolysis techniques for detecting antibodies to influenza A and B viruses, *J. Hyg. (Camb.)* **91**:157-162

Masihi, K. N., and Lange, W., 1980, Enzyme-linked immunosorbent assay for the detection of influenza type-specific antibodies, *J Immunol Methods* **36**:173-179.

McLean D. M., and Wong, K. K., 1984, *Same-Day Diagnosis of Human Virus Infections,* CRC Press, Boca Raton, Fl., p 57.

McQuillin, J., Gardner, P. S , and McGuckin, R., 1970, Rapid diagnosis of influenza by immunofluorescent techniques, *Lancet* **2**:690-695.

McQuillin, J., Madeley, C. R., and Kendal, A. P., 1985, Monoclonal antibodies for the rapid diagnosis of influenza A and B virus infections by immunofluorescence, *Lancet* **2**:911-914

Meguro, H , Bryant, J D , Torrence, A E., et al , 1979, Canine kidney cell line for isolation of respiratory viruses, *J. Clin. Microbiol* **9**:175-179.

Monto, A. S., Maassab, H. F., and Bryan, E R , 1981, Relative efficacy of embryonated eggs and cell culture for isolation of contemporary influenza viruses, *J. Clin. Microbiol* **13**:233-235.

Murphy, B R., Phelan, M A., Nelson, D L , et al., 1981, Hemagglutinin-specific enzyme-linked immunosorbent assay for antibodies to influenza A and B viruses, *J Clin Microbiol* **13**:554-560.

Nerome, K., and Ishida, M , 1978, The multiplication of an influenza C virus in an established line of canine kidney (MDCK) cells, *J. Gen. Virol.* **39**:179.

Pettersson, R. F., 1980, Correlation between antibody titers to influenza A virus subtypes H1N1 and H3N2 as measured by hemagglutination-inhibition and hemolysis-in-gel tests, *J. Infect. Dis.* **142**:450.

Pathier, P., Denayel, G. A., Prudhomme de Saint Maur, G., et. al., 1986, Rapid diagnosis of influenza B virus infections by immunofluorescence with monoclonal antibodies in nasopharyngeal aspirates, *Ann. Inst. Pasteur* [Virol.] **137E**:215–219.

Quash, G. A., Aymard, M., Fougerouze, J., et al., 1982, A fetuin latex agglutination test for detecting the neuraminidases of myxoviruses in allantoic fluid, *J. Biol. Stand.* **10**:115–124.

Racaniello, R. H., 1979, *Biochemical and Genetic Studies of Influenza B and C Viruses*, doctoral dissertation, City University of New York, New York.

Robertson, J. S., Naeve, C. W , Webster, R. G., et al., 1985, Alterations in the hemagglutinin associated with adaptation of influenza B virus to growth in eggs, *Virology* **143**:166–174.

Rytel, M. W., and Sedmak, G. V., 1973, Plaque formation by clinical isolates of influenza A_2/Hong Kong, *Am. J. Clin. Pathol.* **59**:251–253.

Sarkkinen, H. K., Halonen, P. E., and Salmi, A. A., 1981, Detection of influenza A virus by radioimmunoassay and enzyme-immunoassay from nasopharyngeal specimens, *J. Med. Virol.* **7**:213–220.

Schild, G. C., Pereira, M. S., and Chakraverty, P., 1975, Single-radial-haemolysis: A new method for the assay of antibody to influenza haemagglutinin, *Bull. WHO* **52**:43–50.

Schild, G. C., Oxford, J. S., de Jong, J. C., et al., 1983, Evidence for host cell selection of influenza virus antigenic variants, *Nature* **303**:706–709.

Schulman, J. L., and Kilbourne, E. D., 1969, Independent variation in nature of the hemagglutinin and neuraminidase antigens of influenza virus: Distinctiveness of the hemagglutinin antigen of Hong Kong/68 virus, *Proc. Natl. Acad. Sci. U.S.A.* **63**:326–333.

Smith, W., 1935, Cultivation of the virus of influenza, *Br. J. Exp. Pathol.* **16**:508–512.

Smith, T. F., and Reichrath, L., 1974, Comparative recovery of 1972–1973 influenza virus isolates in embryonated eggs and primary rhesus monkey kidney cell cultures after one freeze–thaw cycle, *Am. J. Clin. Pathol.* **61**:579–584.

Styk, B., Russ, G., and Blaskovic, D., 1971, Use of virus recombinants in the analysis of influenza virus antigens, *Acta Virol.* **15**:433–445.

Tateno, I., Suzuki, S., Kawamura, A., Jr., et al., 1962, Diagnosis of influenza by means of fluorescent antibody technique, *Jpn. J. Exp. Med.* **32**:531–559.

Tateno, I., Suzuki, S., Nakamura, S., et al., 1965, Rapid diagnosis of influenza by means of fluorescent antibody technic. I. Some basic informations. II. Relation between immunocytologic, serological and clinical findings. III. Some new aspects of influenza virus infection revealed by fluorescent antibody technic, *Jpn. J. Exp. Med.* **35**:383–417.

Tateno-Ebisawa, I., Makino, M., and Takeuchi, Y., 1966a, Rapid diagnosis of influenza B with fluorescent antibody, *Jpn. J. Exp. Med.* **36**:301–320.

Tateno, I., Kitamoto, O., and Kawamura, A., Jr, 1966b, Diverse immunocytologic findings of nasal smears in influenza, *N. Engl. J. Med.* **274**:237–242.

Taylor, R. M., 1940, Detection of human influenza virus in throat washings by immunity response in Syrian hamster (*Cricetus auratus*), *Proc. Soc. Exp. Biol. Med.* **43**:541–542.

Thomson, D., Stanford, C. F., and Connolly, J. H., 1972, Fluorescent-antibody diagnosis of influenza-A infection, *Lancet* **2**:826.

Tobita, K., 1975, Permanent canine kidney (MDCK) cells for isolation and plaque assay of influenza B viruses, *Med. Microbiol. Immunol.* **162**:23–27.

Tobita, K., Sugiura, A., Enomoto, C., et al., 1975, Plaque assay and primary isolation of influenza A viruses in an established line of canine kidney cells (MDCK) in the presence of trypsin, *Med. Microbiol. Immunol.* **162**:9–14.

Van Voris, L. P., Betts, R. F., Menegus, M. A., et al., 1985, Serological diagnosis of influenza A/USSR/77 H1N1 infection: Value of ELISA compared to other antibody techniques, *J. med. Virol.* **16**:315–320.

Walls, H. H., Harmon, M. W., Slagle, J. J., et al., 1986, Characterization and evaluation of monoclonal antibodies developed for typing influenza A and influenza B viruses, *J. Clin. Microbiol.* **23**:240–245.

Watanabe, H., and Mackenzie, J. S., 1981, The detection of influenza A virus antigens in cultured cells by enzyme-linked immunosorbent assay, *Arch. Virol.* **67**:31–43.

Zakay-Rones, Z., Margalith, E., Levy, R., et al., 1980, A sensitive plaque inhibition technique for assay of antibodies to influenza virus: Use to detect previous antigenic priming with influenza viruses, *J. Virol. Methods* **1**:355–360.

9

Animal Influenza

Ecology and Disease

INTRODUCTION

Before influenza viruses were linked to human disease in 1933, viruses later recognized to be influenza viruses had been identified as the cause of fowl plague and swine influenza. Recovery of the virus from man depended on the susceptibility of the ferret to infection with human virus, although later the chick embryo became the principal experimental host. Thus, from the first, there was evidence that influenza A viruses could infect both avian and mammalian species. However, fowl plague virus appeared to be different not only in antigenic but in pathogenic character from swine and human viruses, so that the concept of host-specific avian and mammalian influenza viruses developed. That this concept is simplistic and overly restrictive has become apparent as ecological studies of the distribution of viruses in nature have been aided by the application of genetic and immunologic methods to their identification and classification. These efforts have been fired by an intense interest in the possible origin of human pandemic viruses from animal sources.

Limited as the ecological studies have been, both in number and in their emphasis on migratory birds, they have already revealed a Pandora's box of influenza A viruses bearing 13 HA and nine NA antigens in various combinations (Table 9-1).

ECOLOGY OF INFLUENZA VIRUSES

Distribution of Influenza A Viruses in Nature

Influenza virus infections are for the most part acute and cytonecrotizing, and the virus is not known to have mechanisms for endogenous persistence in in-

Table 9-1. Subtypes of Hemagglutinin and Neuraminidase Antigens of Influenza A Viruses[a]

Hemagglutinin subtype	Former designation(s)	Neuraminidase subtype	Former designation
H1	H0,H1,Hsw1	N1	N1
H2	H2	N2	N2
H3	H3,Heq2,Hav7	N3	Nav2,Nav3
H4	Hav4	N4	Nav4
H5	Hav5	N5	Nav5
H6	Hav6	N6	Nav1
H7	Heq1,Hav1	N7	Neq1
H8	Hav8	N8	Neq2
H9	Hav9	N9	Nav6
H10	Hav2		
H11	Hav3		
H12	Hav10		
H13	—[b]		

[a]From WHO Memorandum (1980)
[b]Proposed as new H subtype, avian source (Hinshaw et al., 1982)

fected individuals. Therefore, the maintenance of the virus in nature probably depends on a chain of serial transmission in species in which congregation occurs. This attribute characterizes all the species in which propagated interspecific infections have been demonstrated (Table 9-2). The striking points made by this tabulation are (1) the wide representation of HA and NA antigenic subtypes in avian species, especially in ducks, (2) conversely, the limited number of subtypes recovered from outbreaks in mammals, including man, and (3) the lack of restriction of any HA or NA subtype antigen to human or other mammalian species. It is also noteworthy that the species listed, except for the seal, are in contact with man as domestic animals. However, virus carriage is also frequent in wild fowl and other migratory birds (not shown in table). That there is an important measure of host specificity in animal influenza is evidenced by the infrequent influenza virus infection of other herd animals in contact with man (see the following sections).

Host Range of Influenza Viruses

Animals in Which Serially Propagated Infections Occur

Epizootic disease has been most notable in swine, horses, chicken, and turkeys. Infection in ducks is most often asymptomatic. The seal is the only wild mammal reported to suffer from epidemic influenza. The severity of epidemic disease is highly variable and is dependent on both the strain of virus and the specific host infected.

Animals Subject to Sporadic Infection

The ascertainment of influenza virus infection in animals not obviously subject to epidemics is usually accomplished by serological surveys employing the

Table 9-2. Host Distribution of Natural Propagated Infections[a] with Influenza A Viruses with Respect to HA and NA Representation in the Infecting Viruses[b]

	HA subtypes													NA subtypes								
	H1	H2	H3	H4	H5	H6	H7	H8	H9	H10	H11	H12	H13	N1	N2	N3	N4	N5	N6	N7	N8	N9
Man	+	+	+											+	+						—[d]	
Swine	+		+											+	+							
Horse			+				+													+	+	
Seal				+			+											+		+		
Turkey	+			+	+	+	+		+	+				+	+	+	+			+	+	
Chicken			+	+	+	+	+			+				+	+					+	+	+
Duck[c]	+	+	+	+	+	+	+	+	+	+	+	+		+	+		+	+	+	+	+	+

[a]In species in which epidemic, epizootic, or enzootic infections commonly occur
[b]Johnson et al , 1977, Hinshaw and Webster, 1979, Halvorson et al , 1979, 1983, 1985, Hinshaw et al , 1981a,b, Webster et al , 1975, Zhezmer et al , 1973, Shortridge, 1979, WHO Memorandum, 1981, Pysina et al , 1978, Bucher et al , 1980
[c]Feral and domestic
[d]Serological evidence of H3N8 exposure, circa 1900, in man (Fedson et al , 1972, Zilske et al , 1981)

HI test. Because this test may yield false positive results because of nonspecific inhibitors in animal sera, titer increases should be demonstrated or additional tests should be used to verify infection. Nakamura and Easterday (1967) differentiated nonspecific from specific inhibition by challenge with live virus aerosols and observing for secondary antibody response. By this means they showed accelerated secondary responses in turkeys and swine known to have been involved in influenza outbreaks in contrast to primary responses in cattle and sheep not primed by natural infection. Kilbourne and Kehoe (1976) verified infection of domestic New York City dogs by H3N2 virus by demonstrating serum antibody against both HA and NA antigens, as was also done in a survey of north German deer (Ehrengut et al., 1980). Lu et al. (1982) emphasize that conventional HI tests performed with intact virus as antigen may not detect antibodies to avian influenza viruses, and, therefore, the detection of infection of mammals by avian viruses may have been underestimated in the past.

Table 9-3 represents a sampling of cases of apparently sporadic influenza virus infections in a variety of animals for which there is credible evidence of infection from virologic and serological studies. Evidence for canine infection, in some cases with respiratory symptoms, is strong. Natural transmission to cats has not been reported, but this species, as well as dogs (Nikitin et al., 1972), can be artifically infected (Paniker and Nair, 1970).

In ungulates, rare instances of influenza in cattle have been reported (Table 9-3). Goats in West Bengal showed serological evidence of infection with H3N2 virus (Graves and Oppenheimer, 1975).

Demonstration of infection of mink and bears has been in captive or domesticated animals. It is notable that bears in which H3N2 virus infections occurred were symptomatic and involved in epizootic disease (Cretescu et al., 1970). Reported isolations of influenza viruses from whales (Lvov et al., 1978, 1979; Hinshaw et al., 1986) and bats are remarkable for being the only recoveries of virus from wild mammals except for the epidemic infections in seals. Serological evidence of infection of wild deer with both H3N2 and H1N1 human viruses also stands out as an unusual observation in the ecology of influenza. Active search for viruses in other feral species is necessary to obtain a balanced ecological picture of the disease.

Animals Susceptible to Laboratory Infection

Virtually every small laboratory mammal with the exception of the rabbit has proved susceptible to pulmonary infection with influenza A viruses. This includes subhuman primates, mice, rats, hamsters, and guinea pigs. Discussion of the nature of infection in the mouse, ferret, and certain primates follows.

Host Specificity of Influenza Viruses

As shown by the older nomenclature in Tables 2-1 and 9-1, influenza viruses were formerly classified as "swine" (SW), "avian," etc., according to the host from which they were originally isolated. Initially, this conventional approach

Table 9-3. Animals Apparently Subject to Sporadic Infection with Influenza Viruses in Nature

Animal species	Virus	Probable source	Virus isolation	Serological evidence	Disease	Reference
Dog	A/H3N2	Man	–	+	—	Nikitin et al (1972)
	A/H3N2	Man	–	+	"Kennel cough"	Bibrack (1975)
	A/H3N2	Man	+	+	Resp symptoms	Romarary et al (1975)
	A/H3N2	Man	+	0	NS[a]	Chang et al (1976)
	A/H3N2	Man	–	+[b]	Unknown	Kilbourne and Kehoe (1976)
	B	Man	+	–	NS	Chang et al (1976)
Cattle	A/H3N2	?	+	–	Resp symptoms	Fatkhuddinova et al (1973)
	A/H3N2	?	–	+	NS	Molibog et al (1975)
	A/H?N2	?	+	–	"Epizootic cough"	Tanyi et al (1974)
	A/SW/H1N1	?	+	?	?	Barb et al (1962)
Other ruminants (see text)						
Bat	A/H3N2	?	+	–	?	Lvov et al (1979)
Bear[c]	A/H3N2	Man	+	+	Resp symptoms (enzootic)	Cretescu et al (1970)
Deer	A/H3N2	?	–	+[b]	NS	Ehrengut et al (1980)
	A/H1N1	?	–	+[b]	NS	
Whale		Birds	+		—	Lvov et al (1978)
Mink[c]	A/H3N2	Man	–	+	NS	Reviewed by
	A/H1N1	Man	–	+	NS	Okazaki et al (1983)

[a]NS, not stated
[b]Antibody to both HA and NA detected
[c]"In nature," here, refers to presumed natural transmission from man to captive or domesticated animals. Also, see text

to classification made sense because "animal viruses" were shown to be antigenically different from human viruses, although the occurrence of some cross reactivity was noted with swine and human strains from the beginning.

As more strains were recovered from more (particularly avian) species and both HA and NA antigens were studied with sequencing techniques, it became clear that the host of origin approach to classification was simplistic and that increased emphasis should be placed on viral gene composition and structure (see Chapter 2). Furthermore, the present capacity to identify all of the viral genes permits the accurate tracing of interspecific transmission of viruses. Thus, in 1976 the Fort Dix epidemic virus of man was found to be indistinguishable from contemporary strains endemic in American pigs; recently viruses recovered from turkeys were pathogenically as well as chemically "swine" viruses (Hinshaw et al., 1983b).

On the other hand, an H5N9 virus highly pathogenic for turkeys was relatively avirulent for four other avian species, demonstrating the host specificity of this virus with respect to virulence (Slemons and Easterday, 1972). (See also the discussion of virulence in Chapter 4).

The broad representation of the H3 HA in human, equine, and avian prototype strains [H3 and the former Heq2 and Hav7 viruses (Table 9-1)] is consistent with the unusual capacity of the 1968 H3N2 virus to spread from man to a variety of domestic and feral animals. But even within the same HA subtype, antigenic subpopulations can be distinguished that appear to define viruses of different host specificity. H4 avian viruses from different hosts were divided into four groups based on reactivity patterns to a panel of monoclonal antibodies (Fukushi et al., 1982). Some correlation of NP genes with host species has also been reported (Bean, 1984), but efforts to find patterns of species-differentiating receptor specificity have thus far been unavailing except for avian, equine, and human strains within the H3 subtype (Rogers and Paulson, 1983).

At present, no molecular handles or biological traits have been defined that are predictive for host specificity. "Avian viruses" with little molecular change can infect mammalian species in epidemic form (see influenza in seals below). Yet only a limited number of HA, NA subtypes have been identified in mammalian disease (Table 9-2), and in ducks, the melting pot of influenza virus reassortants (Desselberger et al., 1978; Hinshaw et al., 1980a), HA, NA gene combinations appear not to be random (Gardner and Shortridge, 1979).

A genetic property of mammalian strains is their reduced capacity to replicate at 41°C (Giesendorf et al., 1984). It is probable that given the right gene combinations or mutations in critical (i.e., receptor binding or cleavage) regions of the HA, all influenza A viruses are potential pathogens for a wide variety of hosts. To some extent the distribution of viruses may be determined more by geography than by host species (von Hoyningen-Huene and Scholtissek, 1983).

Interspecific Transmission

The collective evidence from other sections of this chapter demonstrates conclusively that transmission of influenza virus infection between species occurs.

At the same time, the frequency of such transmission is unknown but is probably low—except among water fowl.

A relatively stronger barrier against transmission from birds to mammals appears to exist on epizootiological evidence, but the potential for such transmission is demonstrable under laboratory conditions of infection (Hinshaw et al., 1981a). In one case, an avian H2N2 virus proved virulent for mice (Lu et al., 1982). Fowl plague virus has been adapted to replication in human cells in culture (Gerber et al., 1973; Israel, 1980). Fowl plague virus was isolated from the blood of a man with hepatitis (DeLay et al., 1967; Campbell et al., 1970), but neither HI or neutralizing antibody response was shown. In northern Europe during the last few years, transfer of "swinelike" influenza viruses to turkeys has occurred (Andral et al., 1985; Aymard et al., 1985).

Man to Animals

Evidence is strong that H1 and H3 human influenza viruses have spread from man to swine (see below). A largely sporadic but unidirectional flow of viruses down the phylogenetic tree from man to other animals has also been observed, most significantly with H3N2 viruses (Table 9-3). Transmission from man to animal has been noted principally during pandemics and could reflect (1) the density of pandemic infection, (2) a unique pantropicity of pandemic viruses, or, (3) in a more restricted sense, the broad pathogenic potential of the H3 HA (Table 9-3), which in variant form has been demonstrated in equine (Heq2) and avian (Hav7) infections. Because so few species seem to harbor influenza viruses as endemic parasites in nature, this evidence has suggested that man may be the principal host and source of influenza viruses (Kilbourne, 1975). But then, search for these viruses has focused on species in immediate proximity to man (see Fig. 10-6).

Animals to Man

For all the ephemeral and chameleonlike nature of influenza viruses in nature, their adaptation to lower animals appears to be sufficiently restrictive to preclude their transmission to man except in rare instances (see swine and seal influenza, below). Transfer of animal virus antigens, on the other hand, is probably facilitated by reassortment of human and animal viruses, most likely in lower animals more permissive for interspecific infection.

Infection of Animals with Influenza B and C Viruses

Isolation of influenza B virus from dogs (Chang et al., 1976) and a single isolation from horses (Compagnucci et al., 1969) have been reported. Piglets have been experimentally infected with an influenza B virus strain, and neutralizing antibody to influenza B viruses has been found in swine subsequent to outbreaks in man (Takátsy et al., 1967).

Influenza C viruses have been demonstrated in pigs in China (Guo et al., 1983), and serological evidence for infection of other animals has been published (Kawano et al., 1978). Genome analysis of Chinese influenza C virus strains has shown no significant or characteristic differences among strains of human or swine origin but has provided evidence of reassortment (Guo and Desselberger, 1984).

Mongrel dogs have been experimentally infected with influenza C virus (Ohwada et al., 1986).

DISEASE IN COMMONLY INFECTED SPECIES

Influenza in Swine

The Viruses

The virus first identified with influenza in swine is an H1N1 virus thought to be the antigenic prototype of the 1918 pandemic strain. Unless stated otherwise, "swine influenza virus" is a closely related descendant of the virus first isolated by Shope in 1930 and persistent with slight antigenic variation in the United States since that time.

Pigs became naturally infected in 1968 with human H3N2 viruses, some of which have persisted beyond single epidemics. Nerome et al. (1982a) have also reported the isolation of a current H1N1 human virus from swine in Thailand in 1979 and found serological evidence of infection in two of 50 swine examined. Soon after its isolation from man, the WS strain of H1N1 virus was experimentally transmitted to swine by Elkeles (1934).

The only other known human influenza A virus subtype (H2N2) infected and caused disease in pigs after experimental inoculation (Patocka et al., 1958). However, this subtype never became established in swine as did the H1N1 and H3N2 subtypes.

Evidence for influenza C virus infection has been cited above.

The Disease

Like human influenza, swine influenza is an acute disease of sudden onset characterized by paroxysmal cough and lethargy or prostration. Other symptoms are labored or "jerky" breathing, fever, anorexia, weight loss, conjunctivitis, and nasal discharge. Illness lasts only 5 to 7 days and is followed by complete recovery in most cases (Easterday, 1975). Disease varies in severity and is apparently influenced by season, stress, or the occurrence of secondary bacterial infection. Milder infections are comparable to the "filtrate disease" described by Shope (1931) and may be virtually asymptomatic (Nakamura et al., 1972).

Pathology is restricted to the respiratory tract. In experimental infection with both classical swine Hsw1N1 (H1N1) and H3N2 viruses, significant differences in illness or pathology were not observed (Fontaine et al., 1983). Disease can be modified by serum antibody, as is the case in newborn pigs that have acquired passively transferred maternal antibody (Renshaw, 1975). But mild disease is also seen in pigs without detectable antibody prior to infection. Transplacental transmission of both H1N1 and H3N2 influenza viruses resulting in stillbirths has been reported (Gourreau et al., 1985).

In pigs submitted alive for necropsy studies, patchy but clearly demarcated plum-red lung lesions were found, and mediastinal and bronchial lymph nodes were enlarged and edematous and often hyperemic. Histology revealed bronchiolitis and bronchitis and a "serocellular" alveolar reaction (Morin, 1981).

Epizootiology

Following the first appearance of swine influenza in the American Middle West in 1918, the disease has been endemic to the present, recurring in seasonal (fall–winter) local epidemics. Because there are no obvious links among these epidemics, e.g., through transfer of pigs from one farm to another, and because epidemics can occur simultaneously in scattered places, an explanation for latent viral persistence has been sought.

Shope (reviewed in 1955) proposed an intriguing and elaborate mechanism for such persistence, postulating the cryptic retention of virus in the swine lung worm with an intermediate stage in the earthworm. Virus was thought to be reactivated by various stressful stimuli including severe environmental conditions typical of fall and winter seasons. Shope, indeed, provoked disease by stressful stimuli under experimental conditions, but neither he nor any subsequent investigator verified the presence of virus in the lung worm. Subsequent studies have shown that influenza virus infection in swine occurs throughout the year, sometimes asymptomatically, and virus has been recovered from such infections in almost every month of the year (Easterday, 1975). Furthermore, pigs have been shown to be contagious for up to 3 months following infection. Therefore, persistence of virus is readily explained as the result of continuing transmission into and through susceptible animals introduced by birth.

Transmission of virus probably occurs by the respiratory route either by aerosol or direct contact with nasal or pharyngeal secretions. The demonstration of viral antigen in alveolar septa within 4 hr of infection (Easterday, 1975) supports the possibility of aerosol transmission by small droplet nuclei, as may be the case in man (see Pathogenesis, Chapter 7).

Swine Influenza Virus Infection Outside the United States

Swine influenza is now endemic in many parts of Europe, and more recently the virus has been found in the Orient. Recognition of the virus in some countries only since 1976 may reflect concern over the virus as a potential human pathogen, ignited by the Fort Dix, New Jersey outbreak in 1976 (Masurel et al., 1983; Gourreau et al., 1981; Sinnecker et al., 1983; Ogawa et al., 1983; Satsuta et al., 1981).

In most instances the source of virus was not documented, but Nerome et al. (1982b) obtained serological evidence to support the hypothesis that swine viruses that appeared in pigs in Japan in 1977 were introduced by pigs imported from North America.

The swine influenza virus may be said to be widely but not pervasively distributed. Serological surveys have not shown evidence of the virus in India between 1969 and 1973 (Singh et al., 1981), and a survey in Bavaria that detected the presence of H3N2 antibody found no evidence of H1N1 swine virus (Bibrack, 1972).

Kundin (1970) provided the first direct evidence of the interspecies transfer of a human virus to swine by isolating A/Hong Kong/68 (H3N2) influenza virus from naturally infected pigs in Taiwan. Extensive serological surveys of pigs since that time have shown no evidence of antibodies to this virus in pigs prior to 1968, the time of initial appearance of H3N2 viruses in man. Distribution of H3N2 viruses in swine has been widespread, often overlapping with endemic infection with H1N1 swine virus (Gagnon et al., 1974; Aymard et al., 1983; Rao et al., 1983).

Infections of pigs in nature with H3N2 viruses have been largely subclinical, and the virus seems to disappear soon after its introduction. However, the possibility of enzootic infection with this virus remains to be evaluated (Kaplan, 1982). Recent evidence from Belgium suggests the survival in swine of an H3N2 virus antigenically similar to the A/Port Chalmers strain from 1973 to 1984 (Haesebrouck et al., 1985).

The simultaneous circulation in some herds of classical swine and H3N2 viruses has resulted in reassortment, producing a virus with mixed surface antigens (Sugimura et al., 1980), deriving all but the neuraminidase gene from swine H1N1 virus (Nerome et al., 1983).

Minor antigenic variation among classical swine strains has been recognized for years, and strains isolated from ducks are distinguishable from those infecting pigs. It is therefore of interest that evidence for the natural transmission of duck strains to swine has been found (Pensaert et al., 1981). However, a strain of H1N4 virus isolated from wild ducks in Japan did not evoke serological response from experimentally inoculated pigs (Arikawa et al., 1983).

Conversely, H1N1 swine influenza viruses were found to produce epidemics in turkeys without undergoing significant genetic changes (Hinshaw et al., 1983b; C. W. Whitaker and E. D. Kilbourne, unpublished data).

When pig and bird HA genes of H1N1 viruses were analyzed by RNA–RNA hybridization, the HA genes of recent isolates from pigs in northern Europe were found to be genetically more closely related to those of isolates from birds in Europe and North America than to those of isolates from pigs in the United States, Taiwan, and Italy, suggesting that two different H1N1 subtypes are currently circulating in the pig population. Scholtissek et al. (1983) felt these results to be consistent with the view that H1N1 viruses can be transmitted from birds to pigs or vice versa.

Influenza in Horses

The Viruses

Epidemic influenza in horses is caused by either of two discrete subtypes of influenza A viruses, H7N7 (formerly Heq1 Neq1) and H3N8 (formerly Heq2 Neq2). The HA antigens of these viruses are antigenically distinguishable from selected avian (H3, H7) and human (H3) viruses but similar enough to suggest derivation from a common viral ancestor (Chapter 6).

Antigenic Variation

Although some antigenic variation has been seen in H7N7 equine viruses (Powell et al., 1974), the equine H3N8 virus appears to be less stable antigenically. Minor but significant variation in H3N8 strains has been noted (Pereira et al., 1972; Konn et al., 1972; Thomson, 1977; Plateau and Gayot, 1977; Klingeborn and Rockborn, 1980), and recently cocirculation of HA antigenic variants has been described (Hinshaw et al., 1983a). Equine H3 strain showed much less variation than human H3 viruses during the same period (Hinshaw et al., 1983a).

The Disease

The sudden onset of dry cough and moderate to high fever characterizes equine influenza. Respiratory tract symptoms predominate in association with lethargy and loss of appetite. As in man, the disease involves the lower respiratory tract more than the upper, although nasal discharge is not uncommon. Illness lasts for less than a week in most cases (Easterday, 1975; Bryan and Gerber, 1972; Powell et al., 1974).

H3N8 viruses tend to greater virulence than H7N7 viruses, causing higher fever (Gerber, 1969) and sometimes severe pulmonary lesions (reviewed by Smith, 1979). Bacterial pneumonia as a sequel to equine influenza rarely occurs in mature horses unless they are stressed before recovery (Smith, 1979; Easterday, 1975).

The low mortality rate of equine influenza has precluded extensive pathological studies. Gerber (1969) has reviewed the pathology of H3N8 virus infection and cites laryngitis and tracheitis of adults and erosive bronchitis and hyaline membranes in the alveoli of foals. In adults, interstitial and bronchopneumonia can occur.

Epizootiology

Influenza in horses is probably an ancient disease (Chapter 1) but was first linked to type A influenza viruses in 1956 with the isolation of the A equine Prague/156/H7N7 virus during a large epizootic in central and northern Europe (Tumova, 1980; Heller et al., 1956; Sovinova et al., 1958). Since its first appearance, the virus has been endemic in horses in Europe and since 1963 in North America (Tumova, 1980).

The second equine subtype virus (prototype A equine Miami/163/H3N8) was isolated in the United States in 1963 (Waddel et al., 1963) and like H7N7 virus now has virtually a worldwide distribution. The cocirculation of the two equine viruses has led on occasion to simultaneous infections in the same stable (Tumova et al., 1972).

It is most likely that influenza in horses is perpetuated by the frequent mixing of animals from different regions at race tracks and shows. Once a population is seeded with the virus, outbreaks are explosive (Fig. 9-1) (McQueen et al., 1966), probably facilitated by the short 1- to 3-day incubation period and the strong cough, which can spread virus by an aerosol over a distance of 35m (Easterday, 1975). Virus may be introduced by partially immune asymptomatic carriers.

The incidence of epizootics is not seasonal but is influenced principally by factors that bring horses together in large numbers under conditions of crowding.

INFLUENZA IN DOMESTIC FOWL

Chicken

The Viruses

The viral etiology of fowl plague was established long before influenza viruses were isolated from men, swine, or horses (Centanni and Savonuzzi, 1900, cited by Stubbs, 1967). It was Schafer (1955), however, who first identified fowl plague virus (H7N7) as an influenza A virus on the basis of its NP antigen. Since that time at least six different influenza A virus HA and five NA subtypes have been implicated in epizootics of chickens (Table 9-2). Although fowl plague in Europe has been associated principally with viruses with H7 (Hav1) HA, a devastating

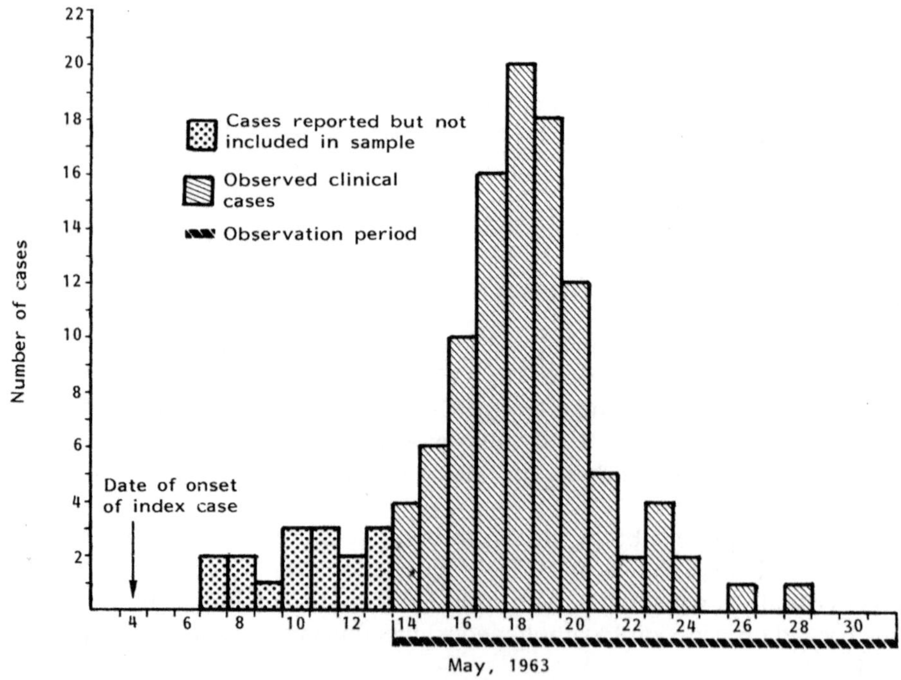

Figure 9-1. An epizootic of H3N8 influenza in horses: 101 of 123 horses were affected. (From McQueen et al., 1966. Reproduced with permission, from *Am. J. Epidemiol.* **83:**283.)

recent epidemic in Pennsylvania was caused by virus of H5N2 subtype (Eckroade et al., 1984). Antigenic type is poorly correlated with virulence in infections with chickens (Allan et al., 1977), although H7 and H5 HA subtypes have been most frequently identified with serious disease. A cleavable HA is a requirement for virulence in infection of fowl (Rott et al., 1984).

The Disease

The extreme variability of disease severity seen with avian influenza in general also characterizes infection of chickens. The malignant epizootic "fowl plague" is attended by extremely high mortality. Much milder disease can occur, as was true at the inception of an outbreak of influenza in chickens in Pennsylvania in 1983. In this outbreak, caused by an H5N2 virus, severe disease later appeared as the virus, presumably derived from wild birds, adapted to chickens on passage (Bean et al., 1985). Because fowl plague is not discretely definable except as a severe epizootic of chickens, the term probably should be abandoned, as has been suggested (Bankowski, cited by Acland, 1984).

Classical fowl plague strikes suddenly, often killing chickens in 2 days without obvious localizing signs or symptoms. Birds are weak and prostrated and stagger if they leave the roost. Cyanosis is evident in comb and wattler, and eyelids are closed with conjunctival edema and redness. Fever may rise to 112°F. Edema affects the head and may extend to the glottis, affecting respiration. Mucous exudate is present in nostrils and pharynx, further compromising breathing. Watery diarrhea also occurs. Convulsions, ataxia, and blindness reflect involvement of the central nervous system (Stubbs, 1967). Virus can be recovered from virtually all organs of the body post-mortem.

Acland et al. (1984) emphasize that there have been no detailed descriptions of pathology in chickens with naturally acquired influenza. In their studies of the 1983 Pennsylvania outbreak, they noted dehydration and shrinkage of breast muscles in broilers (young chickens), diffuse petechiae, and atrophy of the thymus and bursa of Fabricius. Grossly, layers (older chickens) were distinguished by edema and cyanosis of the comb (see above). The distribution of histological lesions was similar in the two age groups, showing evidence of encephalitis, myocarditis, pancreatitis, myositis, sinusitis, and tracheitis, but encephalitis was more severe and comb inflammation less extreme in young chickens.

Epizootiology

The severe infection is often contained by a self-extinguishing high mortality. The origin of epizootics and mechanism of spread are not understood, although virus is demonstrable in ocular and nasal secretions, pharnyx, urine, and feces. Viral contamination of feed, drinking water, and soil as well as clothing of attendants occurs, but survival of virus is probably brief.

Wild birds might serve as sources and vectors of viruses through close association with domestic flocks. At present, control of epizootics requires the "depopulation" or total eradication of infected birds and the imposition of rigid quarantine.

Turkey

The Viruses

Seven HA and six NA subtypes have been associated with influenza infection of turkeys (Table 9-2).

The Disease

Influenza virus can cause a variety of symptoms in turkeys in various combinations, including decreased activity, anorexia, weight loss, respiratory signs or symptoms, cyanosis, nervous disorders, diarrhea, and decreased egg production (Skeeles et al., 1981). Infection of turkeys with H1N1 swine virus was remarkable in its restriction to the upper respiratory tract (Hinshaw et al., 1983b).

Epizootiology

Virus was recoverable from experimentally infected turkeys up to 29 days after inoculation of virus by establishing tracheal cultures from infected birds but was demonstrable for only 7 days by tracheal swabbings (Robinson and Easterday, 1978).

In recent years, avian influenza has been more of a problem in turkeys than in chickens in North America, possibly reflecting their greater contact with wild avian species. The causation of turkey influenza by multiple antigenic types rather than with a single endogenous virus supports this supposition, as does the reduction of turkey influenza in Canada by measures taken to restrict access of wild birds to turkey flocks (Lang and Ferguson, 1981). Turkeys can be naturally infected with swine influenza viruses, either the antigenic variants resembling A/NJ/76 strains or the A/SW/Fin/2899/82 prototype typical of Northern European strains (Hinshaw et al., 1983b; Andral et al., 1985). Disease was cryptic and evidenced principally by a drop in egg production.

Ducks

The Viruses

Twelve of 13 HA subtypes and all nine NA subtypes have been found in viruses recovered from domestic and feral ducks.

The Disease

The apparent immunity of ducks and geese to severe influenza in associated chicken or turkey flocks has long been noted. Ducks are affected by influenza viruses, however, experiencing catarrhal tracheitis and air sacculitis (Hwang et al., 1970). Severe epizootics are far less common than with other domestic avian species, however, although endemic infections are highly prevalent in ducks. Experimental infection of ducks with duck-derived H3N2 and H3N6 viruses led to asymptomatic fecal shedding of virus in high titer, marginal serum antibody response, and brief immunity to reinfection. Viruses of human origin infected ducks but did not replicate in the intestinal track (Kida et al., 1980; Webster et al., 1978).

Of three challenge viruses virulent for chickens and turkeys, only the Rostock strain of H7N7 virus caused any disease in ducks: facial edema and death in five of 17 birds (Alexander et al., 1978).

Epizootiology

More than any other known source in nature, the cloaca of the duck yields influenza viruses in large numbers from apparently undiseased fowl. At a poultry-dressing plant in Hong Kong, swabs from ducks contained six HA subtypes in various combinations with NA subtypes (Shortridge et al., 1977). The extent of genetic diversity among avian influenza viruses was revealed by examination of isolates of the single antigenic subtype H3N8 by competitive hybridization of viral RNA segments. Gene differences suggesting frequent gene reassortment were found even among concurrent isolates from one geographical area (Sriram et al., 1980).

ANIMALS SUBJECT TO SPORADIC INFECTION

Of the animals closely associated with man, only a few appear to be susceptible to human influenza viruses (Fig. 10-6). Dogs (but not cats), ranch mink, cattle, and zoo animals have also been sporadically infected (Table 9-3). Infection of feral bats and whales (Lvov et al., 1978, 1979) and deer (Ehrengut et al., 1980) has been reported. The epizootiology of these infections of wild species is obscure.

It is not clear whether the preponderance of H3N2 virus infections documented in Table 9-3 reflects the unique pantropicity of this viral subtype or increasing study of influenza in animals. However, concurrent inoculation of H2N2 and H3N2 viruses in gibbons resulted in spread only of the H3N2 virus throughout the colony (Johnsen et al., 1971).

EPIZOOTIC AND ENZOOTIC INFLUENZA IN WILD MAMMALS AND BIRDS

Influenza in Migratory and Other Birds

Attention was directed to wild birds as potential sources of influenza virus by a severe epizootic in South African terns in 1961 (Becker, 1963). Since then, scores of isolates with a variety of HA,NA serotypes have been obtained from migratory wild fowl, especially ducks.

Intensive studies in the Soviet Union (reviewed by Lvov and Zhdanov, 1983) and North America (reviewed by Hinshaw et al., 1985) have shown that apparently healthy water fowl carry a large variety of influenza A virus subtypes in their lower intestinal tracts (see ducks, Table 9-2) and that these viruses are recoverable from waters in which they congregate. Therefore, these birds, some of which mingle with domestic flocks, provide a migratory gene pool of viruses available for transmission to barnyard fowl or other species (Webster and Laver, 1972). A 3-year survey in Alberta, Canada demonstrated 27 different combinations of HA and NA subtypes, chiefly in mallards, pintails, and bluewinged teals

(Hinshaw et al., 1980c). Predominant viral subtypes differed from year to year and in different flyways in the same year (Hinshaw et al., 1985). Dual infection of ducks occurs in the natural setting (Hinshaw et al., 1980b; King and Shortridge, 1982) and can result in viral genetic reassortment in vivo (Desselberger et al., 1978; Hinshaw et al., 1980b).

The extent of infection in other avian species is unknown, but influenza viruses have been recovered from imported pet birds, including parakeets and mynahs, that were inapparently infected (Nishikawa et al., 1977) or dead on arrival following shipment (Nerome et al., 1978). Whereas infection of water fowl involves both respiratory and intestinal tracts, experimental infection of budgerigars showed respiratory tract infection only (Kawano et al., 1979). Isolation of an influenza A virus from a bird embryo in Czechoslovakia suggests the possibility of vertical transmission of infection (Gresikova et al., 1979).

Epizootic Influenza in Seals

Two recent epizootics of influenza in harbor seals have been of great interest because they suggest that not only can transmission of "avian" viruses to mammals occur, but that it can lead to propagated infection.

A severe outbreak with pneumonia and significant mortality occurred in New England in 1979–80 from which an H7N7 virus was recovered (Geraci et al., 1982). This virus was similar antigenically and in RNA migration patterns to the classical A/fowl plague/Dutch/77 H7N7 virus but differed in HA gene sequence from A/fowl plague/Rostock/34 (H7N1) at three sites (Naeve and Webster, 1983). Although interesting, this comparison of viruses so different in passage history and year of isolation does not allow conclusions about the relevance of the molecular changes to the pathogenicity of the viruses. The seal H7N7 virus replicated better in mammals than in domestic fowl in experimental infection and caused conjunctivitis in man (Geraci et al., 1982; Webster et al., 1981a). Notably, virus was not shed in the feces of experimentally infected birds and therefore was biologically dissimilar from typical avian viruses (Webster et al., 1981b).

From another outbreak in Massachusetts in 1982 and 1983, another virus previously found only in birds (H4N5) was isolated. This virus also replicated in laboratory mammals but did infect the intestinal tract of the duck, identifying it as an avian virus in pathogenic potential.

It is not clear whether or not the seal viruses represent an intermediate step in the evolution of new mammalian strains (Hinshaw et al., 1984) or are enzootic in seals. Intermingling of marine animals and sea birds occurs commonly, and interspecific infection could occur from direct contact or from guano-contaminated water (Lvov and Zhdanov, 1983). Transmission from contaminated water might explain reported infections of whales with influenza viruses (Lvov et al., 1978; Hinshaw et al., 1986).

INFLUENZA IN LABORATORY ANIMALS

The affinity of influenza viruses for tissues of the respiratory tract of their natural hosts is found also in laboratory-induced infection in other species. Thus, ferrets and mice, not natural hosts of influenza viruses, have provided excellent models for studies of the pathogenesis of human influenza. Many aspects of these studies, as they complement studies in man, are discussed in Chapter 7.

In inbred mice, studies of influenza virus infection have implications for the basic understanding of immunologic mechanisms that go far beyond the elucidation of influenza. The basis of genetic interferon-mediated resistance to the virus in a single mouse strain has been intensively studied (Haller et al., 1979; Staeheli et al., 1984). Study of MHC-mediated recognition of influenza virus-infected cells has increased understanding of viral pathogenesis in general. Most recently, the influenza virus system in mice has been used in the analysis of X-linked genetic defects in humoral immunologic response (Reale et al., 1985).

Of the smaller laboratory animals, guinea pigs (Wetherbee, 1973; Azoulay-Dupuis et al., 1984), hamsters (Taylor, 1940; Kilbourne et al., 1951; Jennings et al., 1976), newborn rats (Michaels et al., 1978), and even chinchillas (Abramson et al., 1981) are susceptible, and in all of these, infection is pneumotropic.

Influenza virus infection of nonhuman primates can occur in nature (O'Brien and Tauraso, 1973; Romváry et al., 1976), and a number of species, including chimpanzees and baboons (Kalter and Heberling, 1978), are subject to contact infection from man. Experimental infections were transmitted to other animals in close proximity (Kalter et al., 1969), notably in gibbons (Johnsen et al., 1971) following introduction of A/Hong Kong (H3N2) virus. Owl, cebus, and squirrel monkeys support replication of H3N2 and H1N1 human viruses. The mild illness manifested by squirrel monkeys simulates human infection, thus providing a useful model for genetic studies of virulence (Murphy et al., 1980).

REFERENCES

Abramson, J S , Giebink, G S , Mills, E. L , et al., 1981, Polymorphonuclear leukocyte dysfunction during influenza virus infection in chinchillas, *J Infect Dis* **143**:836–845

Acland, H M , Silverman-Bachin, L. A , and Eckroade, R J., 1984, Lesions in broiler and layer chickens in an outbreak of highly pathogenic avian influenza virus infection, *Vet. Pathol.* **21** 564–569.

Alexander, D J , Allan, W. H , Parsons, D. G , et al., 1978, The pathogenicity of four avian influenza viruses for fowls, turkeys and ducks, *Res. Vet Sci.* **24** 242–247

Allan, W. H , Alexander, D J , Pomeroy, B S, et al., 1977, Use of virulence index tests for avian influenza viruses, *Avian Dis* **21**:359–363

Andral, B., Toquin, D , Madec, F., et al , 1985, Disease in turkeys associated with H1N1 influenza virus following an outbreak of the disease in pigs, *Vet. Rec.* **116**:616–618.

Arikawa, J., Yamane, N., Totsukawa, K., et al., 1983, Antigenic and genetic characterization of H1 influenza viruses isolated from feral ducks and swine in Japan, *Arch Virol* **78** 19–27.

Aymard, M , Douglas, A -R , Fontaine, M., et al , 1983, Epizootie de grippe porcine dans les élevages de Bretagne, *Rev Epidemiol Santé Pub* **31**:311–327.

Aymard, M , Douglas, A. R., Fontaine, M , et al., 1985, Antigenic characterization of influenza A (H1N1) viruses recently isolated from pigs and turkeys in France, *Bull WHO* **63** 537–542

Azoulay-Dupuis, E., Lambre, C R , Soler, P , et al., 1984, Lung alterations in guinea-pigs infected with influenza virus, *J Comp Pathol* **94**:273–282.

Bankowski, R A , 1981, Introduction and objectives of the symposium, in: *Proceedings of the First International Symposium on Avian Influenza,* The United States Animal Health Association, Richmond, Va , pp vii–xiv

Barb, K , Farkas, E , Romváry, J , et al , 1962, Comparative study of influenza virus strains isolated from domestic animals in Hungary, *Acta Virol* **6**:207–213

Bean, W. J., 1984, Correlation of influenza A virus nucleoprotein genes with host species, *Virology* **133:**438–442.

Bean, W. J., Kawaoka, Y., Wood, J. M., et al., 1985, Characterization of virulent and avirulent A/chicken/Pennsylvania/83 influenza A viruses: Potential role of defective interfering RNAs in nature, *J. Virol.* **54:**151–160.

Becker, W. B., 1963, The morphology of tern virus, *Virology* **20:**318.

Bibrack, B., 1972, Vergleichende serologische Untersuchungen über das Vorkommen von Schweineinfluenza und Influenza A2/Hongkong-Infektionen bei Schweinen in Bayern, *Zentralbl. Vet. Med. B* **19:**397–405.

Bibrack, B., 1975, Serologische Untersuchungen über die Mitbeteiligung von Influenza A2/Hongkong-Infektionen beim Zwingerhusten des Hundes, *Zentralbl. Vet. Med. B* **22:**28–35.

Bryan, J. T., and Gerber, H., 1972, Equine influenza in: *Equine Medicine and Surgery*, 2nd ed. (E. J. Catcott, and J. F. Smithcors, eds.), American Veterinary Publishers, Wheaton, IL, pp. 17–22.

Bucher, D. J., Kharitonenkov, I. G., Lvov, D. K., et al., 1980, Comparative study of influenza virus H2 (avian) hemagglutinins isolated from human and avian sources, *Intervirology* **14:**69–77.

Campbell, C. H., Webster, R. G., and Breese, S. S., Jr., 1970, Fowl plague virus from man, *J. Infect. Dis.* **122:**513–516.

Campbell, C. H., Easterday, B. C., and Webster, R. G., 1977, Strains of Hong Kong influenza virus in calves, *J. Infect. Dis.* **135:**678–680.

Chang, C. P., New, A. E., Taylor, J. F., et al., 1976, Influenza virus isolations from dogs during a human epidemic in Taiwan, *Int. J. Zoon.* **3:**61–64.

Compagnucci, M., Martone, F., and Bonaduce, A., 1969, Influence of influenza myxovirus type B on the horse. Report 1. Characteristics of the isolated strains, *Boll. Inst. Sier. (Milan)* **48:**305–316.

Cretescu, L., Zilisteanu, E., Niculescu, I., et al., 1970, Influenza-Enzootie bei Bären des zoologischen Gartens Bukarest durch ein A2/Hong-Kong/1/68-ähnliches Virus. II. Untersuchungen des isolierten Virusstammes, in: *Verhandlungsbericht des XI. Internationalen Symposiums uber die Erkankungen der Zootiere, Budapest,* Akademie-Verlag, Berlin, pp. 197–199.

DeLay, P. D., Casey, H. L., and Tubiash, H. S., 1967, Comparative study of fowl plague virus and a virus isolated from man, *Public Health Rep.* **82:**615–620.

Desselberger, U., Nakajima, K., Alfino, P., et al., 1978, Biochemical evidence that "new" influenza virus strains in nature may arise by recombination (reassortment), *Proc. Natl. Acad. Sci U.S.A.* **75:**3341–3345.

Easterday, B. C., 1975, Animal influenza, in: *The Influenza Viruses and Influenza* (E. D. Kilbourne, ed.), Academic Press, New York, pp. 449–481.

Eckroade, R. J., Silberman, L. A., and Acland, H. M., 1984, Avian influenza in Pennsylvania in: *Proceedings 33rd Western Poultry Disease Conference,* Davis, Ca, pp. 1–2.

Ehrengut, W., Sarateanu, D. E., and Rutter, G., 1980, Influenza A antibodies in cervine animals, *Infection* **8:**66–69.

Elkeles, G., 1934, Mededeelingen uit het Inst. Praev. Geneesk., Leiden.

Fatkhuddinova, M. F., Kiryanova, A. I., Isachenko, V. A., et al., 1973, Isolation and identification of influenza A virus in respiratory diseases of cattle, *Vop. Virusol.* **4:**474–478.

Fedson, D. S., Huber, M. A., Kasel, J. A., et al., 1972, Presence of A/Equi-2 hemagglutinin and neuraminidase antibodies in man (36245), *Proc. Soc. Exp. Biol. Med.* **139:**825–826.

Fontaine, M., Fontaine, M., Gastellu, J., et. al., 1983, Grippe porcine experimentale. Comparaison du pouvoir pathogène entre souches sauvages de virus grippal de type swine et souches de type humain H3 N2 dont le porc est réservoir, *Ann. Rech. Vet.* **14:**79–88.

Fukushi, H., Yanagawa, R., and Kida, H., 1982, Host species-related antigenic groups of avian influenza viruses possessing H4 hemagglutinin revealed by monoclonal antibodies, *Arch. Virol.* **72:**217–221.

Gagnon, A. N., Boudreault, A., Marsolais, G., et al., 1974, Maladies procines à étiologie virale dans la province de Québec. I. Influenza, *Can. Vet. J.* **15:**312–315.

Gardner, I. D., and Shortridge, K. F., 1979, Recombination as a mechanism in the evolution of influenza viruses: A two-year study of ducks in Hong Kong, *Rev. Infect. Dis.* **1:**885–890.

Geraci, J. R., St. Aubin, D. J., Barker, I. K., et al., 1982, Mass mortality of harbor seals: Pneumonia associated with influenza A virus, *Science* **215:**1129–1131.

Gerber, A., Sauter, C., and Lindenmann, J., 1973, Fowl plague virus adapted to human epithelial tumor

cells and human myeloblasts *in vitro*. II. Replication of human leukemic myeloblast cultures, *Arch. Ges Virusforsch.* **40**:255–264.
Gerber, H., 1969, Clinical features, sequelae, and epidemiology of equine influenza, in *Proceedings International Conference on Equine Infectious Diseases, 2nd, 1969,* S Karger, Basel, pp 63–79
Giesendorf, B., Bosch, F X., Wahn, K , et al , 1984, Temperature sensitivity in maturation of mammalian influenza A viruses, *Virus Res* **1**:655–667.
Gourreau, J. M , Hannoun, C , and Kaiser, C , 1981, Diffusion du virus de la grippe du porc (H1N1 = Hsw1N1) en France, *Ann. Virol* **132 E**:287–294
Gourreau, J. M., Kaiser, C., Madec, F , et al , 1985, Transplacental passage of influenza virus in the sow under natural conditions, *Ann Inst Pasteur* [Virol.] **136E**:55–63
Graves, I L., and Oppenheimer, J R., 1975, Human viruses in animals in West Bengal An ecological analysis, *Hum Ecol* **3**:105–130
Grešíková, M. Sekeyová, M , Tumová, B , et al , 1979, Isolation of an influenza A virus strain from a bird embryo (*Larus ridibundus*) collected in Slovakia, *Acta Virol* **23**:89–92
Guo, Y., and Desselberger, U., 1984, Genome analysis of influenza C viruses isolated in 1981/82 from pigs in China, *J Gen Virol* **65**:1857–1872
Guo, Y J., Jin, F G , Wang, P , et al , 1983, Isolation of influenza C virus from pigs and experimental infection of pigs with influenza C virus, *J Gen. Virol,* **64**:177–182
Haesebrouck, F , Biront, P , Pensaert, M. B et al , 1985, Epizootics of respiratory tract disease in swine in Belgium due to H_3N_2 influenza virus and experimental reproduction of disease, *Am J Vet Res* **46**:1926–1928
Haller, O , Arnheiter, H , Gresser, I , et al , 1979, Genetically determined interferon-dependent resistance to influenza virus in mice, *J Exp Med* **149**:601–612
Halvorson, D A., Karunakaran, D , and Newman, J. A., 1979, Avian influenza in caged laying chickens, *Avian Dis* **24**:288–294.
Halvorson, D , Karunakaran, D , Senne, D., et al., 1983, Epizootiology of avian influenza—simultaneous monitoring of sentinel ducks and turkeys in Minnesota, *Avian Dis* **27**:77–85
Halvorson, D A , Kelleher, C. J , and Senne, D. A , 1985, Epizootiology of avian influenza: Effect of season on incidence in sentinel ducks and domestic turkeys in Minnesota, *Appl Environ. Microbiol* **49**:914–919
Heller, L , Espmark, A., and Viridén, P., 1956, Between two influenzas, *Arch Ges. Virusforsch* **7**:120
Hinshaw, V S., and Webster, R. G , 1979, Characterization of a new avian influenza virus subtype and proposed designation of this haemagglutinin as Hav10, *J Gen Virol.* **45**:751–754.
Hinshaw, V S , Bean, W. J., Webster, R. G , et al , 1980a, Genetic reassortment of influenza A viruses in the intestinal tract of ducks, *Virology* **102**:412–419.
Hinshaw, V S , Webster, R G., Bean, W. J., et al., 1980b, The ecology of influenza viruses in ducks and analysis of influenza viruses with monoclonal antibodies, *Comp Immunol Microbiol Infect Dis* **3**:155–164
Hinshaw, V. S , Webster, R G , and Turner, B , 1980c, The perpetuation of orthomyxoviruses and paramyxoviruses in Canadian waterfowl, *Can. J. Microbiol* **26**:622–629.
Hinshaw, V. S., Webster, R. G., Easterday, B. C., et al., 1981a, Replication of avian influenza A viruses in mammals, *Infect. Immun* **34**:354–361.
Hinshaw, V S., Webster, R. G , and Rodriguez, R. J., 1981b, Influenza A viruses: Combinations of hemagglutinin and neuraminidase subtypes isolated from animals and other sources, *Arch Virol* **67**:191–201.
Hinshaw, V. S., Air, G M., Gibbs, A. J , et al , 1982, Antigenic and genetic characterization of a novel hemagglutinin subtype of influenza A viruses from gulls, *J. Virol* **42**:865–872
Hinshaw, V.S , Naeve, C W , Webster, R. G., et al., 1983a, Analysis of antigenic variation in equine 2 influenza A viruses, *Bull. WHO* **61**:153–158.
Hinshaw, V. S., Webster, R G , Bean, W. J., et al., 1983b, Swine influenza-like viruses in turkeys Potential source of virus for humans? *Science* **220**:206–208
Hinshaw, V S , Bean, W. J , Webster, R G , et al., 1984, Are seals frequently infected with avian influenza viruses? *J Virol.* **51**:863–865
Hinshaw, V S , Bean, W J , Geraci, J., et al , 1986, Characterization of two influenza A viruses from a pilot whale, *J Virol* **58**:655–656
Hinshaw, V. S , Wood, J M , Webster, R. G , et al., 1985, Circulation of influenza viruses and paramyx-

oviruses in waterfowl: Comparison of different migratory flyways in North America, *Bull. WHO* **63**:711–719.

Hwang, J., Lief, F. S., Miller, C. W., et al., 1970, An epornitic of type A influenza virus infection in ducks, *J. Am. Vet. Med. Assoc.* **157**:2106–2108.

Israël, A., 1980, Genotypic and phenotypic characterization of a mammalian cell-adapted mutant of fowl plague virus (FPV), *J. Gen. Virol.* **51**:33–44.

Jennings, R., Denton, M. D., and Potter, C. W., 1976, The hamster as an experimental animal for the study of influenza. I. The role of antibody in protection, *Med. Microbiol. Immunol.* **162**:217–226.

Johnsen, D. O., Wooding, W. L., Tanticharoenyos, P., et al., 1971, An epizootic of A_2/Hong Kong/68 influenza in gibbons, *J. Infect. Dis.* **123**:365–370.

Johnson, D. C., Maxfield, B. G., and Moulthrop, J. I., 1977, Epidemiologic studies of the 1975 avian influenza outbreak in chickens in Alabama, *Avian Dis.* **21**:167–177.

Kalter, S. S., and Heberling, R. L., 1978, Serologic responses of primates to influenza viruses, *Proc Soc. Exp. Biol. Med.* **159**:414–417.

Kalter, S. S., Heberling, R. L., Vice, T. E., et al., 1969, Influenza (A2/Hong Kong/68) in the baboon (*Papio* sp.), *Proc. Soc. Exp. Biol. Med.* **132**:357–361.

Kaplan, M. M., 1982, The epidemiology of influenza as a zoonosis, *Vet. Rec.* **110**:395–399.

Kawano, J., Onta, T., Kida, H., et al., 1978, Distribution of antibodies in animals against influenza B and C viruses, *Jpn. J. Vet. Res.* **26**:74–80.

Kawano, J., Yanagawa, R., and Kida, H., 1979, Site of replication of influenza virus A/budgerigar/Hokkaido/1/77 (Hav4 Nav1) in budgerigars, *Zentralbl Bakteriol. Hyg.* [*A*] **245**:1–7.

Kida, H., Yanagawa, R., and Matsuoka, Y., 1980, Duck influenza lacking evidence of disease signs and immune response, *Infect. Immun.* **30**:547–553.

Kilbourne, E. D., 1975, Epidemiology of influenza, in: *The Influenza Viruses and Influenza* (E. D. Kilbourne, ed.), Academic Press, New York, pp. 483–538.

Kilbourne, E. D., and Kehoe, J. M., 1976, Demonstration of antibodies to both hemagglutinin and neuraminidase antigens of H3N2 influenza A virus in domestic dogs, *Intervirology* **6**:315–318.

Kilbourne, E. D., Anderson, H. C., and Horsfall, F. L., Jr., 1951, Concurrent infection with influenza A and B viruses in a single epidemic, *J. Immunol.* **67**:547–558.

King, A. P., and Shortridge, K. F., 1982, Multiple avian influenza infection: Selection of a non-avid virus by a heterologous avian host, *Res. Vet. Sci.* **33**:127–129.

Klingeborn, B., and Rockborn, G., 1980, Significant antigenic drift within the influenza equi 2 subtype in Sweden, *Vet. Res.* **106**:363–364.

Kono, Y., Ishikawa, K., Fukunaga, Y., et al., 1972, The first outbreak of equine influenza in Japan, *Natl. Inst. Anim. Health Q.* **12**:183–187.

Kundin, W. D., 1970, Hong Kong A2 influenza virus infection among swine during a human epidemic in Taiwan, *Nature* **228**:957–958.

Lang, G., and Ferguson, A. E., 1981, The extent and control of avian influenza in Canada, *Can. Vet J.* **22**:377–381.

Lu, B.-L., Webster, R. G., and Hinshaw, V. S., 1982, Failure to detect hemagglutination-inhibiting antibodies with intact avian influenza virions, *Infect. Immun.* **38**:530–535.

Lvov, D. K., and Zhdanov, V. M., 1983, Persistence of genes of epidemical influenza viruses in natural populations in the USSR, *Med. Biol.* **61**:83–91.

Lvov, D. K., Zhdanov, V. M., Sazonov, A. A., et al., 1978, Comparison of influenza viruses isolated from man and from whales, *Bull. WHO* **56**:923–930.

Lvov, D. K., Easterday, B., Hinshow, W., et al., 1979, Isolation of influenza virus strains of the Hong Kong complex (H3N2) from *Nyctalus noctula* bats in Kazakhstan, *Vopr. Virusol* **4**:338–341.

Masurel, N., de Boer, G. F., Anker, W. J. J., et al., 1983, Prevalence of influenza viruses A-H1N1 and A-H3N2 in swine in The Netherlands, *Comp. Immunol. Microbiol. Infect. Dis.* **6**:141–149.

McQueen, J. L., Davenport, F. M., Keeran, R. J., et al., 1966, Studies of equine influenza in Michigan, 1963. II. Epizootiology, *Am. J. Epidemiol.* **83**:280–286.

Michaels, R. H., Mahmud, M. I. A., Coup, A. J., et al., 1978, Influenza virus infection in newborn rats: A possible marker of attenuation for man, *J. Med. Virol.* **2**:253–264.

Molibog, E. V., Smetanin, M. A., and Yakhno, M. A., 1975, Study of influenza A virus isolated from cattle in the Udmurt A.S.S.R., *Ecol. Viruses* **3**:63–67.

Morin, M., Phaneuf, J. B., Sauvageau, R., et al., 1981, An epizootic of swine influenza in Quebec, *Can. Vet. J* **22**:204–205.
Murphy, B. R., Sly, D. L., Hosier, N. T., et al., 1980, Evaluation of three strains of influenza A virus in humans and in owl, cebus, and squirrel monkeys, *Infect Immun* **28**:688–691
Naeve, C. W., and Webster, R. G., 1983, Sequence of the hemagglutinin gene from influenza virus A/Seal/Mass/1/80, *Virology* **129**:298–308.
Nakamura, R. M., and Easterday, B. C., 1967, Serological studies of influenza in animals, *Bull WHO* **37**:559–567
Nakamura, R. M., Easterday, B. C., Pawlisch, R., et al., 1972, Swine influenza Epizootiological and serological studies, *Bull. WHO* **47**:481–487
Nerome, K., Nakayama, M., Ishida, M., et al., 1978, Isolation and serological characterization of influenza A viruses from birds that were dead on arrival at Tokyo airport, *Arch Virol* **57**:261–270
Nerome, K., Ishida, M., Oya, A., et al., 1982a, Isolation of an influenza H1N1 virus from a pig, *Virology* **117**:485–489
Nerome, K., Ishida, M., Oya, A., et al., 1982b, The possible origin of H1N1 (Hsw1N1) virus in the swine population of Japan and antigenic analysis of the isolates, *J Gen Virol* **62**:171–175
Nerome, K., Sakamoto, S., Yano, N., et al., 1983, Antigenic characteristics and genome composition of a naturally occurring recombinant influenza virus isolated from a pig in Japan, *J. Gen Virol* **64**:2611–2620
Nikitin, T., Cohen, D., Todd, J. D., et al., 1972, Epidemiological studies of A/Hong Kong/68 virus infection in dogs, *Bull WHO* **47**:471–479
Nishikawa, F., Sugiyama, T., Yamamoto, K., et al., 1977, Isolation of influenza type A viruses from imported pet birds, *Jpn J Med Sci Biol* **30**:31–36.
O'Brien, T. C., and Tauraso, N. M., 1973, Antibodies to type A influenza viruses in sera from nonhuman primates, *Arch Ges Virusforsch.* **40**:359–365.
Ogawa, Y., Goto, H., Hirano, T., et al., 1983, Sero-epizootiological study on swine influenza in a prefecture of Japan, *J Hyg (Camb)* **90**:403–406.
Ohwada, K., Kitame, F., and Homma, M., 1986, Experimental infections of dogs with Type C influenza virus, *Microbiol Immunol* **30**:451–460
Okazaki, K., Yanagawa, R., Kida, H., et al., 1983, Human influenza virus infection in mink: Serological evidence of infection in summer and autumn, *Vet. Microbiol* **8**:251–257
Paniker, C. K. J., and Nair, C. M. G., 1970, Infection with A2 Hong Kong influenza virus in domestic cats, *Bull. WHO* **43**:859–862.
Patocka, F., Schreiber, E., Kubelka, V., et al., 1958, An attempt to transmit the human influenza virus strain A-Sing 57 to swine (preliminary report), *J. Hyg. Epidemiol Mikrobiol Immunol* **2**:9–15
Pensaert, M., Ottis, K., Vandeputte, J., et al., 1981, Evidence for the natural transmission of influenza A virus from wild ducks to swine and its potential importance for man, *Bull WHO* **59**:75–78
Pereira, H. G., Takimoto, S., Piegas, N. S., et al., 1972, Antigenic variation of equine (Heq2Neq2) influenza virus, *Bull. WHO* **47**:465–469
Plateau, E., Crucière, C., and Gayot, G., 1983, Antigenic drift of a strain of influenza equi virus isolated in France during winter 1978–1979, *Ann. Rech Vet* **14**:71–77
Powell, D. G., Thomson, G. R., Spooner, P., et al., 1974, The outbreak of equine influenza in England April/May 1973, *Vet. Rec* **94**:282–287
Pysina, T. V., Lvov, D. K., Braude, N. A., et al., 1978, Study of influenza A/*Anas acuta*/Primorie/695/76 virus isolated from wild ducks in the U.S.S.R., *Vopr. Virusol* **3**:300–304
Rao, B. L., Khan, F. U., Bhat, H. R., et al., 1983, Zoonotic studies on influenza in pigs and birds, India, 1980–81, *Int. J. Zoon.* **10**:40–44.
Reale, M. A., Bona, C. A., and Schulman, J. L., 1985, Isotype profiles of anti-influenza antibodies in mice bearing the *xid* defect, *J Virol* **53**:425–429.
Renshaw, H. W., 1975, Influence of antibody-mediated immune suppression on clinical, viral, and immune responses to swine influenza infection, *Am J Vet Res* **36**:5–13
Robinson, J. H., and Easterday, B. C., 1978, Detection of persisting influenza virus with turkey trachea cultures, *Avian Dis.* **23**:354–356.
Rogers, G. N., and Paulson J. C., 1983, Receptor determinants of human and animal influenza virus isolates Differences in receptor specificity of the H3 hemagglutinin based on species of origin,

Virology **127:**361–373.
Romváry, J., Rózsa, J., and Farkas, E., 1975, Infection of dogs and cats with the Hong Kong influenza A(H3N2) virus during an epidemic period in Hungary, *Acta Vet. Acad. Sci. Hung.* **25:**255–259.
Romváry, J., Mészáros, J., Tanyi, J., et al., 1976, Spreading of virus infection among wild birds and monkeys during the influenza epidemic caused by the Victoria(3)75 variant of A(H3N2) virus, *Acta Vet. Acad. Sci. Hung.* **26:**369–376.
Rott, R., Klenk, H.-D., and Scholtissek, C., 1984, Significance of the hemagglutinin for the pathogenicity of avian influenza viruses, *Zentralbl. Bakteriol. Hyg.* **258:**337–349.
Satsuta, K., Endo, Y., Ueyonahara, K., et al., 1981, Serological studies on an epidemic prevalence of swine type influenza virus among swine and human beings, *Jpn. J. Vet. Sci.* **43:**531–538.
Schafer, W., 1955, Vergleichende sero-immunologische Untersuchungen neber die Viren de Influenza und klassichen Gefluegelpest, *Zentrabl. Naturforsh.* **106:**81–91.
Scholtissek, C., Bürger, H., Bachmann, P. A., et al., 1983, Genetic relatedness of hemagglutinins of the H1 subtype of influenza A viruses isolated from swine and birds, *Virology* **129:**521–523.
Shope, R. E., 1931, Swine influenza. III. Filtration experiments and etiology, *J. Exp. Med.* **54:**373–385.
Shortridge, K. F., 1979, H2N2 influenza viruses in domestic ducks, *Lancet* **1:**439.
Shortridge, K. F., Butterfield, W. K., Webster, R. G., et al., 1977, Isolation and characterization of influenza A viruses from avian species in Hong Kong, *Bull. WHO* **55:**15–20.
Singh, P. B., Rodrigues, F. M., and Gupta, N. P., 1981, A serological survey of pigs in India for swine influenza, *Indian J. Med. Res.* **74:**153–155.
Sinnecker, H., Sinnecker, R., Zilske, E., et al., 1983, Influenza virus A/swine—outbreaks in domestic pigs and antibody findings in human sera, *Zentralbl. Bakteriol. Hyg.* [*A*] **255:**209–213.
Skeeles, J. K., Beasley, J. N., Blore, P., et al., 1981, Severe egg-production drops in turkey breeders in southcentral Missouri, *Avian Dis.* **25:**764–767.
Slemons, R. D., and Easterday, B. C., 1972, Host response differences among 5 avian species to an influenzavirus—A/turkey/Ontario/77-32/66 (Hav5N?), *Bull. WHO* **47:**521.
Smith, B. P., 1979, Influenza in foals, *J. Am. Vet. Med. Assoc.* **174:**289–290.
Sovinova, O., Tumova, B., Poustka, F., et al., 1958, Isolation of a virus causing respiratory diseases in horses, *Acta Virol.* **2:**52.
Sriram, G., Bean, W. J., Jr., Hinshaw, V. S., et al., 1980, Genetic diversity among avian influenza viruses, *Virology* **105:**592–599.
Staeheli, P., Horisberger, M. A., and Haller, O., 1984, Mx-dependent resistance to influenza viruses is induced by mouse interferons α and β but not γ, *Virology* **132:**456–461.
Stubbs, E. L., 1967, Fowl plague, in: *Diseases of Poultry* (A. Biester and L. H Schwartz, eds.). University of Iowa Press, Ames, pp. 813–822.
Sugimura, T., Yonemochi, H., Ogawa, T., et al., 1980, Isolation of a recombinant influenza virus (Hsw1N2) from swine in Japan, *Arch. Virol.* **66:**271–274.
Takátsy, G., Romváry, J., and Farkas, E., 1967, Susceptibility of the domestic swine to influenza B virus, *Acta Microbiol. Acad. Sci. Hung.* **14:**309–315.
Tanyi, J., Romváry, J., Áldásy, P., et al., 1974, Isolation of influenza A virus strains from cattle. Preliminary report, *Acta Vet. Acad. Sci. Hung.* **24:**341–343.
Taylor, N. M., 1940, Detection of human influenza virus in throat washings by immunity response in Syrian hamster (*Cricetus auratus*), *Proc. Soc. Exp. Biol. Med.* **43:**541–542.
Thomson, G. R., Mumford, J. A., Spooner, P. R., et al., 1977, The outbreak of equine influenza in England: January 1976, *Vet. Rec.* **100:**465–468.
Tumova, B., 1980, Equine influenza—a segment in influenza virus ecology, *Comp. Immunol. Microbiol Infect. Dis.* **3:**45–59.
Tumova, B., Easterday, B. C., and Stumpa, A., 1972, Simultaneous occurrence of A/equi-1 and A/equi-2 influenza viruses in a small group of horses, *Am. J. Epidemiol.* **95:**80–87.
von Hoyningen-Huene, V., and Scholtissek, C., 1983, Low genetic mixing between avian influenza viruses of different geographic regions. Brief report, *Arch. Virol.* **76:**63–67.
Waddell, G. H., Tiegland, M. B., and Sigel, M. M., 1963, A new influenza virus associated with equine respiratory disease, *J. Am. Vet. Med. Assoc.* **143:**587.
Webster, R. G., and Laver, W. G., 1972, The origin of pandemic influenza, *Bull. WHO* **47:**449–452.
Webster, R. G., Laver, W. G., and Tumova, B., 1975, Studies on the origin of pandemic influenza

viruses V Persistence of Asian influenza virus hemagglutinin (H2) antigen in nature? *Virology* **67**:534–543

Webster, R G , Yakhno, M , Hinshaw, V. S., et al., 1978, Intestinal influenza· Replication and characterization of influenza viruses in ducks, *Virology* **84**:268–278.

Webster, R G , Geraci, J , and Petursson, G., 1981a, Conjunctivitis in human beings caused by influenza A virus of seals, *N Engl J Med* **304**:911.

Webster, R G., Hinshaw, V. S , Bean, W. J , et al , 1981b, Characterization of an influenza A virus from seals, *Virology* **113**:712–724.

Wetherbee, R. E , 1973, Induction of systemic delayed hypersensitivity during experimental viral infection of the respiratory tract with a myxovirus or paramyxovirus, *J Immunol.* **111**:157–163

World Health Organization Memorandum, 1980, A revision of the system of nomenclature for influenza viruses, *Bull WHO* **58**:585–591

World Health Organization Memorandum, 1981, The ecology of influenza viruses, *Bull WHO* **59**:869–873.

Zhezmer, V Yu , Lvov, D K., Isachenko, V. A., et al., 1973, Isolation of Hong Kong variant of influenza type A virus from sick chicken in Kamchatka region, *Vopr. Virusol* **1**:94–99.

Zilske, E , Sinnecker, R , and Sinnecker, H , 1981, Neuraminidase-inhibiting antibodies to avian influenza virus subtypes in human sera, *Zentralbl Bakteriol. Hyg* [*A*] **250**:34–41

IV

Epidemiology, Surveillance, and Control

10

The Epidemiology of Influenza

INTRODUCTION

If the diagnosis of influenza in the individual patient is at times ambiguous, the identification of influenza epidemics is not. In the north temperate zone, the abrupt appearance of febrile respiratory disease affecting adults as well as children in the early winter can be assumed to be influenza pending virologic confirmation. Characteristically, sporadic cases are seen in late November and December; then, following the Christmas holidays, the disease becomes epidemic. School and industrial absenteeism and clinic visits increase, and somewhat later excess mortality related to pneumonia occurs.

Depending on the degree of antigenic novelty of the invading virus and probably other factors, 5–40% of the community may be affected by the disease, and in pandemic years, severe disruption of community services occurs (Rosenstock et al., 1960). In other years, the annual visitation of influenza viruses may be cryptic and revealed only by virus isolations from patients with pneumonia or by antibody rises in community serum samples (Kilbourne, 1961).

INFLUENZA SURVEILLANCE AND THE ASCERTAINMENT OF INFECTION

Influenza lacks the pathognomonic stigmata of poliomyelitis and measles, so its presence in the community is reflected by a variety of indirect measures of its epidemic impact (Table 10-1). For all its importance, influenza is not a reportable or notifiable disease, and few countries have developed systems for estimating its morbidity and mortality on a regular basis. However, much has been done through the cooperating network of WHO regional laboratories to detect and identify influenza viruses so that at least in parts of the world where these laboratories exist, the paths of epidemics can be charted (Assaad et al., 1977).

Table 10-1. Surveillance for Influenza

I. Laboratory (specific) surveillance
   Virus isolation
   Serological studies for specific antibody
II. Estimates of morbidity
   A Direct
      Family studies
      Clinic and emergency room visits
      Physician reporting
      Hospitalized respiratory infection
      Health interview surveys
   B. Indirect
      Absenteeism from school and workplace
      Sickness benefits claims (UK)
      Excess morbidity
      Sales of cold remedies
III Estimates of mortality
   Total excess mortality
   Excess pneumonia and influenza deaths
   Percentage of pneumonia and influenza deaths

Laboratory Surveillance

Virus Isolation and Identification

There is, of course, no substitute for virus isolation in the definitive diagnosis of influenza and the identification of epidemics. Collated data from 61 collaborating state, county, city, or military laboratories in the United States on the testing of specimens from patients with respiratory tract disease have provided a reasonable, if unsystematic, sampling of influenza viruses during the past seven winters (Fig. 10-1). Submission of isolates to the Centers for Disease Control (CDC) not only provides a measure of the relative prevalence of virus type and subtype but affords the opportunity for detection of antigenic variants within a subtype. Such sampling is critical in looking for significant antigenic change that may herald forthcoming epidemics in the following year (Glezen et al., 1982 a,b).

Isolation of virus from specific and limited groups of patients has been rewarding in the early detection of influenza in the community. Both pediatric (Levine et al., 1974) and adult (Goldstein and Block, 1976; Marine et al., 1976; Foy et al., 1979) outpatients have been sampled. Sampling of febrile adult patients detected influenza earlier (Foy et al., 1979) and was more discriminating (Goldstein and Block, 1976) than sampling of children with respiratory disease.

Recently, cocirculation of influenza A viruses has required prompt antigenic characterization of isolates in order to determine periods of prevalence of H1N1 and H3N2 strains, their relative contribution to infection in different age groups, and their antigenic variation (Kendal et al., 1979). Surveillance by viral isolation is of particular importance in assessing the incidence of H1N1 virus infections because the usual indices of influenza occurrence, especially mortality, are inadequate with the young age group (Pereira and Chakraverty, 1982).

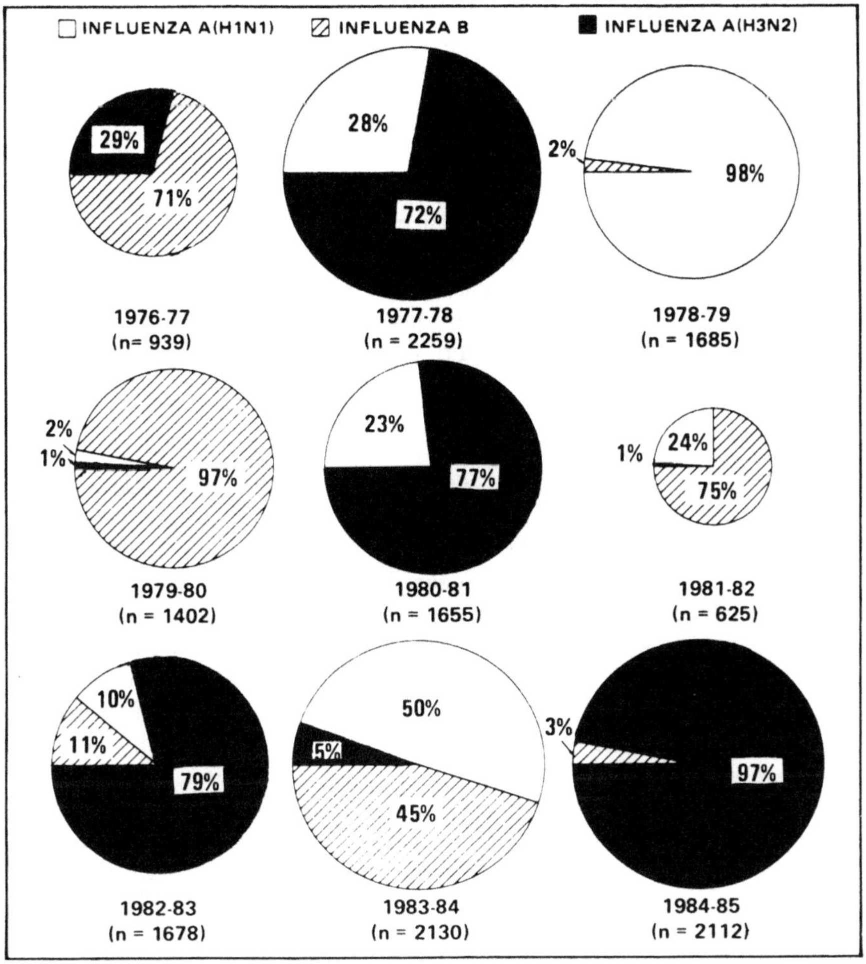

Figure 10-1. Isolation of influenza viruses reported to CDC by collaborating civilian and military laboratories, United States, 1976–1985. (From Centers for Disease Control, 1985.)

Serological Surveillance

Detection of increases of antibody to influenza viruses in various segments of the population provides a definitive picture of the progress of influenza virus in the community as it measures response to both clinical and subclinical infections. Serological studies of New York City's population for the period 1957–1964 utilized sera obtained in routine premarital testing for syphilis in various age groups and demonstrated a progressive rise in HI antibody titer to the H2N2 viruses introduced in 1957 (Widelock et al., 1965). Isolated populations in third-world countries have been monitored by the use of transportable agarose gel plates containing influenza virus antigens (Schild et al., 1977).

Serological surveillance can be important not only in documenting the population's experience with influenza viruses but in ascertaining lack of immunity that may be predictive of an epidemic.

Measurements of Morbidity

Various approaches have been taken in the assessment of the incidence of influenza as a nonreportable disease in the community. These may be classified as direct, in which the extent of disease is estimated from the counting of those affected with typical symptoms in samples of the community, or indirect, in which the community impact of infection is measured.

Direct Methods

Intensive longitudinal studies of families for the detection and identification of acute respiratory infections have provided valuable information on the epidemiology of influenza and also have proved to be a sensitive barometer of its occurrence. Studies of this sort, pioneered by Dingle and his colleagues (1964) and later extended by Fox and Hall (1980) and other groups, combined laboratory and clinical estimates of specific illness. Combined serological and clinical surveillance has shown that unsuspected influenza virus infections can occur "out of season" (Hall et al., 1973; Monto and Kioumehr, 1957). Family studies have contributed to estimation of the burden imposed by influenza and to understanding of disease transmission.

Because the fever and disability of influenza exceed that associated with other respiratory tract infections yet may not require hospitalization of the patient, epidemics cause a sudden increase in visits to physicians, hospital emergency rooms, and outpatient clinics. Increased attendance at such facilities has been correlated positively so often with specific virus isolations (Fig. 10-2) that it is a valid indication of influenza in the community, especially in adult populations (Goldstein and Block, 1976). This method is most effective in urban centers (Rubin and Gregg, 1975) and when high fever is noted among patients (Shen et al., 1983). Similarly, selected general practitioners functioning as "spotters" (Pereira et al., 1978) have been used in the United Kingdom and elsewhere to detect increases in influenzalike illness (Knox and Anderson, 1982). In 1976, 34 of the United States included influenzalike illnesses reported by physicians in morbidity reports to CDC (Poe and Massey, 1977). In the same year, the intensive program of the National Center for Health Statistics was used to collect influenza activity data in the Health Interview Survey (Poe and Massey, 1977). Telephone follow-up interviews after postcard reporting of school absenteeism proved inexpensive and highly specific when utilized in Minnesota (Peterson et al., 1979).

Hospitalization for pneumonia has not been systematically used as an indicator of influenza but is useful, particularly, if adult admissions are sampled (Glezen and Couch, 1978; Kilbourne, 1961) (Fig. 10-3).

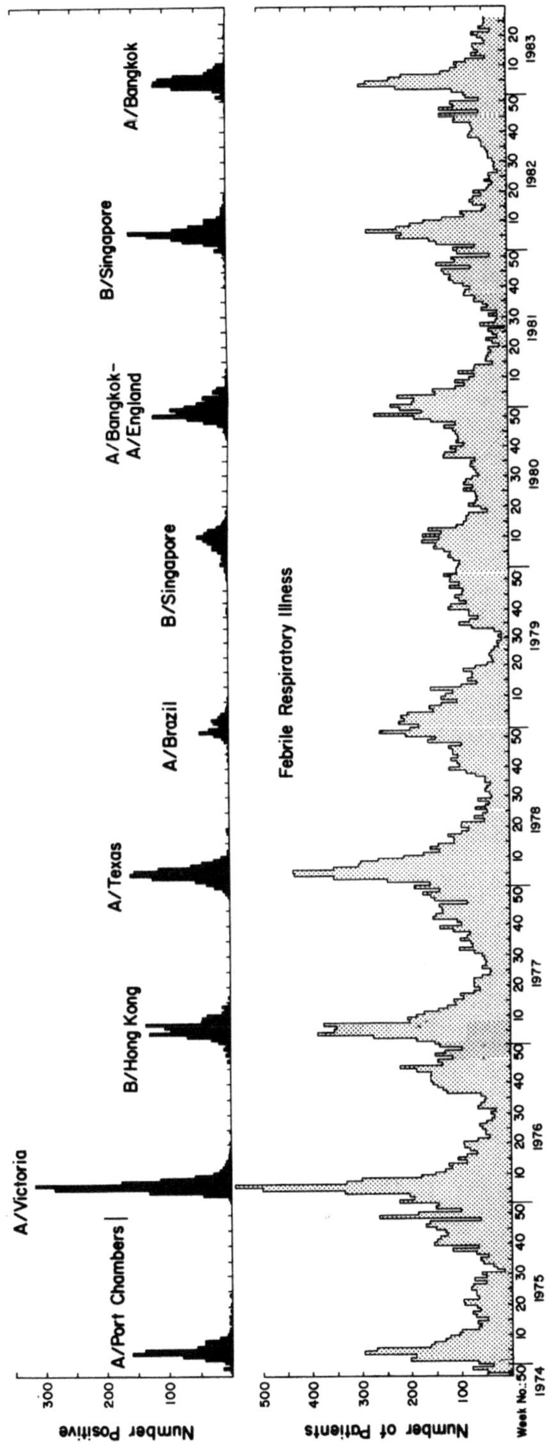

Figure 10-2. Number of persons with acute respiratory illness presenting to sentinel surveillance sites and number with positive cultures for influenza viruses by week, Houston, Texas, 1974–83. (From Glezen et al., 1984. Reproduced with permission, from *The Molecular Virology and Epidemiology of Influenza*, C. Stuart-Harris and C. W. Potter (eds.), p. 19, Academic Press, London.)

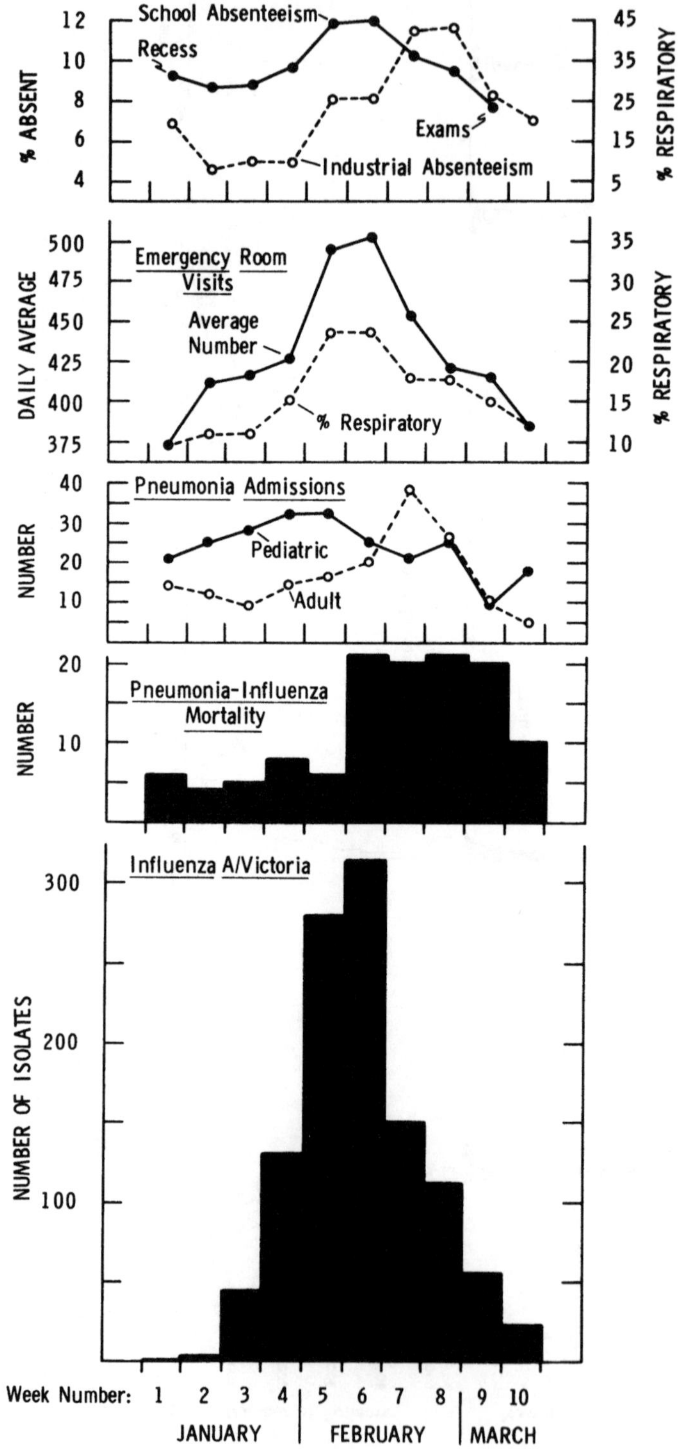

School Absenteeism
Recess
Exams
Industrial Absenteeism
% ABSENT
% RESPIRATORY
Emergency Room Visits
Average Number
% Respiratory
DAILY AVERAGE
Pneumonia Admissions
Pediatric
Adult
NUMBER
Pneumonia-Influenza Mortality
Influenza A/Victoria
NUMBER OF ISOLATES
Week Number: 1 2 3 4 5 6 7 8 9 10
JANUARY
FEBRUARY
MARCH

Indirect Methods

Survey of school and industrial absenteeism, although lacking in specificity, is an effective indicator of the presence of influenza, emphasizing the unique impact of the disease on community life.

In England and Wales, increases in sickness benefits claims in the insured working populations have proved useful because a doctor's certificate is required for absence of more than 3 days.

As a method of assessing excess morbidity against a background of cumulative data from previous years, accumulative summation or "cusum" plots have been used in a manner similar to measurement of excess mortality (see following section) (Tillett and Spencer, 1982). A similar method for comparing the actual daily morbidity of influenza and other acute respiratory diseases with the upper tolerance of the morbidity in nonepidemic years has been used to detect the beginning of an influenza epidemic in Moscow (Peretjagina et al., 1977).

An ingenious but very indirect method tested in Los Angeles was the plotting of a supermarket chain's sales in cold remedies. These sales, and not sales of aspirin or other antipyretics, peaked with influenza activity measured by viral isolation (Welliver et al., 1979).

Impact on Mortality Rates

Although influenza is a disease of "high morbidity and low mortality" (Eickhoff et al., 1961), so many cases occur than normal rates of mortality during epidemics may be perceptibly affected, an observation first attributed to William Farr (1885).

Excess mortality, total or categorical (Housworth and Langmuir, 1974), or mortality attributed to pneumonia and influenza (Collins et al., 1930) have been used as an indicator of influenza in the community (Fig. 1-1). "Excess" implies rates or numbers of fatalities over an "expected" rate or number observed in nonepidemic years. As refined by Serfling (1963) and modified by the CDC, deviation is monitored of the observed number of deaths from an expected number forecast with a regression equation (Serfling, 1963). An epidemic "threshold" is defined in relation to an arbitrarily defined number (1.64) of standard deviations (Choi and Thacker, 1981). Recently, methods using a stepwise regression model (Clifford et al., 1977) or time series analysis of historical mortality data (Choi and Thacker, 1982) have been used without notable increase in accuracy of estimates of pneumonia and influenza deaths. Currently, influenza pneumonia deaths are reported by CDC as percentage of total deaths.

It is probable that current mortality data underestimate the effects of influenza on mortality. Based on virologic surveillance, methods used to predict the

Figure 10-3. Correlation of the nonvirologic indices of epidemic influenza with the number of isolates of influenza A/Victoria virus according to week, Houston, 1976. Industrial absenteeism is indicated by percentage with respiratory complaints. (From Glezen and Couch, 1978. Reprinted by permission of *New England Journal of Medicine* **298**:589, 1978.)

base-line mortality may be inaccurate (i.e., too high) (Glezen et al., 1982b), and some influenza deaths may not be assigned as pneumonia and influenza associated (Barker and Mullooly, 1981). Excess deaths from cardiac causes might most appropriately be considered as part of the excess mortality characteristic of influenza epidemics (Barker and Mullooly, 1982).

INFLUENZA AS A BURDEN ON THE PUBLIC HEALTH

Mortality is only one small measure of the impact of influenza. The incidence and prevalence of influenza as a self-limited but significant disease are more often underestimated than overestimated, particularly in later postpandemic periods when expression of disease is modified in the partially immune. Ongoing surveillance of families with specific virologic methods opens a window on the frequency of infection and also demonstrates its variability related to virus, host, age, and the community.

The overall frequency of infection with A/Victoria/75 H3N2 virus in 37 low-income-group families in Houston in 1976 was 28%, with attack rates inversely related to age. Almost half of the families had at least one infected member (Taber et al., 1981).

In the semiisolated community of Port Chalmers, New Zealand, 46% of selected families had clinical or laboratory evidence of infection with the Port Chalmers H3N2 variant of 1973. Infection was most frequent in school-age children (Jennings and Miles, 1978).

In a 4-year period of surveillance (1965–1969), Hall et al. (1973) found rates per 100 person years just under 20 for both A and B influenza virus infection. In continuing observation of the Seattle population (1975–1979), the same investigators (Fox et al., 1982) found 17–24% overall infection rates and invasion of 38–53% of households during one B and two A H3N2 epidemics.

Although the studies cited above provide an estimate based on serology of the rate of infection, they do not permit exact conclusions about the level of morbidity. In a British study of widely scattered families under surveillance by the Royal College of General Practitioners and the Public Health Laboratory Service, adults experienced twice as many asymptomatic as symptomatic infections (Mann et al., 1981). Although these figures relate to "probable" influenza, they provide a rough correction factor in estimating the disease burden of influenza in nonpandemic years in adults in noninstitutional settings. In pandemic years more than three-fourths of infections are associated with disease (Carey et al., 1958; Clarke et al., 1958; Lim et al., 1957), and serological attack rates in the young may reach 75% (Carey et al., 1958).

Respiratory disease morbidity severe enough to require hospitalization is a less equivocal measure of the impact of influenza. Depending on the population studied and the indications for hospitalization, excess rates for adults of 79–86 per 100,000 (Barker and Mullooly, 1980) or 270 per 100,000 (Glezen and Couch, 1982) have been found during times of influenza prevalence. The impact of three influenza epidemics on hospitalization for acute respiratory disease in one Texas county is shown graphically in Fig. 10-4.

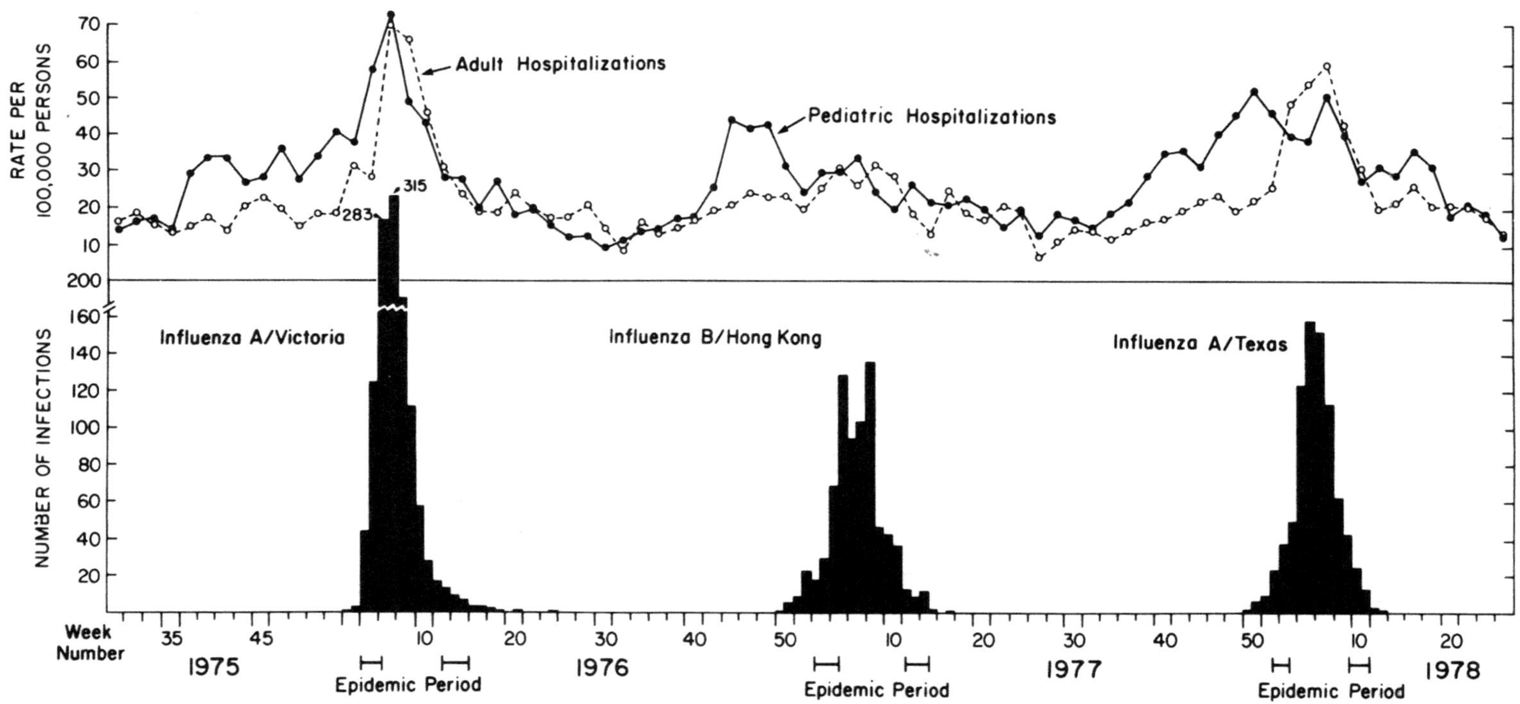

Figure 10-4. Pediatric and adult hospitalization rates for acute respiratory disease by 2-week periods estimated for Harris County, Texas (upper histogram) and their temporal relationship to influenza virus epidemics defined by virologic surveillance at the Influenza Research Center, Houston, Texas (lower histogram), 1975–1978. (From Glezen, 1982. Reproduced with permission, from *Epidemiol. Rev.* **4**:39.)

A study of five American cities in the 1957 pandemic assessed the impact of the epidemic on many aspects of community life (Rosenstock et al., 1960). In two cities, respiratory illness affected nearly one-third of the population and 40% of families. Illness was more frequent in nonwhites and in large families.

The impact of influenza as an epidemic disease in which most individual cases are not serious is perhaps greater than the sum of its parts (Kilbourne, 1975). Preparation for the pandemic of 1957 by public health agencies revealed divergent reactions of public health workers and physicians. Public health officials viewed influenza as "a threat to the community," but influenza was considered by practicing physicians as "not constituting a threat to the individual" (Rosenstock et al., 1960). Stated another way, "a lot of ill people can affect community life and resources, but if few seen by the doctor are seriously ill, he may not be concerned."

The economic impact of influenza has been studied in the prepandemic years 1963 and 1966 and in the pandemic years 1968–1969. In these years total costs, including loss of earnings, ranged between $1.7 and $3.9 billion per year. Direct medically related costs were less than 20% of this amount. Services provided by Medicare to those 65 years of age or older resulted in additional disbursements in 1968–1969 of $86.6 million for physician services, in- and outpatient services, and diagnostic procedures. The total direct cost to the Medicare program was $228 million, and the average cost was $98 per Medicare case. Such cost estimates are incomplete, as they omit the costs of outpatient drugs, deductibles, insurance, and other indirect costs of morbidity and mortality (Kavet, cited in Fox and Kilbourne, 1973; Kavet, 1977). But total costs related to medical care may comprise less than 4% of total economic losses related to reduced productivity of the work force (Plesnik et al., 1978). Costs of a putative uncontrolled pandemic in the United States in 1976 were estimated at $6 billion (Schoenbaum et al., 1976).

PANDEMIC, EPIDEMIC, AND ENDEMIC DISEASE

Depending on the level and distribution of specific immunity in the population, influenza is manifest in global pandemics, regional epidemics, or in cryptic, endemic infections. The different epidemiology of pandemic and interpandemic infections is considered in sections on Pandemic, Interpandemic, and Endemic Influenza. The following epidemiologic determinants apply to each of these epidemic types.

EPIDEMIOLOGIC DETERMINANTS OF INFLUENZA

Viral

Antigenic Variation

Although reinfection with homotypic strains seems to occur fairly frequently (Frank et al., 1979), the survival of influenza viruses in man requires the continual evolution of antigenic mutants in the face of continuing reciprocal change in the immunophenotype of the population (Fig. 10-5). However, it should be appreciated that immunoselection operates against multiple cocirculating antigenic variants, one of which predominates each year (Glezen et al., 1984) (Table 10-2), but not to the total exclusion of others. Pereira and Chakraverty (1982) noted the circulation of A/Victoria/3/75 H3N2 virus in three consecutive winters, but in the third year, it constituted only 4% of the H3N2 viruses identified. Some strains present in a small proportion of infections may presage the predominant strain the following year as a small "herald wave" of infection (Glezen et al., 1982a). At times, careful serological analysis has revealed codominance of two intrasubtypic variants in the same region (Barb and Takatsy, 1973; Pereira and Chakraverty, 1977; Kendal et al., 1978), again emphasizing the antigenic heterogeneity of field isolates of virus (Beyer and Masurel, 1985). Isolates apparently homogeneous when analyzed with polyclonal antisera may show extensive microheterogeneity when studied with monoclonal antibody (Six et al., 1983).

It has been suggested that multiple amino acid substitutions in the HA might be necessary for a virus to acquire epidemic potential (Wiley et al., 1981; see also Chapter 6), but fewer or more changes might be required, depending on the capacity of the virus to replicate efficiently and to be transmitted. Probably of equal importance are the immune status of the host population, population density and environmental factors that foster multiple generations of virus, and the accumulation of mutations in other than the HA gene. That antigenic differences per se may be insufficient for the serial propagation of virus is attested to by the inability of A/Hong Kong/5/72 H3N2 to spread beyond limited regions (Schild et al., 1973) and by the failure of the novel swine influenza virus subtype A/New Jersey/11/76

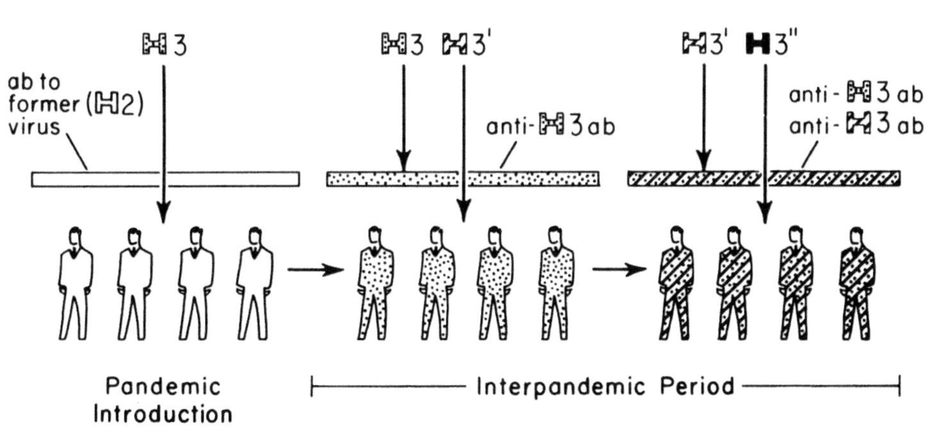

Figure 10-5. Selection of antigenic mutants as a function of population antibody. New pandemic viral subtype H3 transcends barrier of antibody to unrelated previously prevalent virus H2 and readily infects the population. When a critical percentage of the population has been infected with H3, survival of H3 is impeded, and antigenically changed mutant H3′ and later H3″ have survival advantage (minor antigenic variation or antigenic "drift"). (From Kilbourne, 1975. Reproduced with permission, from *The Influenza Viruses and Influenza,* E. D. Kilbourne (ed.), p. 522, Academic Press, New York.)

Table 10-2. Distribution of Major Antigenic Variants of Subtype Influenza A (H3N2) by Year, Houston, 1976–83[a]

	Percentage of H3N2 isolates by year				
Antigenic variant	1976 (N = 1197)	1976–77 (N = 48)	1977–78 (N = 690)	1980–81 (N = 416)	1982–83 (N = 506)
A/Victoria/3/75	97.8	68.0	30.0		
A/Texas/1/77	2.2[b]	32.0[b]	70.0[b]	9.6	5.4
Intermediate strains				47.9[c]	53.6[d]
A/Bangkok/1/79				42.5	30.0
Others[e]					10.9[e]

[a]Reprinted, with permission, from Glezen et al. (1984)
[b]Includes strains sharing antigens with A/Victoria/3/75; many similar to A/Brazil/53/76 (H3N2)
[c]Heterologous group sharing antigens of A/Texas/1/77 and A/Bangkok/1/79, many similar to A/Taiwan/1/79 and A/Arizona/2/80
[d]Heterologous group sharing antigen of A/Bangkok/1/79 and A/Texas/1/77; many similar to A/Oregon/4/80
[e]Includes A/Philippines/2/82 (6 4%), A/Bangkok/2/79 (2 7%), and unclassified (1 8%)

H1N1 to spread beyond Fort Dix (Kilbourne, 1980). But, given the retention of other "human" viral genes, the minimal requirement for the evolution of a pandemic virus appears to be a major antigenic change in the HA.

Variation in Other Viral Properties

Such properties as intrinsic virulence and transmissibility of influenza viruses in human epidemic circumstances are not readily analyzed (see Chapter 7). Even in experimental animal models, these viral properties are difficult to study (see Experimental Epidemiology, below).

No doubt, as field isolates are subjected to closer scrutiny, differences in both genotype and phenotype will be found that may ultimately be correlated with their epidemiologic properties. Thus, reassortants of H3N2 and H1N1 virus unexpected by antigenic analysis have been found, and pleiotropic replication mutants have been identified in contemporary swine influenza viruses isolated from man (see Molecular Epidemiology, below).

Host Factors

Age

Family studies have shown not only that children are more susceptible to influenza but that they are cardinally important in the transmission of the virus within households. Introduction of H3N2 influenza virus into the Isle of Lewis in 1969 affected 44% of households in Stornoway, where the presence of school-age children in the household almost doubled the chance of influenza occurring in that household (Buchan and Reid, 1972). Continuing surveillance of the Michigan community of Tecumseh also has shown the predilection of influenza A and B for households containing children (Longini et al., 1984a).

Immune Phenotype

That community immunity to influenza is an important epidemiologic determinant hardly requires documentation. However, some of the complexities introduced by the interplay of virus and host in different immunologic substrates deserve mention.

First, immunity from natural infection in adults in one winter can confer complete protection against infection with influenza virus with or without illness in the subsequent year. This occurred in the period immediately after the introduction of the H3N2 subtype virus (Miller et al., 1973). Furthermore, as the virus adapts to the new immunologic environment (Fig. 10-5) by intrasubtypic mutation, it does not necessarily escape immunity acquired through infection by antecedent variants. In an English boys' school, infection with a 1972 H3N2 variant conferred almost complete protection against challenge with a 1976 A/Victoria/3/75 H3N2 virus (Hoskins et al., 1979). In second outbreaks of both H3N2 and H1N1 virus infection, asymptomatic infections outnumbered cases with illness. Reinfection with the H1N1 virus was more common after its introduction in the English school in 1978 than was H3N2 reinfection in the 1975–1976 period, possibly related to a background of neuraminidase antibody in the latter case (Davies et al., 1984).

A legacy of antibody to the H2 HA acquired by the elderly (more than 71 years of age) in the 1889–1890 pandemic (Mulder and Masurel, 1958) reduced their susceptibility to the virus in 1957.

Even closely spaced epidemics of different subtypes appear to be modified by heterosubtypic immunity (Sonoguchi et al., 1985), but infection with two subtypes in the same season has been reported (Frank et al., 1983).

Host Genotype

One suspects from animal studies that human genetic determinants of resistance or susceptibility to influenza will be found, but present evidence is inconclusive. Attempts to link ABO blood groups with susceptibility have been contradictory. Part of the problem is in the interpretation of the serological evidence; i.e., is the finding of higher antibody titers in a single blood group indicative of better host response or more severe infection?

Higher incidence of antibody has been found in association with blood group O (Potter and Schild, 1967), with group AB (Aho et al., 1980), and with groups A and B (Lebiush et al., 1981). Mackenzie and Fimmel (1978) found a higher incidence in subjects with blood group B. Others have found no correlation of antibody incidence with blood groups (Watkins et al., 1975; von Heise et al., 1976). The negative studies of Evans et al. (1970) are notable for having been prospective in nature. A decreased antibody response to live influenza viruses was seen in subjects of HLA type W16 (B38, B39), but these subjects had normal response to inactivated virus vaccine (Spencer et al., 1976).

Other Host Factors

The incidence of influenza was higher in cigarette-smoking male college students (Finklea et al., 1969; Kark et al., 1982) and in seronegative smokers (MacKenzie et al., 1976). Illness was reported to be more severe in one study of smokers (Kark et al., 1982).

A significantly higher attack rate of influenza in girls in the premenstrual period (Belbin and Smithson, 1979) is in accord with observations on other viral infections (Dalton, 1977).

Environmental Factors

Season and Climate

In the north temperate zone, the effects of climate and season are not easily divorced, but influenza has a global distribution, and epidemics also occur in the hot weather of the tropics. At times of pandemic introduction, influenza outbreaks have occurred in the summer months (Langmuir, 1961; Dunn, 1958; Tateno et al., 1963; Dowdle et al., 1974). Therefore, cold weather is not required for the occurrence of influenza epidemics, although increased morbidity has been noted during colder winters (Davey and Reid, 1972; Macfarlane, 1977).

In north and south temperate zones, influenza is usually a seasonal wintertime disease; in tropical zones incidence is less focused but is increased during monsoons or rainy seasons. These observations suggest that a principal determinant in the generation of epidemics may be indoor crowding induced by inclement weather—cold in temperate zones and rain in the tropics. In East Africa, as in subtropical or temperate areas, the highest frequency of infection is in populous urban areas (Montefiore et al., 1970).

The low relative humidity characteristic of the winter months in temperate zones is conducive to survival of influenza viruses in aerosols (Hemmes et al., 1960) and facilitates transmission of influenza in experimental infections of mice (Schulman and Kilbourne, 1963b). Season also influences the carrier rate for pneumococci and other potential bacterial invaders, which in turn will affect the incidence of bacterial pneumonia secondary to influenza and hence will influence mortality.

Hope-Simpson (1981) has proposed an all-encompassing role for season in the epidemiology of influenza, postulating "a direct effect of variation in some component of solar radiation on virus or human host." This hypothesis assumes latent carriage of the virus in human tissue, its seasonal reactivation in altered antigenic form, and its subsequent transmission to susceptibles to produce epidemics. This provocative hypothesis attempts to explain the apparent simultaneous eruption of epidemics caused by the same viral antigenic variant in widely scattered areas of the world. In the absence of any demonstrated biological basis for influenza virus latency, and given the ubiquity of subclinical infections and the pervasive nature of human intercourse, current evidence favors the widespread seeding of epidemic viruses and their rapid propagation when ecological conditions are right. The seasonal patterns of influenza are not invariant, and springtime epidemics of influenza have been noted (*British Medical Journal*, 1974).

But the idea of seasonal effects on host resistance cannot be discounted. In experimental infections of mice, transmission of virus was influenced by relative humidity but occurred more frequently in winter than in summer even when environmental controls provided identical conditions of temperature and humidity (Schulman and Kilbourne, 1966). In man, reactions to live virus vaccines have been reported to be more frequent in winter than in summer (Zhdanov and Ritova, 1962; Beare et al., 1968; Shadrin et al., 1977). The concurrence of influenza A and B (Kilbourne et al., 1951; Spelman and McHardy, 1985) emphasizes the importance of factors beyond the nature of the virus in the initiation of epidemics (Fig. 10-6).

Crowding and Infection Density

Any doubt about the communicability of influenza from those ill with the disease is dispelled by studies in crowded, confined, or isolated populations. The

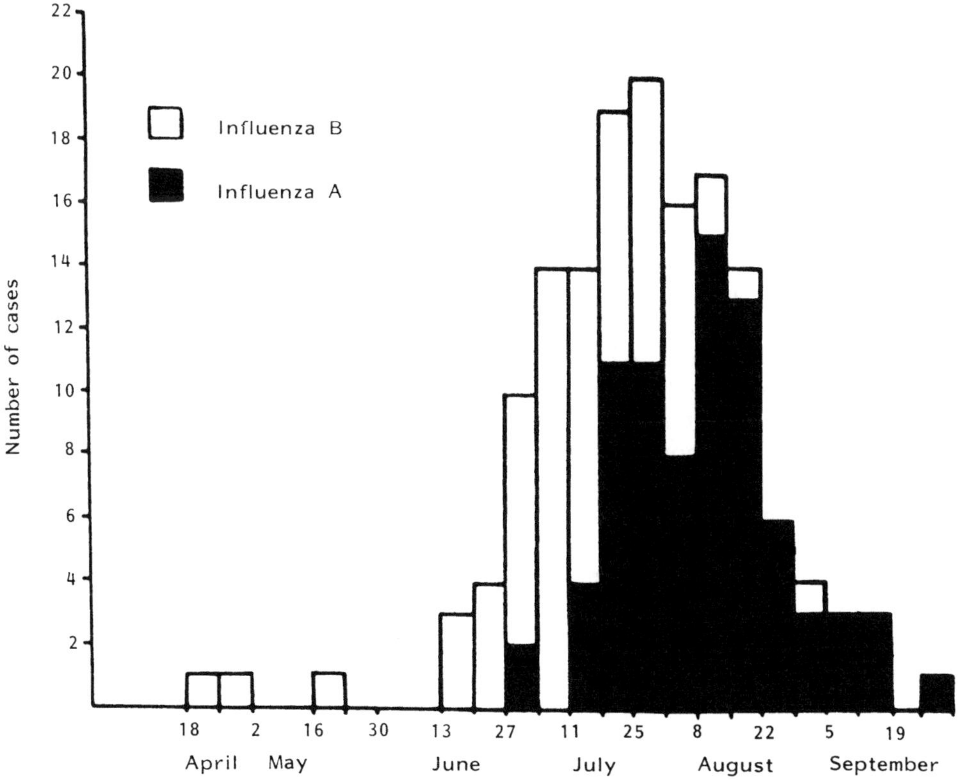

Figure 10-6. Concurrence of influenza A and B in Victoria, Australia. (July and August are winter months.) (From Spelman and McHardy, 1985. Reproduced from *J. Hyg. (Camb.)*, **94**:333. By copyright permission of Cambridge University Press.)

extraordinarily high clinical attack rate aboard Navy ships is notorious and in one outbreak ranged from 27 to 100% in inverse relation to age (Olson and Irving, 1979). Two index cases from New Zealand were sufficient to start a 1-month epidemic involving 41% of the Pacific Island country of Niue (Taylor et al., 1985).

From one single index case, 72% of 54 passengers stranded aboard a grounded airliner without ventilation acquired clinical influenza within 72 hr after exposure (Moser et al., 1979). Outbreaks of H1N1 virus infection on college campuses have seen attack rates of 62–73% in those less than 23 years of age (Layde et al., 1980; Pons et al., 1980).

PANDEMIC INFLUENZA

For all the impact that the persistent annual revisitations of influenza have on our society, it is as the "last great plague" (Kilbourne, 1963; Beveridge, 1977) that the disease holds unique interest. No other disease has the potential to thrust us back into the previous century, as did the pandemic of 1918 (to paraphrase Crosby, 1976).

In Chapter 1 the history of pandemics and their attributes are reviewed in some detail, and Chapters 6, 7, and 9 consider, respectively, aspects of viral genetics, disease, and animal reservoirs that bear on the origin and severity of pandemics. Therefore, the emphasis here is on the unique characteristics and epidemiology of influenza in pandemic form.

Pandemics Defined

Taken literally, the term "pandemic" defines an epidemic that involves all people. In ordinary context the term also implies that the epidemic is pervasive within a relatively brief period (1 to 2 years) and that the disease is more severe than usual. On this latter point there is disagreement. Influenza is a profoundly incapacitating disease from which complete recovery is the rule. The case fatality rate rarely exceeds 0.01%, yet, because of the high morbidity, large numbers of people are killed, recorded as "excess mortality" over usual or preexisting rates. Indeed, excess mortality is a hallmark of influenza. The relatively low mortality associated with the 1947 epidemic [less than that of the 1943–44 epidemic (Sartwell and Long, 1948)] has caused some to question its categorization as a pandemic.

Other characteristics disputably typical of influenza pandemics include cyclical occurrence, a tendency to invade the population in successive waves, and a greater independence of season (Francis, 1958). In fact, early summer outbreaks of influenza were documented in the nonpandemic years of 1943 and 1950–51 (Andrewes, 1951). Thus, these criteria are neither absolute nor restrictive in defining pandemics.

When the H1N1 ("Russian") virus in the early 1950s returned in 1977 bearing recycled HA and NA antigens, its distribution was global, and it produced high morbidity. In that sense it was pandemic—but only pandemic in the young. The failure of the virus to penetrate the firm immunity of those more than 25 years of age precluded the usual emergence of excess mortality that follows influenza in older people. Perhaps the same immune barrier prevented the H1N1 virus from replacing the H3N2 subtype, although H3N2 variants continued to flourish in young people in whom no barrier to H1N1 virus existed.

Pandemic Determinants

The term "pandemic" carries with it the connotation of a ubiquitous infection affecting those in all walks of life and has highly significant implications for public health policy (Chapter 11). The best-studied epidemics of 1957 and 1968 were characterized by (1) major change in the HA antigen of the new virus (Table 10-3), (2) rapid replacement of the preexisting influenza subtype virus variants, (3) susceptibility of all members of the population to the new virus, and (4) significant excess mortality.

In the brief history of influenza accessible to the probes of modern virology (Table 1-3), the "new" pandemic viruses have appeared not to be new at all but previously associated with major epidemics in man (Fig. 10-7). It is then necessary that our pandemic viruses derive their HAs from H1, H2, or H3 subtype viruses (Fig. 10-8), or is the entire arsenal of 13 HA subtypes available for the assembly of human pandemic viruses? (In this connection, constraints on interspecific infection with influenza A virus subtypes are discussed in Chapter 9.)

Constraints seem to exist even on the reentry of "human" HA subtypes into the population. Witness the abortive epidemic of swine influenza virus at Fort Dix, New Jersey in 1976 and the limited incursion of the 1950 H1N1 virus in the juvenile pandemic of 1977–1978. If a human lifetime must span the interval between reappearances of major variants (Chapter 1), both viruses were ahead of schedule.

But the predictions must be tempered by the continuing revelation of the natural history of human influenza. The schema presented in Fig. 10-9 reflects somewhat simplistically the epidemiologic knowledge of a decade ago but is still valid as a representation of the events that follow introduction of a major antigenic variant virus. We now appreciate that there is no constant pattern to the emergence of major variants and that influenza A viruses possessing major antigenic differ-

Table 10-3. Comparative Magnitude of Antigenic Difference between Extreme Interpandemic (Intrasubtypic) "Drift" and Pandemic (Intersubtypic) "Shift" Variants (Plaque Neutralization)[a]

	Virus[b]		
Antiserum	H2('57)	H2('67)	H3
H2N2('57)	*10,240*[c]	2,560	<10
H2N2('67)	320	*10,240*	<10
H3N2	<10	<10	*5,120*

[a]Revised from Kilbourne (1975) Homologous titers are italicized
[b]Test viruses are reassortants with N1 (irrelevant) NA
[c]Highest dilution of antiserum causing 50% reduction in plaque number

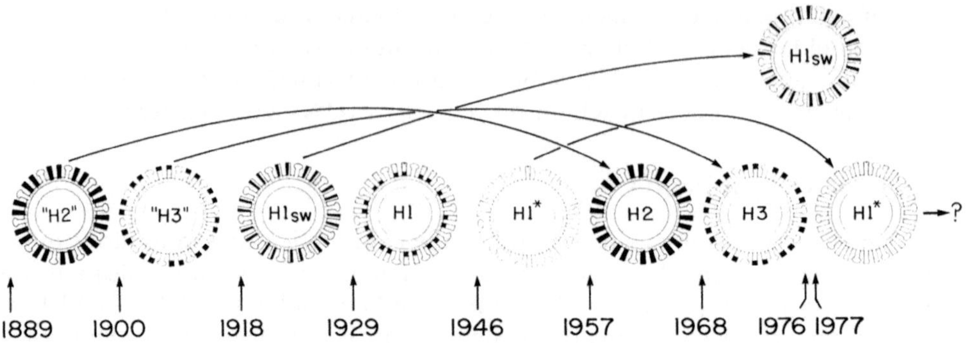

Figure 10-7. Periods of prevalence of major hemagglutinin subtypes of influenza A viruses. The years of introduction of the new subtypes are indicated by arrows. The designation H1, H2, etc. refer to the hemagglutinin subtypes of the viruses without reference to their neuraminidase antigens. Note that the H1 intrasubtypic variants, $H1_{sw}$, H1, and H1*, were associated with significant antigenic change and with major epidemics and are therefore represented. Identification of viral subtypes prior to 1933 (marking the first isolation of a human influenza virus) is deduced from studies of human antibodies. $H1_{sw}$ (swine influenza virus) appeared abortively in the human population in 1976. H3 and H1* subtypes continue to circulate concordantly.

ences can cocirculate. But whether two such variants could cocirculate in a population equally lacking in immunity to each is not yet known. By analogy with the short tenure of the H2N2 subtype (1957–1968), the H3N2 virus should have been replaced in 1979. But by analogy with the H1N1 virus (1918–1957), persistence of the H3N2 virus until the year 2007 might occur. The analogy is imperfect because the former H1N1 viruses had no competition for young respiratory tracts such as that posed to H3N2 viruses by the current "Russian" strains.

Pandemicity and the Phenomenon of Viral Disappearance

Do influenza A viruses disappear because they are forced out by a new more aggressive strain, because they exhaust their capacity for significant antigenic change in the face of rising population immunity, or for both reasons?

One can postulate (Kilbourne, 1957) that

> the periodic disappearance of influenza A viruses suggests that they live in rather precarious balance with man as intensely obligate parasites, and that if survival is dependent on a large number of susceptibles in the population. Pandemic infection results in rapid attainment of high infection density.

The consequent increase in the population of specific antibody response to the virus and broadened immunity to its antigenic variants decreases chances for virus survival. This effect is mediated directly, by decreasing infection of the immune, and indirectly, by decreasing the density and hence transmissibility of infection (Fig. 10-10).

Figure 10-8. Variants of "human" influenza virus subtypes (H1, H2, and H3) are found in domestic animals (shaded figures). See text.

It has been suggested that cumulative antigenic changes at the periphery of the viral HA might compromise the binding and hence replication efficiency of virus at the end of its tenure and thus hasten its disappearance (Whittaker and Underwood, 1980). There is some evidence that viral disappearance is not simultaneous in all areas and that formerly dominant subtype viruses can be isolated several years after new pandemic introductions (Isaacs et al., 1962; Klimov and Ghendon, 1981.)

An influenza virus isolated from an Australian child in 1979 was antigenically and biochemically similar to the prototype H3N2 virus of 1968, A/Aichi/2/68. Laboratory contamination appeared highly improbable, and the NP of the virus differed from 1968 virus but was indistinguishable from that of later H3N2 viruses (Moore et al., 1981). Monto and Maassab (1981) have proposed on the basis of serological evidence that viral subtypes that have "disappeared" continue to circulate in man at low prevalence and that their reemergence is heralded by increased antibody response to them immediately prior to their reappearance as the predominant type.

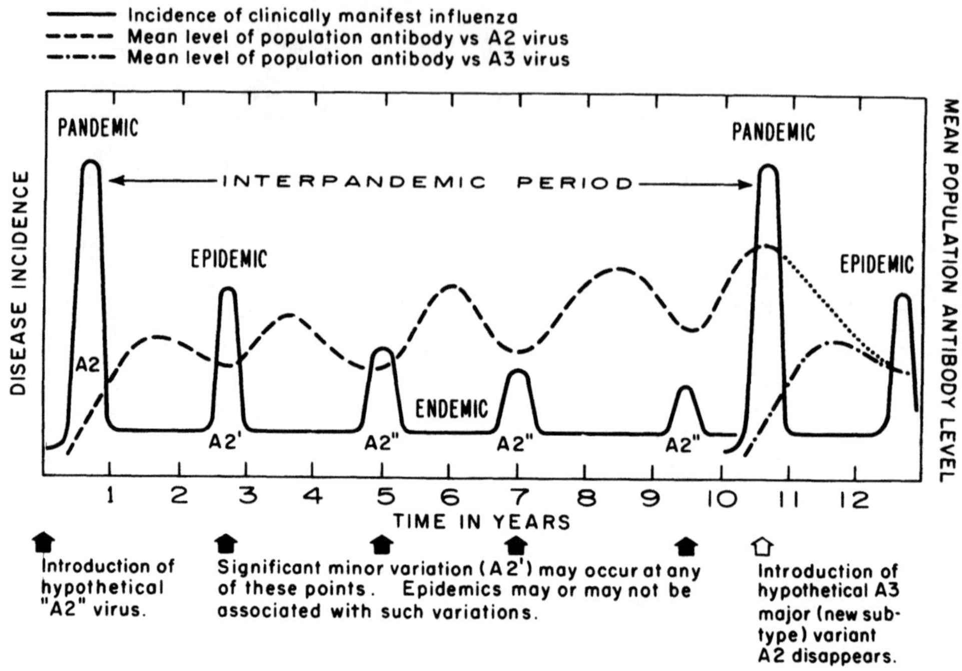

Figure 10-9. Schematic diagram showing the effects of major (pandemic) and minor (interpandemic) influenza virus antigenic variation (global trends). (From Kilbourne, 1975. Reproduced with permission, from *The Influenza Viruses and Influenza,* E. D. Kilbourne (ed.), p. 496, Academic Press, New York.)

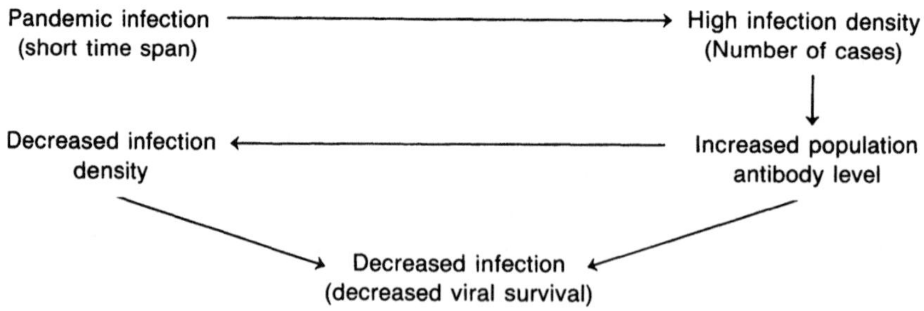

Figure 10-10. Pandemicity may be a factor in influenza virus disappearance. Pandemic infection produces rapid attainment of high infection density in population. Consequent increase in population of specific antibody to the virus decreases chances for virus survival directly by decreasing infection of the immune and indirectly by decreasing infection density (number of cases) and therefore dissemination of virus (herd immunity). (From Kilbourne, 1975. Reproduced with permission, from *The Influenza Viruses and Influenza,* E. D. Kilbourne (ed.), p. 517, Academic Press, New York.)

That viral disappearance need not be extinction is dramatically illustrated by the resurrection of the H1N1 virus in 1977 and by the persistence of vanished human subtypes in lower animals (e.g., swine, Haesebrouck et al., 1985).; i.e., "the graveyard, of old (human) influenza viruses may be the barnyard" (Kilbourne, 1973) (Fig. 10-8).

Conditions for the Entry of Pandemic Viruses

The perpetuation of influenza demands availability of susceptibile human respiratory tracts and the presence in the environment of antigenically novel viruses with human pathogenic potential. At the end of a pandemic era, decreasing circulation of the responsible viral subtype reduces the likelihood of viral interference with invasion by a new virus. The most logical candidate for succession is a virus already well adapted to man but of sufficient antigenic novelty to allow its replication in the immunized population. We are ignorant of the tolerable limits of cumulative point mutations that will allow retention of HA structure necessary for its replication and function but permit antigenic change of sufficient degree for the virus to elude immunologic defenses. At this writing, viruses of the H3N2 subtype continue to flourish. Is it necessary to believe that the seemingly endless cycle of repeated introductions or reintroductions of major variants must continue? Indeed, what is the pathogenic potential for man of HA subtypes H4–H13 no matter how reinforced with human viral genes? Is the human respiratory tract a vacuum ready to engulf any antigenically acceptable errant particle, or is the environment a positive force loaded with viral reassortants ready to assume their place in human history at the slightest faltering of H3N2?

Crowding out is suggested by the persistence or delayed appearance of older strains in isolated populations (Napiorkowski and Black, 1974) beyond the date of introduction of their pandemic successor. Such populations of susceptibles may serve as sensitive indicators of the presence of virus in the larger population at a time when infection is not recognizable in that immune community.

The successor to H3N2 and H1N1 viruses, although it must surmount the barriers posed by their cocirculation, has available the gene pools of both subtypes for combination with the novel antigen, HX.

An exquisite balance of ecological factors must operate in the selection of the viral candidate—no less so than with the political one.

INTERPANDEMIC INFLUENZA

The interpandemic period is a time of complex mutual adaptation of man and virus following the initial saturation of the community with the newly emergent pandemic virus. Minor but significant changes in the virus are occurring as man is responding immunologically to its presence (Kilbourne, 1975) (Fig. 10-5). The gradual acquisition of immunity to the H2N2 virus subtype in a group of Navajo schoolchildren followed for four successive winters is shown in Table 10-4. During this period, significant antigenic changes in the virus were not observed, but

Table 10-4. A Serial Study of Respiratory Illness and Specific Antibody Response to H2N2 Influenza Virus in 77 Navajo School Children[a]

Clinical diagnosis	1956–1957	1957–1958	1958–1959	1959–1960
Influenza	0	18	3	0
Upper respiratory infection[b]	9	16	15	9
Total illnesses	9	34	18	9
Percent illness	(12)	(44)	(22)	(12)
Incidence of antibody increases to H2N2 virus (%)	(0)	(58)	(14)	(26)

[a]Data from Hayslett et al (1962)
[b]For which children sought medical attention

reinfection of this small population, as well as an urban medical student group under surveillance, was shown to occur (Hayslett et al., 1962).

Increasing immunity in the schoolchildren was evidenced by a marked decrease in the proportion of symptomatic infections in 1959–1960 compared to the high clinical attack rate in the first winter. Mean population antibody levels rose during this period (Hayslett et al., 1962). Another measure of the priming of the population through continuing exposure to virus in the postpandemic period is shown by response to vaccine administration. Injection of the same vaccine in the same dose in the pandemic year 1957 and in 1962 resulted in a doubling of the average response in the fifth year of virus prevalence (see Fig. 11-1).

Regional epidemics of influenza A in the interpandemic period tend to occur with approximate biennial frequency (Fig. 10-1) but the pattern is quite variable, and the magnitude of later epidemics may exceed that of earlier ones despite the gradual increase in immunologic experience in the community and a general trend to decreased morbidity (Fig. 10-9). In Houston, six epidemics in an 8-year period were recognizable from excess mortality in the wake of the H3N2 pandemic. The A/Victoria/75 strain produced twice the morbidity and mortality of its immediate predecessor (Glezen and Couch, 1978).

Pandemiclike invasions can occur in isolated populations within interpandemic periods. Almost half the population of the Pacific Island country of Niue developed influenza within a month of the introduction of a 1979 H3N2 virus from New Zealand (Taylor et al., 1985). This experience points to the probable importance of continual viral circulation in large open communities in the maintenance of population immunity. In large communities in the north temperate zone, influenza viruses circulate and produce some measure of disease each winter.

Adaptation of the virus after its initial introduction probably requires more than antigenic change. The 1977 H1N1 virus apparently spared young children in initial outbreaks but affected them as well as previously spared pregnant women in its 1980–81 appearance (Pyhala and Aho, 1982). A much higher incidence in young soldiers than in pregnant women of the same age (Pyhala et al., 1979) can be explained on the basis of differences in intensity of exposure. However, as virus persisted, it began to cause disease in older people in the 1978–79 season (Mathur et al., 1981; Haaheim, 1980). Some H1N1 strains circulating in that winter are known to have been reassortants of H1N1 with H3N2 virus (Young et al., 1979; Bean et al., 1980), which may have affected their virulence. However, reassortant strains have not superseded native H1N1 viruses and have not prevailed in subsequent years.

Although interpandemic disease is more sporadic and less dramatic in its exhibition than pandemic influenza, it constitutes an enormous cumulative burden of disease.

ENDEMIC INFLUENZA

The expression of influenza virus infection as apparent disease or inapparent infection depends on the threshold of clinical perception. Clinical recognition of the individual case is influenced by its observation in the context of an epidemic. Influenza virus infection can occur without symptoms even in those lacking specific antibody to the virus. Such silent infections are part of the epidemiologic chain and, as they increase in proportion to rising population immunity, may facilitate virus spread because they do not immobilize the infected person (Kilbourne, 1975).

Silent or isolated cases of influenza are not detected by the usual indices of community-wide morbidity. However, the endemic presence of virus, even in nonseasonal months, is documented by serological studies. In nonepidemic winters, silent or sporadic infections may be revealed by their sequel of secondary bacterial pneumonia (Maxwell et al., 1949; Stuart-Harris et al., 1949; Kilbourne, 1961).

Thus, the survival of virus between epidemics can be explained as the result of its continuing sequential transmission from person to person at a level below the epidemic threshold.

EPIDEMIOLOGY OF INFLUENZA B AND C

Influenza B

With the important exception that it has not been seen in pandemic form, influenza B is remarkably similar to influenza A in its epidemiology. In seasonal occurrence, high clinical attack rate (in the young), and impact on excess mortality (Nolan et al., 1980), influenza B resembles influenza A on a smaller scale. The concurrence of epidemics of influenza A and B in small populations (reviewed by Kilbourne, 1951) points to common ecological determinants in their causation. Influenza B virus undergoes antigenic drift but not antigenic shift. Its slower antigenic evolution probably accounts for the lower frequency of significant epidemics of influenza B. Perhaps because viral antigenic change is of lesser magnitude and frequency, influenza B affects primarily young adults, adolescents, and children (Retailliau et al., 1979; Wright et al., 1980; La Montagne, 1980).

But severe influenza B with and without bacterial complications has been reported in adults (Baine et al., 1980), and outbreaks in the elderly (Hall et al.,

1981; Van Voris et al., 1982), in whom vaccine failure has been reported (Hall et al., 1981), have caused concern. Attack rates in nosocomial infections of the elderly may be as high as 20% (Van Voris et al., 1982). In another outbreak, influenza B cases were present among hospitalized patients in the absence of recognizable clusters of disease (Bean et al., 1983).

Influenza C

Well-defined epidemics of influenza C have rarely been reported, yet serological evidence demonstrates that influenza C virus infection is very common (see Chapter 7). Influenza C virus has been found in pigs in China (see Chapter 9) but not in other animal hosts. The virus appears to be able to survive with only slight antigenic variation; mutations in its HA are not sequential (Buonagurio et al., 1985), suggesting that the virus is not subject to antigenic drift. Reinfection is common (Dykes et al., 1980; Homma, 1985), so virus survival may not require significant antigenic change. Nevertheless, serological surveillance indicates year-to-year variation in infection frequency similar to that seen with influenza A and B viruses (Troisi and Monto, 1981).

EXPERIMENTAL AND THEORETICAL EPIDEMIOLOGY OF INFLUENZA

Theoretical Epidemiology

Mathematical models have been applied to the prediction of influenza (Baroyan et al., 1971; reviewed by Selby, 1982). When applied retrospectively to actual field observations, the classical epidemic theory of Kermack and McKendrick (1927) proved adequately descriptive, and the estimated values of its parameters could be plausibly related to the influenza deaths in Greater London in the years 1950–1978 (Spicer and Lawrence, 1984).

A stochastic simulation epidemic model developed by Elveback et al. (1976) permits the introduction of family, school, and other subgroup mixing into the structure of the model. The model has been applied to an evaluation of secondary attack rates within households and has suggested that household transmission has an unexpectedly large effect on transmission in the community (Longini et al., 1984b).

A full discussion of epidemic simulation is beyond the scope of this chapter. The present state of the art is well summarized by Selby (1982) and more recently by Longini et al. (1984a,b). To be useful, models must have sufficient flexibility and generality to be appropriate for study of a virus that depends on person-to-person spread in communities with varied populations and social structure (Elvebach et al., 1976).

Experimental Epidemiology

Epidemic conditions can be simulated in animal models (see Chapter 9), and the multiple factors involved can be individually controlled and manipulated.

Transmission experiments in mice, dependent on precise environmental controls, demonstrated aerosol transmission of virus before it was shown to occur in man (Schulman and Kilbourne, 1963a). The demonstration that individual mice differ in their ability to spread virus to their neighbors (Schulman and Kilbourne, 1963b) suggests the possibility of a similar phenomenon in man that may have epidemiologic significance. A hamster model of a human epidemic caused jointly by both influenza A and B viruses showed that viral interference could occur between the two epidemic strains (Kilbourne et al., 1951). Countless other examples of the application of animal model systems to the advancement of knowledge of influenza could be cited, including the development of ts mutant vaccines (see Chapter 11) and the study of age-related susceptibility to infection (Effros and Walford, 1983; Coates et al., 1984).

Experimental infection of human volunteers has also provided valuable information about the incubation period, viral shedding, and secondary spread, although the infecting virus is often attenuated by laboratory passage.

MOLECULAR EPIDEMIOLOGY AND EPIZOOTIOLOGY OF INFLUENZA

Just as understanding of influenza began with the antigenic definition of its causative viruses as A, B, and C, it has been further enhanced by a technology increasingly capable of precise identification of viral variants and analysis of their genetic structure. The definitive epidemiology of influenza is the epidemiology of its molecular determinants—the epidemiology of viral genes. Thus, appreciation of the molecular epidemiology of influenza (Kilbourne, 1973) began with the recognition that the cross reactivity of H2N2 and H3N2 influenza A virus subtypes was mediated by a common NA, antibody to which probably tempered the onslaught of the new Hong Kong H3N2 virus in 1968. Later, molecular genetics demonstrated the new virus to contain all the genes of its H2N2 antecedent save for that coding for its (H3) HA (Scholtissek et al., 1978a). Of course, the reduction of epidemiologic events to molecular terms demands molecular definition not only of the virus but also of the host and even of the environment in which virus and host interact (Kilbourne, 1979). At present we have far more knowledge of the virus than of these other determinants, but studies of receptor specificity, specificity of host proteases in HA cleavage, and the genetics of the immune response should prove rewarding.

Table 10-5 reviews some of the contributions of molecular virology to the understanding of influenza epidemiology and the origin and evolution of influenza viruses. Not shown are contributions of molecular studies to the knowledge of pathogenesis and viral virulence that might have significant indirect effects on influenza epidemiology (see Chapter 7). Rather, the surprising epidemiologic revelations from viral genotyping and more definitive viral characterization are summarized. Gene nucleotide sequencing has opened new perspectives on viral evolution as it affects the genesis and nature of epidemics. Substantive examples of molecular epidemiology are presented briefly in the discussion that follows.

Table 10-5. Contributions of Molecular Virology to Epidemiology of Influenza

Procedure	Epidemiologic contribution
A. Precise viral characterization	Recognition of viral variation (surveillance)
Viral genotyping	Revelation of intraepidemic viral heterogeneity
Monoclonal antibodies	Revelation of viral genetic interaction (reassortment) in man
Oligonucleotide mapping	Tracing of epidemic viruses
Nucleotide and amino acid sequencing	Tracing of viral genes and their origins
	Revelation of virus and gene recycling
	Distribution in nature of viral genes
	Revelation of interspecific gene transfer
B. Analysis of mutations	Definition of rates of nucleotide and amino acid substitutions for viral genes; relation to antigenic variation and epidemiologic events
Nucleotide sequencing	Relation of differential rates of gene evolution to viral adaptation or selection pressure; primary effect on external antigens
	Indication of different evolution of influenza A, B, and C viruses
	Evidence of viral fixation (nonevolution) and reappearance of 1977 H1N1 virus

Intraepidemic Viral Heterogeneity

While antigenic heterogeneity among influenza virus isolates within an epidemic has long been appreciated (Jensen, 1957; Velazco et al., 1983), viral genome analysis (by RNA electrophoresis) has shown that during an epidemic viruses of the same serotype can differ in genes other than that coding for the hemagglutinin (Ghendon et al., 1981). Similar genotyping revealed cocirculation of serologically indistinguishable reassortants and nonreassortant H1N1 virus throughout the world in 1978–80 (Nakajima et al., 1981). Kawaoka and Webster (1985) established that cocirculation occurred during a severe outbreak of virulent and avirulent H5N2 chicken viruses (distinguished by two amino acid differences in their HAs). Cocirculation of antigenic variants of influenza B virus has also been described (Webster and Berton, 1981).

Reassortment of Influenza A H1N1 and H3N2 Viruses in Humans

RNA genotyping of viral isolates after the cocirculation of H1N1 and H3N2 subtype viruses in 1978 revealed that some viruses of H1N1 serotype were genotypically mixed, deriving four or five genes from H3N2 virus (Palese and Young, 1983). Continuing surveillance was able to monitor the relative prevalence of parental and reassortant viruses, and oligonucleotide mapping suggested a similar rate of genetic change of reassortant and nonreassortant viruses (Cox et al., 1983). Double infection of man has been documented (Yamane et al., 1978), as well as direct plaque isolation of an H1N2 reassortant virus from a patient simultaneously infected with H1N1 and H3N2 influenza viruses (Nishikawa and

Sugiyama, 1983). Lvov et al. (1983) reported isolation from a sick child of an H1N3 reassortant of unknown origin.

Tracing of an Epidemic (Epizootic) Virus

Eight months after the termination of a severe epizootic in chickens in Pennsylvania, a direct descendent of the virus was identified in Washington, D.C. by comparative nucleotide sequencing of HA genes (Kawaoka and Webster, 1985).

Molecular Surveillance

The newly reintroduced H1N1 subtype virus is under unprecedented surveillance of its HA molecule as the virus circulates in the human population. Combined with antigenic analysis, this molecular surveillance provides a complete description of the 1982 and 1983 HAs and allows correlation of specific amino acid sequence differences with antigenic changes (Daniels et al., 1985).

Significant correlations of epidemic severity with molecular changes have not yet been made. In the case of H3N2 viruses, the HA of A/Victoria/3/75, a distinctly aggressive strain, uniquely contains an inserted additional asparagine at position 8 (Verhoeyen et al., 1980). Although this insertional substitution may be completely unrelated to the biological or antigenic activity of the variant, further monitoring for change of this site might be instructive if positive correlations with epidemiologic activity can be made.

Evidence of Viral Fixation and Reappearance

When H1N1 virus antigenically similar to 1947–1957 viruses emerged in 1977, it was rapidly demonstrated that it was remarkably similar in all genes to viruses of circa 1950 (Nakajima et al., 1978; Scholtissek et al., 1978b; Bean et al., 1980). This molecular analysis led to startling epidemiologic conclusions: a virus genetically unchanged after more than 20 years could not have replicated continuously during its apparent absence from man without significant mutations; therefore, it must have been maintained in the latent state or have been introduced following laboratory storage. Biological persistence of influenza virus is not known to occur, but this finding prompts reexamination of the possibility.

The genetic relatedness of some 1978–1979 H1N1 strains to a 1953 H1N1 virus (Nakajima et al., 1979) cannot be explained as a recapitulation of the same evolutionary pathway but again suggests the stable evolutionary fixation of virus for more than two decades, although the postulation of two separate reintroductions of H1N1 viruses strains credulity.

Implications of Molecular Evolution for Influenza Epidemiology: The Case of Influenza C Virus

The HA of influenza C virus lacks primary sequence homology with the HAs of influenza A and B viruses (Nakada et al., 1984a; Pfeifer and Compans, 1984), but common structural features suggest the derivation of the three viral types from a common ancestor (Nakada et al., 1984a). A common origin is further suggested by the partial sequence homology of the influenza C virus NP gene with that of influenza A and B viruses (Nakada et al., 1984b).

The lesser genomic diversity among type C strains (Palese et al., 1981) is concordant with their lesser antigenic variation and epidemicity. The previously cited evidence of noncumulative mutations in influenza C virus HA (Buonagurio et al., 1985) can be correlated with the different epidemiology of influenza C, in which sharply defined epidemics of influenza associated with new variants are uncommon. Genome diversity has been found in contemporaneously circulating influenza C viruses (Kawamura et al., 1986).

Finally, molecular virology indicates that influenza C virus strains from pigs are similar to but distinguishable from human strains (Guo and Desselberger, 1984). The viral proteins of human influenza C viruses have been conserved over a 35-year period and differ little from those of contemporary swine isolates (Elliott et al., 1984). These findings do not resolve current epidemiologic questions about influenza C as a zoonosis but offer a valuable base line for further study.

The Limitations and Promise of Molecular Epidemiology

The yield of gene sequencing, as with other basic techniques, will depend more on the quality of the questions asked than the number of genes analyzed. The accumulating data base is of undeniable value but in essence represents descriptive science using advanced technology. One sees the beginnings now of gene manipulation and site-specific mutagenesis that can approach the more profound epidemiologic questions. For example, did H2N2 virus disappear from man after only a decade of prevalence because it had exhausted its capability of evolving new HA epitopes compatible with maintenance of its structural integrity? Or, perhaps more likely, did the virus simply suffer from bad luck in not finding the right mutation in the right host at the right time carry it on?

REFERENCES

Aho, K., Pyhälä, R., and Visakorpi, R., 1980, ABO associated genetic determinant in H1N1 influenza, *Tissue Antigens* **16:**310–313.

Andrewes, C. H., 1951, Epidemiology of influenza in the light of the 1951 outbreak, *Proc. R. Soc. Med.* **44:**803–804.

Assaad, F., Brès, P., and Delon, P., 1977, Surveillance internationale de la grippe, *Rev. Epidemiol. Sante Publ.* **25:**441–445

Baine, W. B., Luby, J. P., and Martin, S. M., 1980, Severe illness with influenza B, *Am. J. Med.* **68:**181.

Barb, K., and Takàtsy, G. Y., 1973, Unusual features of the 1971 influenza epidemic in Hungary and the antigen analysis of the isolates by a serum adsorption test, *Bull. WHO* **49:**21–24.

Barker, W. H., and Mullooly, J. P., 1980, Impact of epidemic type A influenza in a defined adult population, *Am. J. Epidemiol.* **112:**798–813.

Barker, W. H., and Mullooly, J. P., 1981, Underestimation of the role of pneumonia and influenza in causing excess mortality, *Am. J. Public Health* **71:**643–645.

Barker, W. H , and Mullooly, J. P., 1982, Re: "A study of excess mortality during influenza epidemics

in the United States, 1968-1976," *Am J Epidemiol* **115**:479-480
Baroyan, O V, Rvachev, L. A., Basilevsky, U. V, et al., 1971, *Adv. Appl Probabil.* **3**:224-226.
Bean, W J., Jr, Cox, N. J., and Kendal, A. P, 1980, Recombination of human influenza A viruses in nature, *Nature* **284**:638-640.
Bean, B, Rhame, F S, Hughes R. S., et al, 1983, Influenza B Hospital activity during a community epidemic, *Diag. Microbiol. Infect Dis* **1**:177-183
Beare, A. S., Bynoe, M L., and Tyrrell, D A J., 1968, Investigation into the attenuation of influenza viruses by serial passage, *Br Med J* **4**:482-484.
Belbin, D J, and Smithson, W H, 1979, "Red 'flu'" A study of an epidemic in a girls' boarding school in February 1978, *J. R Coll Gen Practit* **29**:151-154
Beveridge, W I B, 1977, *Influenza The Last Great Plague An Unfinished Story of Discovery* Heinemann, London
Beyer, W E P, and Masurel, N, 1985, Antigenic heterogeneity among influenza A(H3N2) field isolates during an outbreak in 1982/83, estimated by methods of numerical taxonomy, *J Hyg (Camb)* **94**:97-109
British Medical Journal, 1974, Epidemiology Seasonal variations in influenza, *Br Med J* **2**:511
Buchan, K A, and Reid, D, 1972, Epidemiological aspects of influenza in Lewis, *Health Bull* **30**:183-186
Buonagurio, D A, Nakada, S, Desselberger, U, et al 1985, Noncumulative sequence changes in the hemagglutinin genes of influenza C virus isolates, *Virology* **146**:221-232
Carey, D L, Dunn, F. L, Robinson, R Q, 1958, Communitywide epidemic of Asian Strain influenza Clinical and subclinical illness among school children, *J A M.A.* **167**:1459-1463
Centers for Disease Control, 1985, Influenza—United States, 1984-1985 season, *Morbid. Mortal Weekly Rep.* **34**:440-443.
Choi, K., and Thacker, S. G., 1981, An evaluation of influenza mortality surveillance, 1962-1979 II. Percentage of pneumonia and influenza deaths as an indicator of influenza activity, *Am J. Epidemiol* **113**:227-235.
Choi, K, and Thacker, S. G., 1982, Mortality during influenza epidemics in the United States, 1967-1978, *Am J Public Health* **72**:1280-1283
Clarke, S K R, Heath, R. B, Sutton, R N. P., et al., 1958, Serological studies with Asian strain influenza A, *Lancet* **1**:814-818.
Clifford, R E, Smith, J. W. G., Tillett, H E, 1977, Excess mortality associated with influenza in England and Wales, *Int J Epidemiol.* **6**:115-128.
Coates, D. M., Husseini, R H., Rushton, D I, et al., 1984, The role of lung development in the age-related susceptibility of ferrets to influenza virus, *Br. J Exp. Pathol.* **65**:543-547.
Collins, S D, Frost, W. H, Gover, M, et al., 1930, Mortality from influenza and pneumonia in 50 large cities of the United States, 1910-1929, *Public Health Rep* **45**:2277-2329.
Cox, N J, Bai, Z S, and Kendal, A P, 1983, Laboratory-based surveillance of influenza A(H1N1) and A(H3N2) viruses in 1980-81 Antigenic and genomic analyses, *Bull. WHO* **61**:143-152
Crosby, A W, 1976, *Epidemic and Peace, 1918,* Greenwood Press, Westport, CT.
Dalton, K, 1977, *The Premenstrual Syndrome and Progesterone Therapy,* Heinemann Medical Books, London
Daniels, R S, Douglas, A R, Skehel, J J, et al, 1985, Antigenic and amino acid sequence analyses of influenza viruses of the H1N1 subtype isolated between 1982 and 1984, *Bull WHO* **63**:273-277.
Davey, M L, and Reid, D, 1972, Relationship of air temperature to outbreaks of influenza, *Br. J Prev Soc Med* **26**:28-32
Davies, J R, Grilli, E A., and Smith, A J., 1984, Influenza A Infection and reinfection, *J Hyg (Camb)* **92**:125-127
Dingle, J H, Badger, G F, and Jordan, W S., 1964, *Illness in the Home,* The Press of Western Reserve University, Cleveland
Dowdle, W R, Coleman, M T., and Gregg, M. B., 1974, Natural history of influenza type A in the United States, 1957-1972, *Prog Med Virol* **17**:91-135
Dunn, F L, 1958, Pandemic influenza in 1957. Review of international spread of new Asian strain, *J A M A* **166**:1140-1148
Dykes, A C, Cherry, J D, and Nolan, C. E., 1980, A clinical, epidemiologic, serologic, and viro-

logic study of influenza C virus infection, *Arch. Intern. Med.* **140**:1295–1298.

Effros, R. B., and Walford, R. L., 1983, The immune response of aged mice to influenza: Diminished T-cell proliferation, interleukin 2 production and cytotoxicity, *Cell. Immunol.* **81**:298–305.

Eickhoff, T. C., Sherman, I. L., and Serfling, R. E., 1961, Excess mortality associated with epidemic influenza, *J.A.M.A* **176**:776.

Elliott, R. M., Guo, Y., and Desselberger, U., 1984, Polypeptide synthesis in MDCK cells infected with human and pig influenza C viruses, *J. Gen. Virol* **65**:1873–1880, 1984.

Elveback, L. R., Fox, J. P., Ackerman, E., et al., 1976, An influenza simulation model for immunization studies, *Am. J. Epidemiol.* **103**:152–165.

Evans, A. S., Shepard, K., and Richards, V., 1970, Blood-groups and influenza, *Lancet* **2**:1364.

Farr, W., 1885, *Vital Statistics,* Office of the Sanitary Institute, London, p 330.

Finklea, J. F., Sandifer, S. H., and Smith, D. D., 1969, Cigarette smoking and epidemic influenza, *Am. J Epidemiol.* **90**:390–399.

Fox, J. P., and Hall, C. E., 1980, *Viruses in Families,* PSG Publishing, Littleton, MA.

Fox, J. P., and Kilbourne, E. D., 1973, Epidemiology of influenza—Summary of influenza workshop IV, *J. Infect. Dis.* **128**:361–386.

Fox, J. P., Hall, C. E., Cooney, M. K., et al., 1982, Influenzavirus infections in Seattle families, 1975–1979. I. Study design, methods and the occurrence of infections by time and age, *Am J Epidemiol* **116**:212–227.

Foy, H. M., Hall, C., Cooney, M.K., et al., 1979, Influenza surveillance by age and target group, *Am. J Epidemiol* **109**:582–587.

Francis, T., Jr., 1958, Influenza. Epidemiology. Influenza as a military problem, in: *Preventive Medicine in World War II.* Vol. IV. *Communicable Diseases* (J B. Coates, Jr., E. D. Hoff, and P. M. Hoff, eds.), The Surgeon General, Department of the Army, Washington, pp. 85–128.

Frank, A. L., Taber, L. H., and Wells, J. M., 1983, Individuals infected with two subtypes of influenza A virus in the same season, *J. Infect. Dis.* **147**:120–124.

Frank, A. L., Taber, L. H., Glezen, W. P., et al., 1979, Reinfection with influenza A (H3N2) virus in young children and their families, *J. Infect. Dis* **140**:829–835.

Ghendon, Y., Klimov. A., Gorodkova, N., et al., 1981, Genome analysis of influenza A virus strains isolated during an epidemic of 1979–1980, *J. Gen. Virol.* **56**:303–313.

Glezen, W. P., 1982, Serious morbidity and mortality associated with influenza epidemics, *Epidemiol. Rev.* **4**:25–44.

Glezen, W. P., and Couch, R. B., 1978, Interpandemic influenza in the Houston area, 1974–76, *N. Engl. J. Med.* **298**:587–592.

Glezen, W. P., Couch, R. B., Six, H. R. 1982a, The influenza herald wave, *Am. J. Epidemiol.* **116**:589–598.

Glezen, W. P., Payne, A. A., Snyder, D. N., et al., 1982b, Mortality and influenza, *J. Infect. Dis.* **146**:313–321.

Glezen, W. P., Six, H. R., Perrotta, D. M., et al., 1984, Epidemics and their causative viruses—Community experience, in: *The Molecular Virology and Epidemiology of Influenza* (C. Stuart-Harris and C. W. Potter, eds.), Academic Press, pp. 17–38.

Goldstein, I. F., and Block, G., 1976, A method for surveillance of influenza epidemics, *Am J. Public Health* **66**:992–993.

Guo, Y., and Desselberger, U., 1984, Genome analysis of influenza C viruses isolated in 1981/82 from pigs in China, *J. Gen. Virol.* **65**:1857–1872.

Haaheim, L. R., 1980, Surveillance of the antibody status to current influenza strains in the Norwegian population. 2: Antibodies to the A/USSR/90/77(H1N1) virus in the various age groups after the second epidemic wave, *Natl. Inst. Public Health Ann.* **3**:49–56.

Haesebrouck, F., Biront, P., Pensaert, M. B., et al., 1985, Epizootics of respiratory tract disease in swine in Belgium due to H_3N_2 influenza virus and experimental reproduction of disease, *Am J Vet. Res.* **46**:1926–1928.

Hall, C. E., Cooney, M. K., and Fox, J. P., 1973, The Seattle virus watch. IV. Comparative epidemiologic observations of infections with influenza A and B viruses, 1965–1969, in families with young children, *Am. J. Epidemiol.* **98**:365–380.

Hall, W. N., Goodman, R. A , Noble, G. R , et al , 1981, An outbreak of influenza B in an elderly population, *J Infect Dis* **144**:297-302

Hayslett, J., McCarroll, J , Brady, E., et al., 1962, Endemic influenza I Serologic evidence of continuing and subclinical infection in disparate populations in the post-pandemic period, *Am. Rev. Respir Dis* **85**:1-8.

Hemmes, J H , Winkler, K. C , and Kool, S. M., 1960, Virus survival as a seasonal factor in influenza and poliomyelitis, *Nature* **188**:430-431.

Homma, M., 1985, Epidemiological characteristics of type C influenza, *J Cell Biochem (Suppl)* **9C**:278

Hope-Simpson, R. E , 1981, The role of season in the epidemiology of influenza, *J Hyg (Camb)* **86**:35-47

Hoskins, T W , Davies, J. R , Smith, A. J., et al , 1979, Assessment of inactivated influenza A vaccine after three outbreaks of influenza A at Christ's Hospital, *Lancet* **1**:33-35

Housworth, J , and Langmuir, A D., 1974, Excess mortality from epidemic influenza, 1957-1966, *Am J Epidemiol.* **100**:40-48.

Isaacs, A , Hart, R J C , and Law, V. G , 1962, Influenza viruses, 1957-1960, *Bull WHO* **26**:253-259

Jennings, L C , and Miles, J A R , 1978, A study of acute respiratory disease in the community of Port Chalmers II Influenza A/Port Chalmers/1/73 intrafamilial spread and the effect of antibodies to the surface antigens, *J Hyg (Camb)* **81**:67-75

Jensen, K E , 1957, Immunologic significance of antigenic differences among strains of influenza A isolated during the same epidemic, *J Immunol* **78**:373-379.

Kark, J D , Lebiush, M , and Rannon, L , 1982, Cigarette smoking as a risk factor for epidemic A(H_1N_1) influenza in young men, *N Engl J Med.* **307**:1042-1046.

Kavet, J., 1977, A perspective on the significance of pandemic influenza, *Am J Public Health* **67**:1063-1070.

Kawamura, H , Tashiro, M , Kitame, F , et al , 1986, Genetic variation among human strains of influenza C virus isolated in Japan, *Virus Res* **4**:275-288

Kawaoka, Y , and Webster, R G , 1985, Evolution of the A/chicken/Pennsylvania/83 (H5N2) influenza virus, *Virology* **146**:130-137

Kendal, A. P , Schieble, J., Cooney, M. K., et al , 1978, Co-circulation of two influenza A (H3N2) antigenic variants detected by virus surveillance in individual communities, *Am J Epidemiol.* **108**:308-311

Kendal, A P , Joseph, J. M., Kobayashi, G., et al., 1979, Laboratory-based surveillance of influenza virus in the United States during the winter of 1977-1978, *Am. J Epidemiol.* **110**:449-461

Kermack, W O., and McKendrick, A G., et al , 1927, A contribution to the mathematical theory of epidemics, *Proc R Soc.* **A115**:700-721

Kilbourne, E D , 1961, Influenza—the cryptic killer, *Am Rev Resp. Dis* **83**:265-266

Kilbourne, E D., 1963, Infections primarily of the respiratory tract, in: *Preventive Medicine and Public Health* (W. G. Smillie and E D Kilbourne, eds), Macmillan, New York, p. 192

Kilbourne, E D , 1973, The molecular epidemiology of influenza *J Infect. Dis* **127**:478-487.

Kilbourne, E D , 1975, Epidemiology of influenza, in: *The Influenza Viruses and Influenza* (E D Kilbourne, ed.), Academic Press, New York, pp 483-538.

Kilbourne, E. D., 1979, Molecular epidemiology—influenza as archetype, *Harvey Lect.* **73**:225-258.

Kilbourne, E D., 1980, Influenza: Viral determinants of the pathogenicity and epidemicity of an invariant disease of variable occurrence, *Phil. Trans. R Soc. Lond [Biol]* **288**:291-297.

Kilbourne, E. D , Anderson, H. C., and Horsfall, F L., Jr , 1951, Concurrent infection with influenza A and B viruses in a single epidemic, *J. Immunol.* **67**:547-558

Klimov, A I., and Ghendon, Y. Z., 1981, Genome analysis of H1N1 influenza virus strains isolated in the U.S.S.R during an epidemic in 1961-1962, *Arch Virol.* **70**:225-232.

Knox, J. D E , and Anderson, R. A., 1982, Influenza-like illness: A monitoring system, *Ecology Dis.* **1**:87-91

LaMontagne, J. R , 1980, Summary of a workshop on influenza B viruses and Reye's syndrome, *J Infect Dis* **142**:452-465

Langmuir, A. D , 1961, Epidemiology of Asian influenza With special emphasis on the United States, *Am Rev Respir Dis* **83**:2-14.

Layde, P. M., Engelberg, A. L., Dobbs, H. I., et al., 1980, Outbreak of influenza A/USSR/77 at Marquette University, *J. Infect. Dis.* **142**:347–352.

Lebiush, M., Rannon, L., and Kark, J. D, 1981, The relationship between epidemic influenza $A(H_1N_1)$ and ABO blood groups, *J. Hyg. (Camb.)* **87**:139–146.

Levine, M. M., Togo, Y., and Wald, E. R., 1974, Practicability of influenza surveillance by virus culture of children, *Am. J. Epidemiol.* **100**:272–276.

Lim, K. A., Smith, A., Hale, J. H., et al., 1957, Influenza outbreak in Singapore, *Lancet* **2**:791–796.

Longini, I. M., Jr., Monto, A. S., and Koopman, J, 1984a, Statistical procedures for estimating the community probability of illness in family studies: Rhinovirus and influenza, *Int. J. Epidemiol* **13**:99–106.

Longini, I. M., Jr., Seaholm, S. K., Ackerman, E., et al., 1984b, Simulation studies of influenza epidemics: Assessement of parameter estimation and sensitivity, *Int. J. Epidemiol.* **13**:496–501.

Lvov, D. K., Zhdanov, V. M., Sadykhova, F., et al., 1983, Isolation of a natural recombinant of influenza A virus (H1N3) from a sick child, *Lancet* **2**:1246–1247.

Macfarlane, A., 1977, Daily mortality and environment in English conurbations. I. Air pollution, low temperature, and influenza in Greater London, *Br. J Prev. Soc. Med.* **31**:54–61.

MacKenzie, J. S., and Fimmel, P. J., 1978, The effect of ABO blood groups on the incidence of epidemic influenza and on the response to live attenuated and detergent split influenza virus vaccines, *J. Hyg (Camb.)* **80**:21–30.

MacKenzie, J. S., MacKenzie, I. H., and Holt, P. G., 1976, The effect of cigarette smoking on susceptibility to epidemic influenza and on serological responses to live attenuated and killed subunit influenza vaccines, *J. Hyg. (Camb.)* **77**:409–417.

Mann, P. G., Pereira, M S., Smith, J. W. G., et al., 1981, A five-year study of influenza in families, *J. Hyg. (Camb.)* **87**:191–200.

Marine, W. M., McGowan, J. E., Jr., and Thomas, J. E., 1976, Influenza detection: A prospective comparison of surveillance methods and analysis of isolates, *Am. J. Epidemiol.* **104**:248–255.

Mathur, U., Bentley, D. W., Hall, C. B., et al., 1981, Influenza A/Brazil/78(H1N1) infection in the elderly, *Am. Rev. Respir. Dis* **123**:633–635.

Maxwell, E. S., Ward, T. G., and Van Metre, T. E., Jr., 1949, The relation of influenza virus and bacteria in the etiology of pneumonia, *J Clin. Invest.*, **28**:307–318.

Miller, D. L, Reid, D., Diamond, J. R, et al., 1973, Hong Kong influenza in the Royal Air Force 1968–70, *J Hyg. (Camb.)* **71**:535–547.

Montefiore, D., Drozdov, S. G., Kafuko, G. W., et al., 1970, Influenza in East Africa, 1969–70, *Bull WHO* **43**:269–273.

Monto, A. S., and Kioumehr, F., 1975, The Tecumseh study of respiratory illness. IX. Occurrence of influenza in the community, 1966–1971, *Am. J. Epidemiol.* **102**:553–563.

Monto, A. S., and Maassab, H. F., 1981, Serologic responses to nonprevalent influenza A viruses during intercyclic periods, *Am. J. Epidemiol.* **113**:236–244.

Moore, B. W., Webster, R. G., Bean, W. J., et al., 1981, Reappearance in 1979 of a 1968 Hong Kong-like influenza virus, *Virology* **109**:219–222.

Moser, M. R., Bender, T. R., Margolis, H. S., et al., 1979, An outbreak of influenza aboard a commercial airliner, *Am. J. Epidemiol.* **110**:1–6.

Mulder, J., and Masurel, N., 1958, Pre-epidemic antibody against 1957 strain of Asiatic influenza, *Lancet* **2**:810–814.

Nakada, S., Creager, R. S., Krystal, M., et al., 1984a, Influenza C virus hemagglutinin: Comparison with influenza A and B virus hemagglutinins, *J. Virol.* **50**:118–124.

Nakada, S., Creager, R. S., Krystal, M., et al., 1984b, Complete nucleotide sequence of the influenza C/California/78 virus nucleoprotein gene, *Virus Res.* **1**:433–441.

Nakajima, K., Desselberger, U., and Palese, P., 1978, Recent human influenza A (H1N1) viruses are closely related genetically to strains isolated in 1950, *Nature* **274**:334–339.

Nakajima, K., Nakajima, S., Nerome, K., et al., 1979, Genetic relatedness of some 1978–1979 influenza H1N1 strains to 1953 H1N1 strain, *Virology* **99**:423–426.

Napiorkowski, P. A., and Black, F. L., 1974, Influenza A in an isolated population in the Amazon, *Lancet* **2**:1390–1391.

Nishikawa, F., and Sugiyama, T., 1983, Direct isolation of H1N2 recombinant virus from a throat swab of a patient simultaneously infected with H1N1 and H3N2 influenza A viruses, *J Clin Microbiol* **18**:425-427.

Nolan, T F , Goodman, R. A., Hinman, A R , et al , 1980, Morbidity and mortality associated with influenza B in the United States, 1979-1980: A report from the Center for Disease Control, *J Infect Dis.* **142**:360-362

Olson, J. G., and Irving, G S , 1979, An explosive outbreak of influenza caused by A/USSR/77-like virus on a United States naval ship, *Mil Med* **144**:743-745

Palese, P., and Young, J. F , 1983, Molecular epidemiology of influenza, in: *Influenza Virus Genetics* (P Palese and D W Kingsbury, eds), Springer-Verlag, New York, pp 321-336

Palese, P , Elliott, R M., Baez, M , et al , 1981, Genome diversity among influenza A, B and C viruses and genetic structure of RNA 7 and RNA 8 of influenza A viruses, in: *Genetic Variation among Influenza Viruses* (D. P. Nayak, ed.), Academic Press, New York, pp 127-140

Pereira, M S , and Chakraverty, P , 1977, The laboratory surveillance of influenza epidemics in the United Kingdom 1968-1976, *J Hyg. (Camb)* **79**:77

Pereira, M S , and Chakraverty, P , 1982, Influenza in the United Kingdom 1977-1981, *J. Hyg. (Camb)* **88**:501-512

Pereira, M , Assaad, F A , and Delon, P. J., 1978, Influenza surveillance, *Bull WHO* **56**:193-203

Peretjagina, N S , Antonova, I V , and Urbah, V. J , 1977, The upper tolerance limits of nonepidemic daily morbidity for influenza and other acute respiratory diseases in the epidemic season, *Bull WHO* **55**:761-763

Peterson, D , Andrews, J S., Jr., Levy, B. S , et al., 1979, An effective school-based influenza surveillance system, *Public Health Rep* **94**:88-92

Pfeifer, J B , and Compans, R. W , 1984, Structure of the influenza C glycoprotein gene as determined from cloned DNA, *Virus Res* **1**:281-296

Plesník, V , Heinz, F , Matuska, J , et al., 1978, Economic losses in influenza epidemics, *Cas Lek Cesk* **117**:53-57

Poe, G S., and Massey, J T , 1977, Estimating influenza cases and vaccinations by means of weekly rapid reporting system Methodological considerations and results obtained in the U S Health Interview Survey, *Public Health Rep.* **92**:299-306

Pons, V G , Canter, J , and Dolin, R., 1980, Influenza A/USSR/77 (H1N1) on a university campus, *Am J Epidemiol.* **111**:23-30.

Potter, C W , and Schild, G C., 1967, The incidence of HI antibody to influenza virus A2/Singapore/1/57 in individuals of blood groups A and O, *J Immunol* **98**:1320-1325

Pyhala, R , and Aho, K , 1982, Seroepidemiology of H1N1 influenza The range of the affected population is widening, *Med Biol* **60**:328-332.

Pyhälä, R , Aho, K , and Visakorpi, R., 1979, Seroepidemiology of H1N1 influenza Striking differences in the attack rate among young people, *Acta Pathol. Microbiol. Scand.* **B87**:161-164.

Retailliau, H. F., Storch, G A , Curtis, A. C., et al., 1979, The epidemiology of influenza B in a rural setting in 1977, *Am J Epidemiol* **109**:639-649.

Rosenstock, I M., Hochbaum, G M., Leventhal, H., et al., 1960, *The Impact of Asian Influenza on Community Life. A Study in Five Cities,* U.S. Dept. of Health, Education, and Welfare, Bethesda

Rubin, R J , and Gregg, M. B., 1975, Influenza surveillance in the United States, 1972-1974, *Am. J. Epidemiol* **102**:225-232

Sartwell, P E., and Long, A P., 1948, The Army experience with influenza, 1946-1947 I Epidemiological aspects, *Am J. Hyg.* **47**:135-141

Schild, G C., Henry-Aymard, M., Pereira, M. S., et al., 1973, Antigenic variation in current human type A influenza viruses: Antigenic characteristics of the variants and their geographic distribution, *Bull WHO* **48**:269-278.

Schild, G C , Newman, R W , McGregor, I A , et al , 1977, The use of transportable single-radial-diffusion immunoplates in seroepidemiological studies of influenza in the Gambia. The occurrence and persistence of antibody to influenza A/Hong Kong/68 (H3N2) virus in selected inhabitants of two rural villages, *Bull WHO* **55**:3-13

Schoenbaum, S C , McNeil, B J., and Kavet, J., 1976, The Swine-influenza decision, *N Engl. J. Med.*

295:759–765.

Scholtissek, C., Rohde, W., von Hoyningen, V., et al., 1978a, On the origin of the human influenza subtypes H2N2 and H3N2, *Virology* **87**:13–20.

Scholtissek, C., von Hoyningen, V., and Rott, R., 1978b, Genetic relatedness between the new 1977 epidemic strains (H1N1) of influenza and human influenza strains isolated between 1947 and 1957 (H1N1), *Virology* **89**:613–617.

Schulman, J. L., and Kilbourne, E. D., 1963a, Experimental transmission of influenza virus infection in mice. I. The period of transmissibility, *J. Exp. Med.* **118**:257–266.

Schulman, J. L., and Kilbourne, E. D., 1963b, Experimental transmission of influenza virus infection in mice. II. Some factors affecting the incidence of transmitted infection, *J. Exp. Med.* **118**:267–275.

Schulman, J. L., and Kilbourne, E. D., 1966, Seasonal variations in the transmission of influenza virus infection in mice, in *Biometeorology II. Proceedings Third International Biometeorology Congress, Pau, S. France, 1–7 Sept., 1963,* Pergamon Press, New York, pp. 83–87.

Selby, P., 1982, Influenza Models. Prospects for Development and Use, MTP Press, Lancaster.

Serfling, R. E., 1963, Methods for current statistical analysis of excess pneumonia–influenza deaths, *Public Health. Rep* **78**:494–506.

Shadrin, A. S., Marinich, I. G., and Taros, L. Y., 1977, Experimental and epidemiological estimation of seasonal and climatogeographical features of non-specific resistance of the organism to influenza, *J. Hyg. Epidemiol. Microbiol. Immunol.* **21**:155–161.

Shen, F.-Z., Zhao, G -F., and Xu, X.-L., 1983, Influenza surveillance, *Chin Med. J.* **96**:349–354.

Six, H. R., Webster, R. G., Kendal, A. P., et al , 1983, Antigenic analysis of H1N1 viruses isolated in the Houston metropolitan area during four successive seasons, *Infect. Immun.* **42**:453–458.

Sonoguchi, T., Naito, H., Hara, M., et al., 1985, Cross-subtype protection in humans during sequential, overlapping, and/or concurrent epidemics caused by H3N2 and H1N1 influenza viruses, *J. Infect. Dis.* **151**:81–88.

Spelman, D. W., and McHardy, C. J., 1985, Concurrent outbreaks of influenza A and influenza B, *J. Hyg. (Camb.)* **94**:331–339

Spencer, M. J., Cherry, J. D., and Teresaki, P. I., 1976, HL-A antigens and antibody response after influenza A vaccination. Decreased response associated with HL-A type W16, *N. Engl. J. Med.* **294**:13–16.

Spicer, C. C., and Lawrence, C. J , 1984, Epidemic influenza in Greater London, *J Hyg. (Camb)* **93**:105–112.

Stuart-Harris, C. H., Laird, J., Tyrrell, D. A., et al., 1949, The relationship between influenza and pneumonia, *J. Hyg.* **47**:434–448.

Taber, L. H., Paredes, A., Glezen, W. P., et al., 1981, Infection with influenza A/Victoria virus in Houston families, 1976, *J. Hyg. (Camb.)* **86**:303–313.

Tateno, I., Suzuki, S., Nakamura, S., et al., 1963, Etiology and seasonal incidence of acute nonbacterial respiratory infections in adults with emphasis on influenza, *Jpn. J. Exp. Med* **33**:159.

Taylor, R., Nemaia, H., Tukuitonga, C., et al., 1985, An epidemic of influenza in the population of Niue, *J. Med. Virol.* **16**:127–136.

Tillett, H. E., and Spencer, I.-L., 1982, Influenza surveillance in England and Wales using routine statistics, *J. Hyg. (Camb.)* **88**:83–94.

Troisi, C. L., and Monto, A. S., 1981, Comparison of enzyme-linked immunosorbent assay and hemagglutination inhibition in a seroepidemiological study of influenza type C infection, *J. Clin Microbiol.* **14**:515–521.

Van Voris, L. P., Belshe, R. B., and Shaffer, J. L., 1982, Nosocomial influenza B virus infection in the elderly, *Ann. Intern. Med.* **96**:153–158.

Velazco, J. G., Couch, R. B., Six, H. R., et al., 1983, Surveillance of influenza in Houston, Texas, USA: Gradual transition from A/Victoria/75 (H3N2) to A/Texas/77 (H3N2) predominance and antigenic characterization of "intermediate" strains, *Bull. WHO* **61**:345–352.

Verhoeyen, M., Fang, R., Min Jou, W., et al., 1980, Antigenic drift between the haemagglutinin of the Hong Kong influenza strains A/Aichi/2/68 and A/Victoria/3/75, *Nature* **286**:771–776

von Heise, R., Stelzner, A., Urbach, H., et al., 1976, Blood group affiliation of positive influenza A2 sera, *Z. Ges. Hyg* **22**:229–231.

Watkins, I J., Tills, D , and Health, R B , 1975, Studies of the genetic susceptibility of individuals to infection with influenza viruses, *Humangenetik* **30:**75-79
Webster, R G , and Berton, M T , 1981, Analysis of antigenic drift in the haemagglutinin molecule of influenza B virus with monoclonal antibodies, *J Gen Virol* **54:**243-251
Wellıver, R C , Cherry, J D , Boyer, K M , et al , 1979, Sales of nonprescription cold remedies· A unique method of influenza surveillance, *Pediatr Res* **13:**1015-1017.
Whittaker, R G , and Underwood, P. A , 1980, A mechanism for influenza subtype disappearance, *Med Hypotheses* **6:**997-1008.
Widelock, D., Schaeffer, M., and Millian, S J , 1965, Surveillance of infectious disease by serologic methods I Asian influenza, 1957-1964, *Am J Public Health.* **55:**578-586
Wiley, D C , Wilson, I A , and Skehel, J J., 1981, Structural identification of the antibody binding sites of Hong Kong influenza hemagglutinin and their involvement in antigenic variation, *Nature* **289:**373-378
Wright, P. F , Bryant, J D , and Karzon, D T , 1980, Comparison of influenza B/Hong Kong virus infections among infants, children, and young adults, *J Infect Dis* **141:**430-435
Yamane, N., Arikawa, J., Odagiri, T., et al , 1978, Isolation of three different influenza A viruses from an individual after probable double infection with H3N2 and H1N1 viruses, *Jpn J Med Sci. Biol* **31:**431-434
Young, J F , and Palese, P , 1979, Evolution of human influenza A viruses in nature Recombination contributes to genetic variation of H1N1 strains, *Proc Natl Acad Sci U S A* **76:**6647-6651
Zhdanov, V M , and Ritova, V V , 1962, Seasonal fluctuations in reactivity to live influenza vaccines, *Vest Akad Med Nauk SSR* **17:**3-5

11

The Control of Influenza

INTRODUCTION

Influenza is essentially an uncontrolled disease. The effect of current vaccination programs on morbidity is insignificant, and that on mortality marginal. Although effective artificial immunization against influenza was demonstrated soon after the first isolation of the virus from man (Chenoweth et al., 1936), the antigenic variability of the virus has hampered vaccine production and precluded effective control (Table 11-1). The transient protection afforded by presently available inactivated virus vaccines is related, in part, to a rapid decline in vaccine-induced homologous antibody but also to vaccine obsolescence as the result of continuous viral mutation to antigens unlike or less like those of viral predecessors contained in the vaccine (Kilbourne, 1978).

But if influenza is primarily a problem in viral genetics and immunology, it also presents, as a major and unpredictable disease burden, problems in preventive medicine, public health policy, economics, and even politics. Even if effective vaccines can be made available in adequate amounts, control will require continual ascertainment and surveillance of infection and constant monitoring of the virus on a global scale to detect significant antigenic change at the earliest possible time. This chapter addresses broadly all these issues, including a review of the immunology of influenza vaccination and new and hypothetical approaches to control of the disease.

VACCINES: LICENSED AND EXPERIMENTAL

The Immunology of Influenza Vaccination

The basic immunology of influenza is reviewed in Chapter 7. However, the artificial induction of the immune state with vaccines introduces special considerations related to the use of attenuated viruses or subvirion antigens that require further discussion.

Table 11-1. A Unique Virus Presents Unique Problems in Vaccination[a]

Viral attribute	Result	Epidemiologic event	Vaccination problem(s)
Mutable RNA genome and antigenic selectability[b]	Sequential antigenic variation as reciprocal of rising host population antibody response	Interpandemic (epidemic) influenza	Need for continual vaccine revision or induction of polyvalent or broadened heterovariant immunity
Segmented genome	Capability for genetic reassortment and introduction of "new" antigens into host population	Pandemic influenza	Need for totally new vaccine or broadened heterosubtypic immunity

[a]From Kilbourne (1985), with permission.
[b]Antigenic mutability is inherent in all RNA viruses, influenza viruses seem unique in the natural emergence and selection of epidemiologically significant antigenic variants

Categorical Problems in the Artificial Presentation of Antigen

Either in man (Chapter 7) or in animal models (Schulman, 1967; Potter et al., 1972), a single infection with unattenuated influenza virus confers lasting and solid homologous immunity. Infection elicits local and humoral antibody responses and cell-mediated immunity as well. This result, perhaps, is unremarkable, except that it emphasizes again the importance of antigenic variation in enabling the virus to escape host immune mechanisms.

Because the primary determinants of immunity to influenza are the HA and NA antigens of the virus, it has been a logical approach to separate these polypeptides, especially the HA, from the virus particle with the goal of reducing vaccine complexity and therefore reactogenicity. In fact, simple disruption of the virus particle with organic solvents or detergents does reduce toxicity but also impairs the immunogenicity (notably in primary immunization) of the external glycoproteins freed from the structural constraints of the intact virion (see also discussion of subvirion components below). However, the price of antigenic purification and simplicity, in general, is reduced immunogenicity (Table 11-2), and this problem must be dealt with in the development of practical vaccines.

Kinetics and Duration of Response to Influenza Vaccines

As early as 7 days after parenteral injection of an inactivated whole-virus vaccine in primed subjects, there is an increase both in HA antibody and in peripheral blood lymphocyte response to in vitro stimulation with the vaccine virus (Mitchell et al., 1982). Even in unprimed subjects, HI antibodies have been detected 7 days after vaccination, with peak titers appearing at 14 days (M. Rosenbaum, cited in Kilbourne et al., 1974). It should be noted, however, that experience with common epitopes among influenza A virus subtypes is widespread in adults and probably conditions most vaccine responses to some degree. Obviously, studies of the kinetics of the immune response are best undertaken in seronegative young children, but in such subjects whole-virus vaccines are unacceptable.

Table 11-2. The Direct Relationship of Immunogenicity and Antigen Complexity[a]

Complexity	Antigen(s)	Adjuvant requires carrier	Immunogenicity
Most ↓	Vaccinia hybrid (live)	0	Most ↓
	Live virus (attenuated)	0	
	Inactivated virus	0–+	
	Disrupted virus	0–+	
	Isolated viral proteins/fusion proteins	0–+	
Least	Oligopeptide	+	Least

[a]Revised from Kilbourne (1985)

The duration of immunity from vaccination is variable and, like the kinetics of response, must reflect prior contact with the viral antigen through vaccination or natural exposure or perhaps cryptic antigenic stimulation by infection subsequent to vaccination. Of children responding to a single dose of swine influenza vaccine, only 58% had demonstrable serum HI antibody 10–13 months after vaccination compared to 91–100% of adults comparably immunized (Lerman et al., 1980). In this situation, prior or subsequent exposure to the swine influenza virus H1N1 antigen was precluded in the children but not in the adults. Similarly, in studies of serum antibody after vaccination with H1N1 USSR vaccine in younger (immunologically inexperienced) adults, decline of antibody 6 months after immunization exceeded that in older adults.

Assessment of actual immunity to infection following vaccination with inactivated viral vaccines is also difficult. The assumption of public health authorities that immunity does not usually exceed 1 year (declining between 15 and 27 months after immunization) is probably justified (Kilbourne et al., 1974). In a comparison of live and inactivated virus vaccines, Clark et al. (1983) found that the incidence of induced infection by the challenge virus 8 months after immunization was directly related to the titer of serum antibody 1 month after vaccination with either type of vaccine.

It is clear, however, that an immunization plateau can be reached by repeated annual vaccinations (Meiklejohn, 1961; Powers et al., 1984), at least with respect to populations in whom the virus is circulating as a boosting agent.

Homologous (Variant-Specific) Immunity

The induction solely of variant-specific immunity probably never occurs because of shared epitopes among intrasubtypic mutants. However, the relative amount of variant-specific and cross-reacting antibody can vary in relation to the immunizing procedure (Kasel et al., 1979).

Homosubtypic (Heterovariant) Immunity

Although less effective than strain or variant-specific vaccines that exactly "match" the challenge virus, both inactivated and live virus vaccines of intrasubtypic variants induce significant heterovariant antibody response and immunity (see Table 7-3).

When young adults and adolescents were vaccinated with A/USSR/77 (H1N1) vaccine, those who had previously received A/NJ/76 (H1N1) swine influenza vaccine were found to have (1) reduced homologous response to A/USSR/77 and (2) a marked boost in A/NJ/76 (heterovariant) antibody titer (Masurel et al., 1981). This study epitomizes the paradoxical effect of cross-reactive anamnestic response to influenza virus glycoproteins (original antigenic sin); i.e., although heterovariant priming and boosting can be achieved by intrasubtypic variant vaccines and hence provide a broader immune response, response to the new antigen in the vaccine may be impaired. This response probably results from cross stimulation of an enlarged population of cross-reactive memory B cells remaining after the primary antigenic stimulation (Schulman, 1975). Such response is not necessarily in the best interests of the host if most of the antigen in the second immunization is preferentially bound to the expanded B-cell population that is committed to the formation of antibody to the original (now heterovariant) antigen (Kilbourne, 1978). However, heterotypic antibody formed in the secondary response is sufficient in quantity (Webster, 1966) and high enough in affinity so that among intrasubtypic variants cross immunity may provide substantial protection. This point has been repeatedly made in studies conducted during the present 1968 H3N2 era.

Vaccines to the original H3N2 1968 variants proved protective against the later A/England/42/72 virus (Table 11-3). In one large carefully controlled study in Air Force cadets, the frequency of influenza was reduced 60% by inactivated heterovariant vaccine (Stiver et al., 1973). Only marginal effects were seen in a girls' boarding school (Dunn and Wormald, 1973), but schoolboys in another study were protected (Hoskins et al., 1973). An attenuated live virus vaccine, A/Scotland/74, provided significant protection against artificial challenge with A/Victoria/75, an intratypic antigenic mutant (Betts et al., 1977). A/Victoria/75, in turn, protected against the later emergent A/Texas/77 variant with 80% efficiency (Meiklejohn et al., 1978).

Vaccine-induced heterovariant immunity is not unique to H3N2 subtype viruses but was noted originally by Meiklejohn (1958) with antigenically dissimilar viruses of the H1N1 (1947–1957) period. Meiklejohn et al. (1978) have emphasized that changes in vaccine formula should be made only after consideration of the length of time that a virus subtype has been prevalent. During later years of prevalence, most of the human population has been primed and, hence, is more responsive to heterovariant immunization. The validity of this point is supported by the demonstrated 80% efficacy of heterovariant immunization in 1977 (Meiklejohn et al., 1978), 9 years after introduction of the H3N2 subtype.

Homotypic (Heterosubtypic) Immunity

As reviewed in Chapter 7, there is some epidemiologic evidence of cross protection among influenza A subtype infections in man. Cross protection between H2N2 and H3N2 subtypes seen in 1957 probably was mediated through their common NA antigens. With vaccination, HA-mediated heterosubtypic antibody response is not uncommon, however, and has been observed between H3 and H7 (Kilbourne, 1976), H1(Sw) and H3 (Evans and Kraus, 1978; Noble et al., 1977), and HI(Sw) and H2 subtypes (Noble et al., 1977). Marine and Thomas (1979), however, noted cross reactivity only between H2 and H3 and not between H1 and H2 or H1 and H3 subtypes. Intersubtypic cross-reacting (non-HI) antibodies have been demonstrated by ELISA in children undergoing primary infection (Burlington et al., 1985). This finding might have implications for live virus vaccines. Prior heterosubtypic infection of hamsters primed them to respond to virus of another influenza A virus subtype (Jennings and Potter, 1973).

Local versus Systemic Immunity

Influenza virus infection, whether naturally occurring or induced with live virus vaccines, results in both humoral and local antibody response. The pres-

Table 11-3. Examples of Heterovariant Immunity Induced with H3N2 Influenza Virus Vaccines

Vaccine strain	Vaccine type	Challenge virus	Antigenic similarity[c]	Result	Reference
A/Hong Kong/68	I[a]	A/England/42/72	50%	"Marginal"	Dunn and Wormald (1973)
A/Aichi/68	I	A/England/42/72	50%	60% reduction of influenza frequency	Stiver et al. (1973)
A/Hong Kong/68	I	A/England/42/72	50%	69% reduction of influenza frequency	Hoskins et al. (1973)
A/Scotland/74	L[b]	A/Victoria/3/75	—	Reduced frequency and severity of illness	Betts et al. (1977)
A/Victoria/3/75	I	A/Texas/1/77	12.5%	80% reduction of influenza frequency	Meiklejohn et al. (1978)

[a]I, inactivated whole-virus vaccine
[b]L, live attenuated vaccine
[c]Defined by hemagglutination inhibition tests (E D Kilbourne, unpublished data)

ence of nasal IgA HA antibody has been correlated with resistance to infection with vaccine virus (Clements et al., 1983).

Topical administration of either live or inactivated viral vaccine led to rises in secretory IgA HA antibody in nasal washings, an effect not observed with intramuscular injection of the inactivated vaccine. However, all three regimens were equal in their capacity to elicit local IgG antibody response. Serum antibody was best stimulated by intramuscular (i.m.) inactivated vaccine (Zahradnik et al., 1983). Inactivated virus by aerosol can stimulate both nasal IgA and serum IgG neutralizing antibody (Waldman et al., 1970). Stimulation of local antibody with nonreplicating antigen is influenced by dose (Ruben et al., 1975), antigen presentation, and host priming (Hruskova et al., 1978).

A comparison of humoral and local antibody responses to whole and subunit virus vaccines demonstrated predominantly strain and subtype specificity in systemic response but broader response to older strains in local antibody formation (Kuwert et al., 1979). With inactivated whole-virus vaccines administered by the parenteral route, a direct correlation between serum HI and nasal neutralizing antibody titers was noted (Wenzel et al., 1973).

Thus, live or inactivated virus administered by the intranasal route will evoke local and systemic antibody response; conversely, inactivated virus administered parenterally will induce both systemic (humoral) and local antibody responses (but usually involving IgG at the local site). In unprimed subjects, the frequency and magnitude of rise in serum antibody were highest (100%) after i.m. inactivated and lowest (29%) after an intranasal live cold-adapted (ca) vaccine (Zahradnik et al., 1983).

Induction of Cellular Immunity by Vaccines

Cytotoxic T-lymphocyte (Tc) response occurs following the administration of live attenuated, inactivated whole-virus, or subunit influenza virus vaccine and may be detectable in the absence of demonstrable antibody response (Ennis et al., 1982). McMichael et al. (1981) found HLA-restricted H1–H3 heterosubtypic Tc response 1 month after injection of inactivated whole-virus (H1N1) vaccine. Subunit vaccine was much less effective in inducing Tc response. The boosting effect on Tc memory of killed whole-virus vaccine was relatively short-lived. In man, individuals with high Tc response subsequently shed little or no virus after challenge with an H1N1 virus (McMichael et al., 1982). In a mouse model, infection induced high levels of Tc memory (Webster and Askonas, 1980). Live-virus vaccines induced cross-reactive T cells and heterosubtypic protection in Balb/C mice (Armerding et al., 1982).

Vaccine-Induced Response to Internal Viral Proteins

Although the HA and NA glycoproteins represent the major inducers of immunity to influenza, humoral or cell-mediated response to the internal proteins of the virus occurs in influenza virus infection (see Table 7-4) and may play a role in immunity. Little data are available on such a response following human vaccination. Immunization with formalin-inactivated vaccine induced a significant increase in serum antibody to purified M protein in ELISA assays. This response was dissociated from the response to HA or NA antigens (Khan et al., 1982).

Host Differences in Immunologic Responses to Vaccine

Host differences in response to influenza vaccines are principally related to individual differences in experience with viral antigens (Fig. 11-1) or to immunodeficiency (see Implementation of Present Vaccination Policy, below). Such differences, related to accumulated priming, must account for most age-related differences in response. However, the proliferative response to influenza vaccines of both B and T cells is reduced in old age (Biro and Beregi, 1980).

Decreased antibody response to live-virus vaccines was associated with HLA type W-16, which reflected increased cellular resistance to infection rather than a suppressed immune response (Spencer et al., 1976).

The Influence of Vaccine Dosage on Response to Inactivated Influenza Vaccines

It was established soon after the initial development of inactivated influenza viral vaccines that (in swine) the relation of log antibody titer to log dose of vaccine per unit weight appeared linear (McLean et al., 1945); in man, a tenfold increase in inactivated vaccine virus concentration induced a two- to threefold increase in geometric mean antibody titer (Hirst et al., 1942; Mostow et al., 1970). Mostow et al. (1970, 1973) found no plateau in response to higher dose levels, as had been observed by Hirst et al. (1942) and by Salk (1948).

In a primed population, an apparent plateau in response (as measured by conventional methods of antibody titration) was observed with injection of 6 μg of HA (Jennings et al., 1985). However, in another study, an accidental tenfold over-

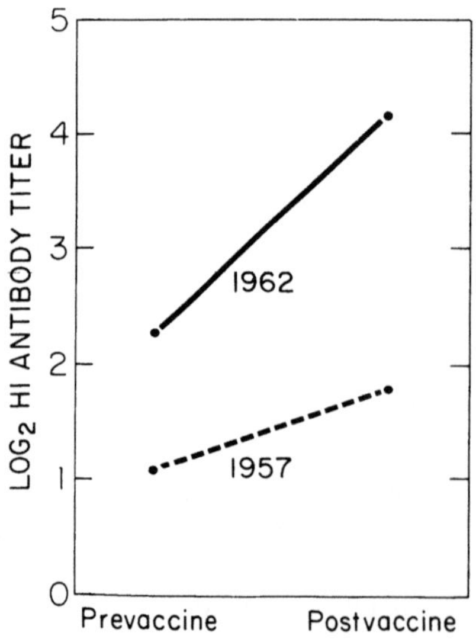

Figure 11-1. Comparative responses to the same vaccine in the first year of prevalence of H2N2 virus (1957) and 5 years later (1962) in the same population primed by natural infection. (E.D. Kilbourne and W.N. Christenson, unpublished data.)

dose of vaccine resulted in significantly greater response, although response to the usual dosage of 7 μg of HA was relatively poor (Matzkin and Nili, 1984). Repeated administration of vaccine at 6-month intervals did not increase the number of subjects with protective levels of antibody except after initial vaccination (Powers et al., 1984).

Obviously, a response plateau may reflect in part diminishing returns in antibody titration on a geometric scale when initial antibody concentrations are high.

Effects of the Route of Vaccine Administration: Nonreplicating Antigen Vaccines

Inactivated whole-virus vaccines elicit humoral antibody response when given by subcutaneous (s.c.), intradermal (i.d.), respiratory tract, or oral route. Unfortunately, few studies have been made of the comparative efficacy of the same vaccine preparation administered by various routes in the same study population. In one such study (Phillips et al., 1970), the same dose of vaccine by intranasal and subcutaneous routes favored the subcutaneous as far as humoral antibody response was concerned, whereas vaccine in lesser dose by the intradermal route resulted in lesser response.

Comparison of subcutaneous and intradermal routes is commonly flawed by use of larger amounts of antigen in the subcutaneous inoculation. Because most studies have been carried out in primed subjects, the limited amounts of vaccine that it is possible to inject intradermally often have appeared equivalent in efficacy to larger amounts given subcutaneously. When equal amounts of antigen have been given by subcutaneous and intradermal routes, they have usually proved to be equally immunogenic (Bruyn et al., 1949; McCarroll and Kilbourne, 1958). In a population primed with H3N2, an H3 variant vaccine was slightly more immunogenic by the intradermal route, whereas in the same population unprimed to $H1_{SW}$ (H1N1), responses to 0.5-ml subcutaneous and 0.1-ml intradermal injections were equivalent (Halperin et al., 1979). This study suggests that the intradermal route might be immunogenically superior in primed subjects. However, a similar study with the same vaccines demonstrated equivalent response by both routes to i.d. and s.c. injection of the H3 vaccine; response to the H1 swine component was greater with the s.c. dose of five times more HA antigen (Herbert et al., 1979). Thus, the limited amounts of vaccine that can be given by i.d. administration are equal in efficacy to larger amounts given by the conventional s.c. route only when administered as a boosting injection in immunologically primed subjects.

The success of aerosol immunization with inactivated vaccines may also depend on previous experience with the vaccine antigens, which may reflect the greater success of aerosol administration of H2N2-containing vaccine in the late H2N2 period (Waldman et al., 1969a) than was seen in the early prevalence of H3N2 viruses (Waldman et al., 1969b). Use of the oral and intranasal routes for administration of inactivated virus has been reported to be successful over many years of experience (Petrescu et al., 1977), although controlled comparisons with parenteral inoculation were not made. Administration of subunit influenza vaccine in enteric-coated capsules resulted in increased secretory IgA in nasal secretions, presumably by stimulation of IgA precursor cells in the small intestine that spread to populate the respiratory tract (Lazzell et al., 1984).

The hypothetical advantage of direct application of inactivated vaccine to the sites later to be challenged by natural infection must be weighed against the problems of inferior humoral antibody response and possible hypersensitization (Baskerville et al., 1975). In patients with chronic bronchopulmonary disease, Shore et al. (1973) found administration of inactivated vaccine by nasal spray to be inferior to the SC route in the induction of either HI serum antibodies or nasal neutralizing antibody.

The Use of Adjuvants

The effect of adjuvants as immunopotentiators of influenza vaccines has received sporadic study for the past 40 years. There is no doubt that adjuvants work in enhancing the magnitude and duration of the immune response to influenza vaccines (Henle and Henle, 1945; Salk et al., 1952; Davenport, 1961) (Fig. 11-2). This effect is probably mediated through B lymphocytes, at least with oil

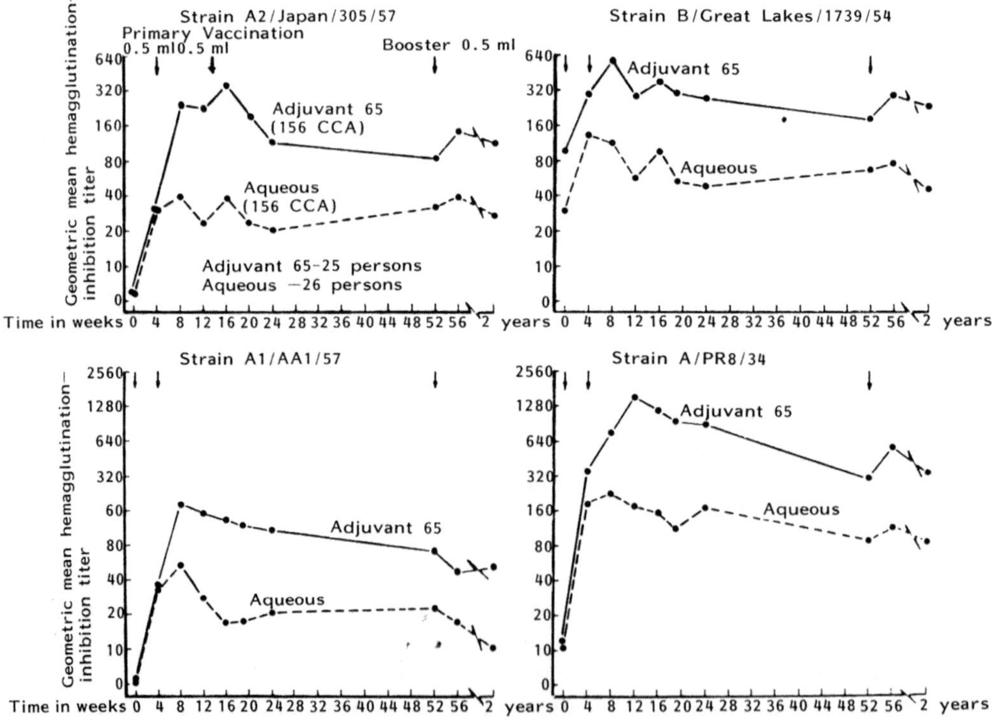

Figure 11-2. Antibody response in human subjects following three injections of aqueous or adjuvant 65 polyvalent influenza virus vaccine given at 0, 4–8, and 52 weeks. (From Hilleman, 1966. Reproduced with permission, from *Progress in Medical Virology,* Volume 8, J. L. Melnick (ed.), p. 161, S. Karger AG, Basel, Switzerland.)

adjuvant stimulation (Cole and Molyneux, 1980). However, the local reactogenicity and the potential carcinogenicity of the most effective ones have deterred general acceptance of adjuvant vaccines. Beebe et al. (1964) studied 44,000 army personnel injected years before with mineral oil adjuvant or aqueous or saline–formalin control vaccines. No association of adjuvant vaccine with chronic or neoplastic disease was found, but persistent local reactions had required medical attention in up to 4% of adjuvant vaccinees. An increased incidence of urticaria in association with adjuvant use was probably related to trace amounts of penicillin in the vaccine. In British studies, an incidence of 3.3 abscesses requiring surgical intervention was found per 10,000 inoculations (Perkins, 1972).

A water-in-oil emulsion of influenza vaccine in peanut oil with mannide monooleate (Arlacel A®), an emulsifier, was judged to be minimally reactogenic on 10-year follow-up, perhaps related to the metabolizable nature of its components (Weibel et al., 1973). Alum has also been used to increase antibody response to influenza vaccine but is inferior to oil adjuvants (Woodhour et al., 1969). Local reactions to an aluminum hydrogel split-virus vaccine occurred in children at an unacceptable rate and without any increase in immunogenicity (Miles et al., 1982).

Recently, purified bacterial cell wall components (peptidoglycans) have aroused interest as immunopotentiators for peptide antigens (Stewart-Tull, 1985) and have increased HI antibody response to subunit influenza vaccines in hamsters (Crawford et al., 1982).

Acrylic acid derivatives also appear to enhance antibody response to inactivated influenza virus preparations in animals (Kreuter et al., 1976; Petrov et al., 1985).

Nonreplicating Antigens

Inactivated Virus

Whole-virus preparations grown in the allantoic sac of the chick embryo and rendered noninfective by various procedures (see Vaccine Standardization, below) have been the standard licensed influenza vaccines throughout the world for four decades. Current vaccines are purified by zonal gradient centrifugation or chromatography and must be free of bacterial endotoxins (Table 11-4).

Most influenza A and B viruses can be isolated directly by inoculation of the chick embryo allantoic sac, but maximal yields of virus are not obtained except after prolonged passage for the selection of genetic variants adapted for chick embryo replication. It was shown over 25 years ago (Kilbourne and Murphy, 1960) that reassortment of a high-yielding laboratory strain [A/PR/8/34 (H1N1)] with low-yielding H2N2 strains freshly recovered from man followed by immunologic selection (Fig. 11-3) produced high-yielding virus bearing surface antigens of the new H2N2 subtype. After the 1968 pandemic, routine use of such reassortants in vaccines was advocated (Kilbourne, 1969). The immunogenicity and efficacy of one such reassortant, X-31, was demonstrated in correlated laboratory, volunteer, and field studies (Kilbourne et al., 1971; Schulman and Kilbourne, 1971; Couch et al., 1971; Leibovitz et al., 1971). The method has been generally successful (Higashihara, 1975; McCahon and Schild, 1972; Mihail et al., 1981;

Table 11-4. Licensed and Experimental Influenza Vaccines Utilizing Nonreplicating Antigens

Inactivated whole virus
Wild type and high-yield reassortant
Subvirion components
Split and subunit vaccines
Individual antigens
Purified or cloned
Individual epitopes
Oligopeptides
Internal image
(antiidiotype)

Downie, 1984), and virtually all contemporary inactivated influenza vaccines contain reassortant viruses. Although high yield is associated primarily with the M gene (RNA 7) (Baez et al., 1980; Klimov et al., 1983), other genes (Schulman and Palese, 1978) including that coding for the HA (Kilbourne, 1978) can influence chick embryo virus yields.

The selection of high-yielding reassortants usually involves cloning of the virus, a procedure not ordinarily carried out with the antigenically heterogeneous parental wild-type virus. Consequently, reassortant viruses may show slight antigenic differences (detectable with monoclonal antibodies) from wild type even if wild-type virus has been subjected to a single cloning (Downie, 1984). Differences in virus passage history may lead to similar minor antigenic variations of reassortant from wild-type virus (Pemberton et al., 1985). The extent of these differences is probably of no practical significance. Selection of the most appropriate antigenic substrain from a new variant obviously must be arbitrary. High-yielding B reassortants have not yet been obtained (E.D. Kilbourne, personal experience; Goodeve et al., 1985).

Subvirion Components: Split and Subunit Vaccines

When influenza virus particles are treated with organic solvents or detergents, the viral membrane is disrupted, and the amphiphilic glycoproteins (HA and NA)

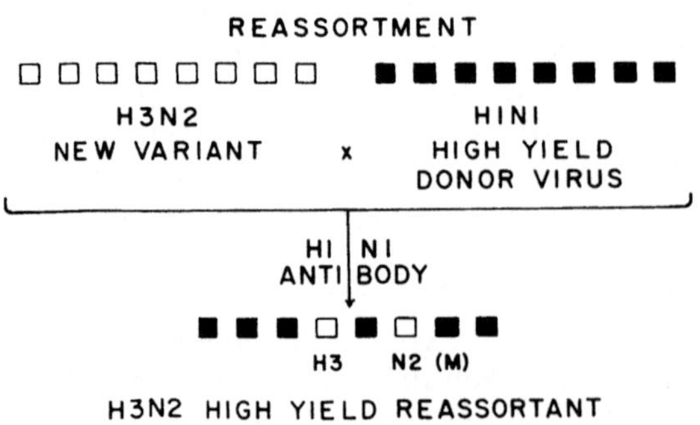

Figure 11-3. Genetic reassortment of high-yield donor virus and new antigenic variant to produce a high-yield reassortant vaccine virus. The reassortant contains H3- and N2-coding RNAs from the new variant and high-yield-determining M (and other) genes from high-yield H1N1 donor virus.

are released and solubilized. Such preparations, when used as vaccines, have been termed "split" or "subunit" and are characterized by low reactogenicity, especially in children. The earliest such vaccines (still in widespread use) were prepared by viral disruption with ether, which, like other organic solvents, is relatively inefficient in dissociation of the viral envelope (Helenius and Simons, 1975). Such split-virus vaccines are heterogeneous in composition, containing fragments of viral membrane in addition to rosettes of HA monomers, viral cores, and aggregates of viral components (Jennings et al., 1984). Detergents such as SDS are far more efficient and solubilize membrane proteins into protein–detergent complexes that can be separated from lipids, forming mixed micelles with the detergent molecules (Morein and Simons, 1985).Vaccines so prepared, which contain principally stellate, aggregated HA monomers (Jennings et al., 1984), after detergent removal, are known as subunit vaccines. Although these vaccines are enriched for HA and NA components, they contain residual internal viral proteins, principally NP (Webster et al., 1976; Jennings et al., 1984; Crawford et al., 1984). In practice, significant differences in immunogenicity and reactogenicity of split and subunit vaccines have not been demonstrated (Bernstein and Cherry, 1983; Gross et al., 1981).

The principal value of split and subunit vaccines is in their low toxicity, especially in children. It is not known whether the lessened reactogenicity reflects reduced lipid or internal viral protein content or alteration in the presentation of the surface glycoproteins. Reduced pyrogenicity of disruptive virus for rabbits is thought to result from disturbances of the spatial arrangement of the viral lipids and occurs even if the lipids are not removed after viral disruption (Grossebauer et al., 1969; Siegert and Braune, 1964).

But the price of lessened toxicity is reduced immunogenicity. In unprimed subjects of all ages, subvirion vaccines are clearly less immunogenic than vaccines containing whole virus (Barry et al., 1976b; Hilleman et al., 1977; Bernstein et al., 1982). In children given 100 CCA units of swine influenza vaccine, response to whole-virus vaccine was 65–93%, whereas that to split-virus vaccine did not exceed 17% (Hilleman, 1977). Immunogenicity of one subunit vaccine was enhanced in unprimed subjects by addition of a small amount of whole virus (Frank et al., 1981). The mechanism of the effect is unknown. In unprimed persons, two or more injections of subunit vaccine may be required.

In primed subjects, subvirion vaccines are equal to whole-virus vaccines in their capacity to evoke humoral antibody response (Couch et al., 1979). Indeed, response to split-virus vaccine in older primed subjects may exceed that induced by whole-virus vaccine (Parkman et al., 1977; Quinnan et al., 1983).

Not only primary antibody response but cell-mediated immunity is less effectively stimulated by HA–NA component vaccines (Webster and Askonas, 1980).

New Approaches to the Formulation of Purified Antigen Vaccines

Although barriers to the purification of isolated HA and NA antigens by chemical extraction methods are considerable, the answer may lie in genetic purification (cloning), in the synthesis of epitopes (oligopeptides), or in internal image-based vaccines (antiidiotype antibodies).

Cloning of the HA and NA genes into bacteria (e.g., Davis et al., 1983) offers promise of obtaining purified antigens in their entirety or perhaps as fusion proteins. The synthesis of oligopeptides representing the four principal antigenic sites on the HA molecule has not thus far fulfilled its promise for specific immunization, although when conjugated with high-molecular-weight proteins, synthetic peptides induce the formation of antibody reactive with hemagglutinin (Green et al., 1982) (see below).

Newly proposed antiidiotype or internal image vaccines (Dreesman et al., 1985), like those incorporating synthetic peptides, would have the potential for low toxicity and antigenic specificity but similarly might require the use of adjuvants.

Submolecular Antigens: Oligopeptides. Definition of major antigenic sites on the HA molecule (Wiley et al., 1981) has opened the door to investigation of submolecular oligopeptides corresponding to these and other sites in immunization against influenza virus infection. Expectations concerning the practical value of this approach must be tempered by the following considerations:

1. Oligopeptides with sequences not contained in the major antigenic sites are immunogenic and produce antibody reactive with the intact HA molecule (Green et al., 1982).
2. Antibody to the intact HA may not react with the synthetic immunogenic oligopeptides (Green et al., 1982), even those representing putative major antigenic sites (Nestorowicz et al., 1985).
3. Antipeptide antibodies that react strongly with the intact HA protein recognize an immunodominant peptide in the subunit interface of the HA trimer. This area represents a previously unrecognized antigenic site and demonstrates the ability of synthetic peptides to generate antibodies that bind to regions of the HA that are not immunogenic or accessible when the HA itself is the immunogen (Wilson et al., 1984).

In summary, the immunogenicity of the HA as an intact protein molecule is not merely the sum of the immunogenicity of its pieces (Green et al., 1982).

As yet, immunization with synthetic oligopeptides of the HA has been only marginally effective in protecting experimental animals (Shapira et al., 1984).

Antigenicity of the HA2 Chain of the HA. Because the amino acid sequence of HA2 is more conserved, and the isolated peptide is antigenic despite its less accessible location (Erickson, 1977; Brown et al., 1980; Becht et al., 1984), its potential as an immunogen has been explored. Subviral influenza virus particles from which the HA1 subunit had been removed induced the formation of HA2-specific antibody that was partially heterotypic, reacting with the HA2 of an influenza virus of a different subtype (Graves et al., 1983). Antibody to the fusion region of HA2 was not neutralizing (Atassi and Webster, 1983).

Synthetic HA Oligopeptides and Cellular Immunity. Human T-cell clones recognize chemically synthesized peptides of the HA (Lamb et al., 1982). In mice, a small synthetic peptide stimulated the induction of Tc cells (Melchizedek et al., 1984). Hackett et al. (1985) found a synthetic decapeptide of HA_1 that elicited Th cells with the same recognition specificity as possessed by cells stimulated by the whole virus.

Potential Advantages and Problems of Submolecular and Synthetic Immunogens. Partial disruption of the virus particle (Graves et al., 1983) or synthesis of peptides buried in the native HA molecule may produce novel immunogens (Wilson et al., 1984) capable of inducing immunity of altered dimension. Such immunity might be broader than immunity engendered by the native virus.

Problems with the practical utilization of submolecular antigens include not only the probable requirement for adjuvants or carrier proteins but also that of carrier-induced epitopic suppression (Schutze et al., 1985).

Replicating Antigens

Mutants

Empirically Selected Attenuated Viral Vaccines. Artificial infection of human subjects with influenza viruses has been employed in experimental influenza vaccines studied since 1937 (Smorodintsev et al., 1937) , and live virus vaccines were used fairly widely in the USSR in the 1960s (Slepushkin et al., 1968; Smorodintsev et al., 1967). These vaccines were attenuated by serial chick embryo passage, as empirically selected host range (hr) mutants and were shown to be generally safe (Beare, 1975), although efficacy was variable and was not always evaluated in controlled field trials. Study of such attenuated vaccines (largely confined to volunteers) proceeded sporadically in Australia (Burnet, 1943), the United States (Francis et al., 1944), the United Kingdom (McDonald et al., 1962), and Japan (Okuno and Nakamura, 1966).

Adaptation of virus by passage in a nonhuman host (e.g., chick embryo) might be expected to ''deadapt'' the virus to man, and, in fact, some reduction in human virulence usually occurs with such manipulation, but often attenuation is minimal (Beare et al., 1968a) except after protracted passage (Okuno and Nakamura, 1966) and is unpredictable. Furthermore, evaluation of vaccine virulence or lack of it must be made against the shifting backdrop of variable and varying human immunity to influenza virus infection. Although many empirically attenuated vaccines have proved acceptable in adults during interpandemic periods, few have been stringently tested in young seronegative children. That problems in adequate attenuation remain is attested to by the absence of licensed live influenza virus vaccines in the United States at this time (1987). But much progress has been made in the genetic manipulation and engineering of live influenza vaccines, which is summarized briefly in the sections that follow.

Marker-Selected Mutants. Specific genetic laboratory markers for virus attenuation in man would have obvious utility in selection of viral strains for use in vaccines. Certain markers such as temperature sensitivity have an intrinsically logical basis, i.e., restriction of virus replication to the cooler upper respiratory, but other viral characteristics demonstrable in vitro are not obvious correlates of avirulence.

Host range (hr) mutants. No deliberate alteration of genes determining host range has yet been practically applied. Chick-embryo-passaged hr mutants (discussed above as high-yield variants) have been empirically derived by serial propagation.

Temperature-sensitive (ts) mutant vaccines. The use of experimental ts mutant vaccines, promoted principally by Murphy and Chanock (Murphy et al., 1980), is ingenious in that the replication of such mutants should be forbidden in the higher temperature of the lower respiratory tract, thus precluding the more serious effects of replication in the lung. The ts mutant strains have not been employed directly as vaccines but have contributed genes with ts defects to mutant-reassortant viruses (group III, Table 11-5). Attenuating genes conferring ts sensitivity have been identified as those coding for PB_1 and PB_2 proteins (Massicot et al., 1980a,b).

Cold-adapted (ca) mutants. Maassab has pioneered in the development of ca mutants for vaccine use. Serial passage of the A/Ann Arbor (H2N2) virus at progressively lowered temperatures in primary chick kidney and chick embryo cultivation resulted in the selection of virus adapted to growth at 25° and 33° but restricted at 39°. Thus, the virus proved to be ts as well as ca, with both phenotypes contributing to attenuation (Maassab et al., 1981; Odagiri et al., 1983). During low-temperature adaptation, mutations have occurred in all genes for internal virus proteins of the Ann Arbor mutant strain and in five genes of a Leningrad mutant (Cox et al., 1985).

Table 11-5. Genetic Categorization of Experimental[a] Influenza Live Virus Vaccines

	Example(s)	Reference
I. Mutants[b]		
hr	Most empirically derived lab-passaged viruses (e.g., A/PR/8/34)	Beare et al (1975)
ts	Lab selected (mutagenized) H2N2 virus	Mills and Chanock (1971) Murphy et al. (1976)
	Naturally occurring H1N2 virus	Tian et al. (1984)
ca	A/Ann Arbor/6/60 (H2N2) adapted to 25°C	Maassab (1967)
HA	HA mutants resistant to serum inhibitors	Soloviev et al. (1961) Beare and Bynoe (1969)
II. Wild-type reassortants		
wt/wt	Avian/human influenza virus reassortants	Murphy et al. (1984)
III. Mutant reassortants		
wt/hr(lab)[c]	H3N2 (wt) and H1N1 (PR8) (hr)	Beare et al. (1975)
wt/ts	H3N2 (wt) and H2N2 (ts)	Murphy et al. (1972)
wt/ca	Various H3N2 (wt)/H2N2 (ca)	Spring et al. (1977)
IV. Reassortant mutants		
wt/hr/HA	Inhibitor-resistant mutant from high-yield H3N2/H1N1 (PR8) reassortant	Schiff et al (1975)

[a]Category IV has been licensed for use in Western Europe
[b]Employed primarily as donor strains for group III vaccines
[c]Empirically selected for replication in chick embryo

Hemagglutinin (HA) inhibitor-resistant mutants. Soloviev et al. (1961) observed that empirically attenuated vaccine viruses that were resistant to normal serum inhibitors appeared to be less virulent for human subjects. Beare and Bynoe (1969) selected and compared H2N2 inhibitor-sensitive and -resistant mutants and showed the latter to be less virulent. The phenotype of inhibitor resistance is mediated by the HA protein so that such mutants are most precisely defined as HA mutants (Table 11-5).

Wild-Type (Avian/Human) Influenza Virus Reassortants

Murphy and his colleagues (Murphy et al., 1984) have mated naturally occurring avian influenza viruses attenuated for replication in the respiratory tract of nonhuman primates with human influenza viruses with the expectation that genetically stable attenuated reassortants would emerge. The expectation of stability is based on evolutionary divergence of some avian virus genes from their human virus counterparts, reflected by significant differences in nucleotide sequence. Attenuation for the squirrel monkey of an H2N2 mallard virus and its reassortants has been shown to be dependent on the NP and M protein genes, with RNA 1 and NS genes playing a supplemental role (Tian et al., 1985).

Mutant Reassortants

Wild-Type Laboratory-Adapted hr Mutants. Reassortment per se should lessen the probability of expression of any polygenic characteristic, such as virulence, dependent on combinations of genes (Kilbourne, 1969), but novel phenotypes could arise with reassortment, so that the possible emergence of virus of increased virulence cannot be excluded (Kilbourne, 1969; Scholtissek et al., 1979). The limited application of simple reassortment of lab-adapted and wild-type parental viruses has thus far led to production only of virus less virulent than the more virulent of the reassortant pair. Reassortment using the laboratory PR8 (H1N1) strain as donor of attenuating genes produced an array of reassortants of H3N2 phenotype from an H3N2–H1N1 virus cross that were intermediate in virulence (Beare et al., 1975). However, no H3N2 recombinants were obtained that, like PR8, were noninfective for man, indicating incomplete segregation of virulence from the HA and NA genes required for immunization.

Wild Type-ts Mutant Reassortants. The transfer of ts attenuating genes for the production of vaccine viruses has received intensive study. The ts mutants exhibit a temperature dependence of replication in vivo, indicating that it is the ts mutations that are responsible for suppression of growth in the respiratory tract (Fig. 11-4) and, therefore, for attenuation (Murphy et al., 1980).

A number of reassortants of various epidemic viruses bearing the PB_1 and PB_2 genes of the donor strain have been studied in human subjects and have been shown to be immunogenic and relatively well tolerated. However, genetic instability of the vaccine viruses has proved to be "an insoluble problem" (Murphy and Chanock, 1985). Change to virulence has occurred both from reversion and from second-site suppressor mutations (Tolpin et al., 1981, 1982).

Wild Type-ca Mutant Reassortants. The principal donor of attenuating genes to experimental ca mutant reassortant vaccine viruses is the A/Ann Arbor/6/60

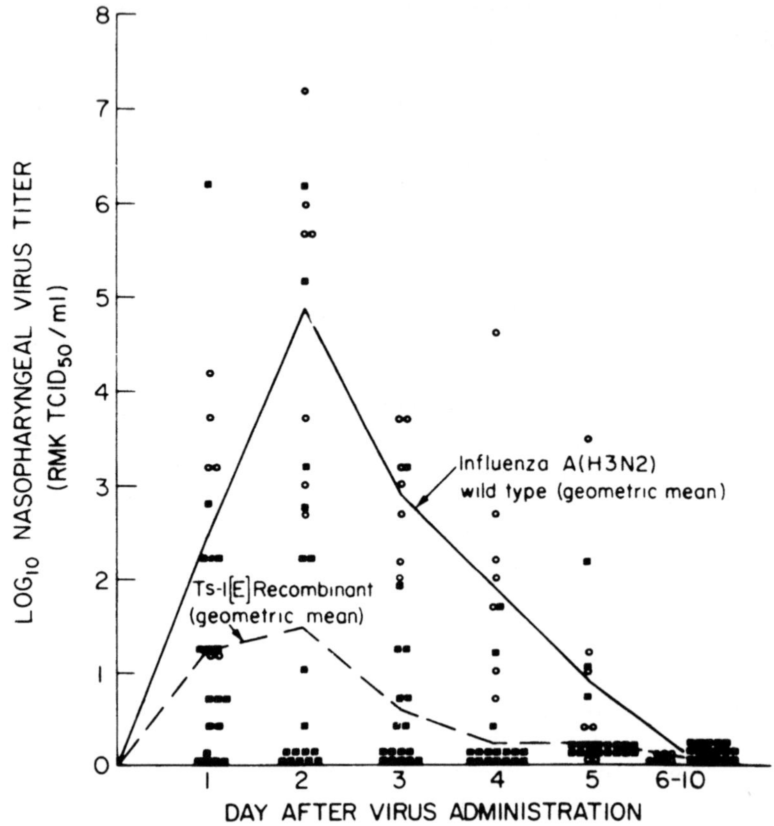

Figure 11-4. Replication of influenza A (H3N2) wild-type ($10^{4.5}$ $TCID_{50}$) and ts-1[E] recombinant (10^6 $TCID_{50}$) viruses in the nasopharynx of volunteers. Circle, wild type; filled square, ts-1[E] recombinant. (From Murphy et al., 1973. Reproduced with permission, from *J. Infect. Dis.* **128:**481. Copyright © 1973 by The University of Chicago Press.)

H2N2 virus of Maassab. This virus possesses separate mutations for ca and ts properties, the former being a more reliable marker for attenuation. Transfer of the six RNAs coding for all but the external glycoproteins to reassortants bearing H1N1 or H3N2 surface antigens produces viruses that possess "(1) a satisfactory level of attenuation, (2) good immunogenicity, (3) lack of transmissibility, and (4) an acceptable level of genetic stability" (Clements et al., 1984b). Markers in the mutant donor strain appear not to change on transfer to recombinant ca vaccine viruses. Furthermore, isolates from vaccinated seronegative children reveal a high degree of genetic stability of the vaccine virus (Cox and Kendal, 1984). It should be borne in mind that attenuation and genetic stability in man of derivatives of the A/Ann Arbor/6/60 virus may reflect not only its ca and ts adaptation (Maassab et al., 1981) but a complex laboratory history (Maassab, 1967) that includes mouse lung passage. However, other ca recombinant vaccines without multiple host passage, including those for influenza B

(Monto et al., 1982; Reeve et al., 1982) and influenza A (Ghendon et al., 1984), also appear promising.

Reassortant Mutants

An H3N2 virus was first recombined with A/PR/8/32 virus for incorporation of high-yield genes. From the H3N2 reassortant produced, an inhibitor-resistant mutant was then selected. The "Alice strain" selected was attenuated and used as a vaccine in several studies (Schiff et al., 1975). (The contribution of inhibitor resistance to attenuation was not demonstrated.)

Live Influenza Virus Vaccines: Summary

Virulence, and therefore its reciprocal, attenuation, are polygenic properties that must ultimately be defined in a single specific host, i.e., man. The identification of specific attenuating genes has been accomplished with certain ts and avian–human virus reassortants and to some extent with ca reassortants (Odagiri et al., 1982). Furthermore, attenuation has been linked to ca and ts phenotypes so that transfer of genetic defects to newly emergent epidemic viruses can be recognized in the laboratory prior to testing of the vaccine candidate viruses in man. However, problems remain with live-virus vaccines. Primary is genetic instability, which may plague even the multiply mutated ca vaccine viruses if five rather than six ca virus genes are transferred (Maassab et al., 1982). Second-site suppressor mutations leading to virulence reversion have been mentioned, as well as the importance of gene combinations (see Chapter 6) in the expression of virulence. But an equally important problem is presented by the need for vaccines balanced for avirulence (reduced replication) and adequate immunogenicity in a population of varied immunologic experience with the vaccine antigens. Thus, a vaccine virus adequately attenuated for adults may cause significant illness in children, or, conversely, one attenuated in seronegatives may be antigenically impotent in experienced adults. Although the use of bivalent (two-virus) live-virus vaccines appears to be feasible (Wright et al., 1985), genetic reassortment demonstrably occurs, enhancing the hypothetical probability of virulence reversion through complementation or new gene combinations.

Cloned Influenza Virus HA Gene Replicated and Expressed In Vaccinia

The cloned HA gene of influenza viruses has been inserted into recombinant vaccinia viruses (Panicali et al., 1983). Expression of the gene during infection with vaccinia stimulates IgA or IgG antibody response depending on the site of inoculation, stimulates HA-specific Tc cells, and induces immunity in experimental animals (Smith et al., 1983; Bennink et al., 1984; World Health Organization, 1985).

Comparison of Replicating versus Nonreplicating Influenza Virus Antigens in Vaccination

Not unexpectedly, live-virus vaccines administered topically in the upper respiratory tract are most effective in inducing formation of local secretory IgA antibody, whereas inactivated vaccines given parenterally stimulate better serum IgG response (Table 11-6). But serum IgA and local (secretory) IgG are raised by both types of vaccine (Johnson et al., 1985), and both live and nonreplicating vaccines induce cell-mediated immunity and can provide heterovariant immunity. Thus, the distinctions between the two types of immunization are not clear cut, particularly since the relative contributions of humoral and local immunity in resistance to influenza have not yet been sorted out. At present, the two different approaches must be judged as equivalent alternatives. However, the greater reduction in viral shedding observed after live-virus vaccines (Clements et al., 1984b, 1986) offers promise in control of the secondary spread of infection.

Infection-Permissive Immunization with NA-Specific Vaccine: A Diphasic Approach to Influenza Immunization

Antibody to influenza virus NA does not directly neutralize virus but is effective in reducing the yields of multiple-cycle infection (Kilbourne et al., 1968): NA-specific immunization has been shown to permit infection but prevent disease in experimental animals (Schulman et al., 1968; Allan et al., 1971; Rott et al., 1974; McLaren et al., 1974) and in man (Couch et al., 1974; Ogra et al., 1977; Beutner et al., 1979).

It is the fundamental premise of NA-specific immunization that partial (i.e., infection-permissive) immunity is desirable and that advantage can be taken of subsequent definitive immunization by natural infection or minimally attenuated live-virus vaccines (see Fig. 11-5). This approach assumes that HA-specific (conventional) immunization with inactivated vaccinia is only briefly effective in preventing infection and merely postpones that event (see Hoskins et al., 1979).

An NA-specific vaccine may prove to be useful (1) as a sole and primary vaccine, preferably administered early in life before initial infection, (2) as a booster vaccine in primed subjects, (3) as a prelude to administration of live-virus vaccine of unknown degree of attenuation, or (4) as an accompaniment to HA oligopeptide vaccines to insure a balanced immune response.

The concept of infection-permissive immunization could also be applied to immunization with other viral epitopes that do not evoke neutralizing antibody response (Kilbourne, 1984).

Vaccine Standardization

Present inactivated whole virus and subunit vaccines are standardized by their HA activity in an agarose gel single radial immunodiffusion (SRD) test (Wood et al., 1977). Vaccine potency is described in terms of micrograms of HA (antigenic) activity per dose. The SRD tests are highly reproducible (Wood et al., 1981), and the potency of assayed vaccines correlates well with antibody response detected in field studies (Ennis et al., 1977b). Current vaccines contain about 7 μg of HA per dose. Quantification of the HA activity of ether-split vaccines is subject to error from aggregate formation during solubilization of the HA—a problem that can be solved by additional SDS solubilization (Johannsen et al., 1983).

Table 11-6. Comparative Immunogenic Effects of Replicating and Nonreplicating[a] Influenza Virus Vaccines

Effect	Vaccine: Nonreplicating	Vaccine: Replicating	Reference (example)
Antibody response			
Serum IgG: anti-HA, anti-NA	++	+	Beare et al (1968b)
Serum IgA	+	+	Clements and Murphy (1986)
Respiratory tract IgA	+	++	Clements and Murphy (1986)
Secretory IgG	+	+	Johnson et al. (1985)
Heterovariant anti-HA	+	+	Yamane et al (1984)
Heterosubtypic anti-HA	+	+[b]	Kilbourne (1976)
			Burlington et al (1985)
Duration of primary antibody response	+	++	Johnson et al (1985)
Cell-mediated immune response	+	+	Ennis et al (1982)
Reduced viral shedding	±	+	Clements et al (1984b, 1986)
Cross-reactive (heterovariant) immunity	+	+	Stiver et al (1973)
			Betts et al (1977)
Duration of immunity	?	?	

[a]Whole virus or subunit vaccines
[b]Demonstrated by ELISA in primary natural infection

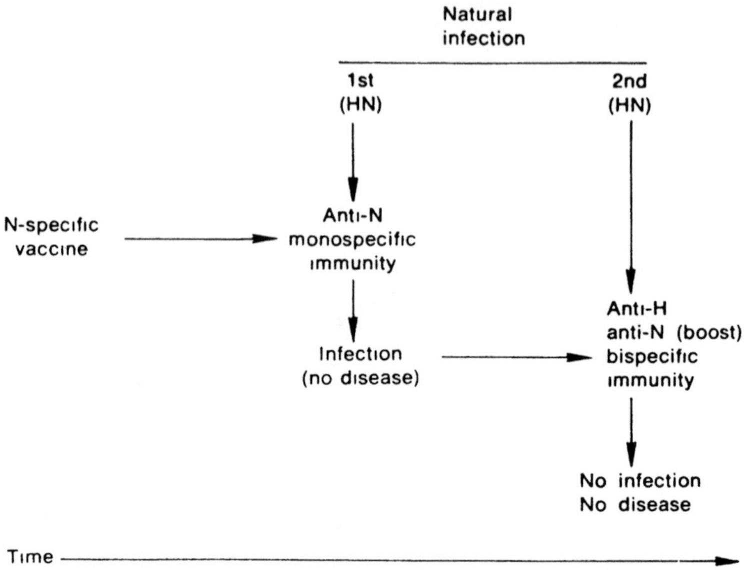

Figure 11-5. Concept of infection-permissive immunization with NA specific (N) vaccine (see text). (From Couch et al., 1974. Reproduced, with permission, from *J. Infect. Dis.* **129:**412. Copyright © 1974 by The University of Chicago Press.)

The validity of the SRD HA assay is dependent on use of homologous virus and antibody in the test and can be altered by small differences between test and reference strains (Rodda et al., 1981). This problem has been addressed by the use of monoclonal antibody broadly reactive with different H3N2 strains (Van de Donk et al., 1984).

At present, vaccines are not assayed for NA, an antigen subject to destruction during formalin inactivation, especially in H1N1 vaccines (Kendal et al., 1977b).

Immunologic response of experimental animals has been employed in efforts to standardize immunogenicity of both inactivated and live virus vaccines, without much success.

Complications of Influenza Vaccination

Early influenza vaccines were relatively crude preparations of chick embryo allantoic fluid virus and were often contaminated with bacterial endotoxin. Present vaccines contain virus purified by zonal centrifugation or chromotography; they rarely evoke febrile responses or systemic reactions except in children. Furthermore, with these modern vaccines there is no correlation between systemic reaction rate and endotoxin content (Ennis et al., 1977a).

Like other vaccines, influenza vaccines cause local systemic reactions in a small proportion of vaccinees. These reactions can be categorized as caused by intrinsic toxicity of purified virus or by hypersensitivity or host idiosyncrasy in response.

Intrinsic Viral Cytotoxicity

As discussed in Chapter 5, there is evidence that influenza viruses devoid of replication capacity can injure cells or impair cell function. Among these effects are depressed phagocytosis, respiratory burst, and superoxide production of polymorphonuclear leukocytes by UV-inactivated virus and stimulation of chemiluminescence by HANA liposomes. The pyrogenic effect of inactivated influenza viruses in rabbits was observed in the absence of contaminating bacterial endotoxin (Wagner et al., 1949). The nature of the toxic factor in influenza virus has not been established but is known to be reduced by disturbances of the spatial arrangement of viral lipid during solubilization of the virus with detergent (see "Subvirion Components," this chapter).

The incidence and severity of vaccine reactions are dose related (Goodeve et al., 1983). Systemic reactions are fewer in vaccinees with preexisting HI antibody to the vaccine virus (Dolin et al., 1977), suggesting participation of viral surface components in toxicity.

Nature of Vaccine Reactions

Local Reactions. Erythema, edema, pain, tenderness, and itching may occur at the site of injection 12 to 24 hr after inoculation. These reactions occasionally proceed to induration but usually do not interfere with normal activity. Adults have higher local reaction rates than children (Barry et al., 1976a). In one study, cutaneous reactions to vaccine were absent in young children but common in adults, supporting the view that local reactions to influenza vaccine are acquired and probably have an immunologic basis. Reactions were characterized by mononuclear cell infiltration and deposition of complement (C3) and IgM (Buttar et al., 1981).

Systemic Reactions. Systemic reactions to inactivated influenza virus vaccine include myalgia, arthralgia, headache, malaise, and fever. In adults fever rarely exceeds 101°F and is briefly sustained. Fever occurs in fewer than 1% of adult subjects injected with purified vaccines (Mostow et al., 1970).

Because these symptoms—lacking only the respiratory component—simulate influenza itself, laymen may suspect the vaccine of causing influenza, which obviously it cannot, as it contains no infective virus.

In seronegative (previously unimmunized) children, systemic reaction rates may be prohibitively high with whole virus vaccines. In infants febrile response in over 50% has been observed with either influenza A or influenza B vaccines (Marine and Stuart-Harris, 1976; Wright et al., 1976). In slightly older children systemic reaction rates are less (Wright et al., 1976), and in school children they approximate 8–9% (Hoskins, 1977). Reduction in dosage of whole-virus vaccine by use of a two-dose schedule or use of split-virus vaccines significantly reduces the incidence of systemic reactions in children 3–18 years old (Gross, 1977). Reaction rates after the second vaccine dose were similar to that with placebo—further indication that vaccine toxicity is primary in the virus particle and susceptible to modulation by specific antiviral immunity. There is some evidence that systemic toxicity may vary with the strain of virus in the vaccine. Certain strains of influenza B virus appear to have been unusually reactogenic (Gwaltney et al., 1977; Wright et al., 1976; Foy et al., 1969).

Hypersensitivity Reactions. Although systemic reactions to vaccine are clearly related to viral toxicity, local reactions to vaccine increase with age and increasing immunologic experience and, therefore, may be presumed to have a hypersensitivity component.

Because influenza vaccines are prepared in eggs, their use in those with allergy to egg proteins has not been recommended routinely. However, allergic children, even those with a history of egg allergy, should not be denied vaccine, even when skin testing with small amounts of vaccine is positive. The majority of "egg-allergic" subjects can be safely immunized (Bierman et al., 1977). This point is important because asthmatic allergic children are particularly vulnerable to influenza and should be vaccinated. Murphy and Strunk (1985) found that even children with a history of egg protein hypersensitivity and positive skin tests to vaccine could be immunized by a rapid divided dose schedule.

Inactivated influenza vaccine had no effect on pulmonary function of young healthy adults 2 and 6 weeks after vaccination (Dull et al., 1980). However, in children with chronic asthma, a transient decrease in peak expiratory flow rate (Bell et al., 1978) or increased airway sensitivity to histamine (Banks et al., 1985) has been observed. These effects do not appear to represent allergic reactions to vaccine antigens but might reflect stimulation of the synthesis and release of interleukin 1 (Banks et al., 1985) with subsequent synthesis of prostaglandins.

Neurological Complications of Influenza Vaccination

As is the case with other vaccines, various neurological syndromes have been sporadically associated with the administration of influenza vaccines during the more than four decades of their use. Of these, credible reports include reversible and irreversible optic neuritis (Perry et al., 1979; Macoul, 1982), brachial and upper extremity neuritis (Hasselbacher, 1977; Weintraub, 1977), cranial nerve palsies (Felix et al., 1976), and Guillain–Barré syndrome. Only in the case of Guillain–Barré syndrome has a statistically significant linkage been established, and only with specific reference to swine influenza vaccine. (In 1976, Felix and colleagues reported that up to that time only 13 neurological complications of influenza vaccination had been reported.)

The Landry-Guillain-Barré-Strohl or (more generally) Guillain–Barré syndrome (Wiederholt et al., 1964) is a subacute, usually symmetrical, paralytic polyneuronitis of unknown cause associated with sensory disturbance. Pathology displays a lymphocytic cellular infiltration of peripheral nerves and segmental demyelination.

During a 10-week period in 1976, 43,327,000 Americans were given monovalent influenza vaccines containing a reassortant virus, X-53a, bearing the surface antigens of a swine influenza virus A/NJ/11/76 (H1N1) recovered from an abortive epidemic in recruits at Fort Dix, New Jersey. Because of reports of Guillain-Barré syndrome in temporal association with "swine flu" vaccination, the National Immunization Program was suspended in mid-December, 1976, and a national surveillance for Guillain-Barré syndrome was initiated.

Initial reports indicated that in persons more than 18 years of age the risk of getting Guillain-Barré syndrome was five to six times greater than in the unvaccinated (Schonberger et al., 1979; Langmuir, 1979). These reports were criticized during subsequent court proceedings in which victims sued for damages because (1) the Schoenberger report did not contain the basic data for calculation of incidence and estimation of risk and (2) the Langmuir report had not corrected for the reduced population at risk as the vaccination program continued.

In response, a definitive study was made of approximately 1300 cases reported by State Health Departments to CDC (Langmuir et al., 1984). This study essentially corroborated the earlier ones but refined the analysis by separately considering severe and milder cases. Vaccine-associated risk (4.9–5.9 per million) was found to last for at least 6 weeks following vaccination but not beyond 8 weeks. A separate statistical analysis by Greenstreet (1984) described the associated risk after 6 weeks as unlikely and after 10 weeks as negative. Less intensive surveillance of subsequent more limited influenza vaccination programs in 1978–79 (Hurwitz et al., 1981) and 1979–80 and 1980–81 (Kaplan et al., 1982) have not shown a similar risk of Guillain-Barré syndrome, and, therefore, it has been proposed that the elicitation of Guillain-Barré syndrome is unique to swine influenza vaccine. It seems much more likely that the revelation of increased risk is unique to the massive swine influenza immunization program and to subsequent intensive surveillance for Guillain-Barré syndrome rather than to the vaccine itself.

Guillain-Barré syndrome may or may not be associated with antecedent immunization (Brooks and Reznik, 1980) or respiratory or gastrointestinal infections. Therefore, triggering appears to be nonspecific and not restricted to specific pathogens. It is probable that massive immunization with any antigen would evoke the same incidence of disease. Thus, the very low risk (five- to sixfold over background or one excess case per 100,000) related to influenza vaccination is difficult to demonstrate during conventional vaccine usage and less intensive surveillance for Guillain-Barré syndrome. No evidence has been provided that swine influenza virus is uniquely neurotropic or that the vaccine alone can elicit experimental Guillain-Barré syndrome in animal models. Evocation of Guillain-Barré syndrome by influenza vaccine or any other entity must have an immunologic basis, given the time lag from vaccination to expression of disease. A cellular and humoral response to peripheral nerve myelin has been described (Caspary et al., 1971), and atypical lymphocytes have been found during the course of the disease (Cook et al., 1970). Guillain-Barré syndrome shares features with experimental allergic neuropathy in animals (Waksman and Adams, 1955).

Although recurrent Guillain-Barré syndrome is rare (Seyal et al., 1978; Poser et al., 1978), it is possible that the definitive basis of Guillain-Barré syndrome is genetic. If so, identification of susceptibles by HLA or genotyping methods not as yet developed will be required to obviate this complication in future immunization programs.

Vasculitis

Among the disorders that have been temporally if not causally related to influenza vaccination is small-vessel vasculitis (Blumberg et al., 1980; Cannata et al., 1981; Houston, 1983), manifested by fever, arthralgia, uveitis, or purpura and appearing 7–10 days after immunization.

Effects of Influenza Vaccine on Drug Metabolism

Influenza vaccines, in common with other interferon-inducing agents, have been reported to depress metabolism of certain drugs by the liver, presumably by depression of cytochrome P-450-dependent hepatic microsomal mixed-function oxidases. These drugs include theophylline (Renton et al., 1980), aminopyrine (Kramer and McClain, 1981), and phenytoin (Levine et al., 1984).

Other studies suggest that clinically apparent adverse reactions to warfarin and theophylline are rare after influenza vaccination and are no more frequent than the reactions in unvaccinated people (Patriarca et al., 1983; Bukowskyz et al., 1984) and are uncommon even in the elderly (Gomolin et al., 1985).

In a comprehensive study, Meredith et al. (1985) found increases in α and γ interferon levels up to 72 hr after vaccination but variable and selective effects on drug metabolism. The degree of depression of theophylline clearance was small and transient. On balance, these studies do not interdict the use of influenza vaccine in high-risk subjects.

Untoward Effects of Live-Virus Vaccines

Because they actually induce respiratory tract infection, the "toxicity" and untoward effects of live-virus vaccines might be expected to parallel those of natural infection. Although they are well tolerated in healthy young adults, it has been of some concern that these vaccines might have unfavorable effects in just those chronically ill subjects who are in greatest need of immunization. Indeed, an increase in respiratory symptoms has been found in subjects with chronic lung disease (Zeck et al., 1976; Fell et al., 1977), also attended by greater antibiotic usage (Fell et al., 1977). Trials of live attenuated influenza vaccines in a Medical Research Council study in patients with chronic obstructive airway disease were generally favorable but did find some adverse effects with one of three vaccines tested (MRC Advisory Group, 1984). Minor functional effects were found in other studies (Dierckx et al., 1976; Rosenzweig et al., 1975). Hall et al. (1976) found no disturbance of pulmonary function with ts reassortant vaccines. Asthmatic subjects appear to tolerate live-virus vaccines without disturbance in small airway function (Storms et al., 1976; Kava and Laitinen, 1985).

Although it is encouraging that immunization with live-virus vaccines can be achieved without significant compromise of respiratory funciton, it must be understood that the reactogenic potential of each vaccine must be defined individually until attenuation can be predictably attained. Furthermore, the ultimate test of attenuation must always be in nonimmune subjects.

CHEMOPROPHYLAXIS

Influenza caused by influenza A viruses can be prevented by the oral administration of L-adamantanamine hydrochloride (amantadine) or methyl-1-adamantanemethylamine hydrochloride (rimantadine). These symmetrical amines interfere with an early step in virus replication (see chapter 6) but at recommended dosage of 100–200 mg/day apparently allow a degree of infection sufficient to induce specific but reduced antibody response (Quarles et al., 1981). The drugs have reduced virus shedding in some studies. Amantadine has shown prophylactic activity in boarding schools (Payler and Purdham, 1984; Muldoon et al., 1976; Rose, 1980), semi-open populations (Smorodintsev et al., 1970), nosocomial infections (O'Donoghue et al., 1973), and general populations (Dolin et al., 1982).

The principal side effects of amantadine are related to its actions on the central nervous system, which may cause insomnia, lightheadedness, irritability, and difficulty in concentration in 5–10% of healthy adults. Rimantadine appears to be better tolerated than amantadine and is comparable in prophylactic efficacy (Dolin et al., 1982). Reduced dosage is recommended in the elderly (Patriarca et al., 1984).

Because these drugs have significant side effects and must be taken continuously to be effective, they are considered to be a supplement to immunization in the control of influenza and are recommended only for certain specifically defined situations. These include (1) short-term institutional prophylaxis during an epidemic in which vaccine is ineffective because of major antigenic change, (2) as an adjunct to late immunization of high-risk individuals, (3) in unprotected health care workers during epidemics, (4) as a supplement to immunization in those with impaired immune response, or (5) in those few who are allergic to influenza vaccine (see above).

Under conditions simulating natural exposure, amantadine and rimantadine prevented experimental infection in chickens with a highly virulent virus, but mortality ensued from emergent drug-resistant variants. However, combined administration of amantadine and vaccination provided protection (Webster et al., 1985).

SPECIFIC RECOMMENDATIONS FOR THE CONTROL OF INFLUENZA BY VACCINE AND CHEMOPROPHYLAXIS*

Use of inactivated influenza vaccine is the single most important measure in preventing and/or attenuating influenza infection. Potency of present vaccines is such that nearly all vaccinated young adults develop hemagglutination-inhibition

*Edited recommendations of the Immunization Practices Advisory Committee (ACIP) of the National Centers for Disease Control for 1985-86 (Centers for Disease Control 1985a)

antibody titers that are likely to protect them against infection by strains like those in the vaccine and, often, by related variants that emerge. The elderly, the very young, and patients with certain chronic diseases may develop lower postvaccination antibody titers than young adults. Under these circumstances, however, influenza vaccine may be more effective in preventing lower respiratory tract involvement or other complications of influenza than in preventing infection and involvement of the upper respiratory tract. Influenza vaccine will not prevent primary illnesses caused by other respiratory pathogens.

Annual vaccination against influenza has been recommended since 1963 for individuals at high risk of lower respiratory tract complications and death following influenza infection, i.e., the elderly and persons with chronic disorders of the cardiovascular, pulmonary, and/or renal systems, metabolic diseases, severe anemia, and/or compromised immune function. These groups have been identified primarily by reviews of death certificate data, supported by hospital-based or population-based studies. Each group encompasses patients along a continuum of underlying general health. In other words, within each broadly defined high-risk category, some persons may be more likely than others to develop severe complications from influenza infection.

Investigations of influenza outbreaks in nursing homes, for example, have demonstrated attack rates as high as 60%, with case-fatality ratios of 30% or more. Chronic diseases and other debilitating conditions are common among nursing home residents, and spread of infection can often be explosive in such relatively crowded and closed environments. Recent retrospective studies of noninstitutionalized patients also suggest that chronic, underlying diseases, particularly those that affect the cardiovascular and pulmonary systems, may contribute more to the severity of illness than age alone. Since influenza infections are also known to invoke abnormalities in gas exchange and peripheral airways dysfunction in adults, children with compromised pulmonary function, including those with cystic fibrosis, chronic asthma, and bronchopulmonary dysplasia, as well as neonates in intensive-care units, may also be at higher risk of severe illness, although firm evidence is lacking. Children with congenital heart disease may also be considered at high risk, since respiratory viruses in general often produce severe infections in this population.

Target Groups for Vaccination

1. Based on the above observations, the previous, broadly defined high-risk group has been further classified on the basis of priority, so special efforts can be directed at providing vaccine to those who may derive the greatest benefit. Groups for which active, targeted vaccination efforts are most necessary are:
   a. Adults and children with chronic disorders of the cardiovascular or pulmonary systems that are severe enough to have required regular medical follow-ups or hospitalization during the preceding year.
   b. Residents of nursing homes and other chronic-care facilities (e.g., institutions housing patients of any age with chronic medical conditions).
2. Although not proven, it is reasonable to believe that medical personnel can transmit influenza infections to their high-risk patients while they are

themselves incubating infection, undergoing subclinical infection, or working despite the existing of mild symptoms. In many winters, nosocomial outbreaks of influenza are reported. The potential for introducing influenza to high-risk groups, such as patients with severely compromised cardiopulmonary or immune systems or infants in neonatal intensive-care units, should be reduced by vaccination programs targeted at medical personnel. Therefore, physicians, nurses, and other personnel who have extensive contact with high-risk patients (e.g., primary-care and certain specialty clinicians and staff of intensive-care units) should receive influenza vaccination annually.

3. After consideration is given to the needs of the above two target groups, high priority should also be given to organizing special programs making vaccine readily available to persons at moderately increased risk of serious illness compared with the general population:
   a. Otherwise healthy individuals over 65 years of age.
   b. Adults and children with chronic metabolic diseases (including diabetes mellitus), renal dysfunction, anemia, immunosuppression, or asthma that are severe enough to require regular medical follow-ups or hospitalization during the preceding year.

Vaccine Recommendations

Influenza vaccine is recommended for high-risk persons 6 months or older, for their medical-care personnel, and for other persons wishing to reduce their chances of acquiring influenza illness. Vaccine composition and doses are given in Table 11-7. Because adequate evaluation of other routes in high-risk persons is lacking, the preferred route of vaccination is at the conventional site, the deltoid muscle.

Antiviral Agent: Amantadine

Amantadine prophylaxis is recommended as short-term prophylaxis during the course of a presumed influenza A outbreak (e.g., in institutions for persons at high risk), particularly when the vaccine may be relatively ineffective (e.g., because of major antigenic changes in the virus). The drug should be given early in the outbreak in an effort to reduce the spread of the infection. Contingency planning for influenza outbreaks in institutions is needed to establish specific steps for rapid administration of amantadine when appropriate, including obtaining physicians' orders at short notice. When the decision to give amantadine for outbreak control is made, it is desirable to administer the drug to all residents of the affected institution. Amantadine dosage is described in Table 11-8.

CONTROL OF ANIMAL RESERVOIRS OF INFECTION

Thus far, only three of the 13 HA influenza A subtypes (H1, H2, and H3 viruses) have been associated with human epidemics. If, indeed, only viruses of these subtypes can produce epidemic disease in man, and if pandemic viruses

Table 11-7. Influenza Vaccine[a] Dosage by Age of Patient: United States, 1985–1986 Season[b]

Age group	Product[c]	Dosage[d]	Number of doses	Route[e]
6–35 months	Split virus only	0 25 ml	2[f]	i m
3–12 years	Split virus only	0.5 ml	2[e]	i m
>12 years	Whole or split virus	0.5 ml	1	i m

[a]Contains 15 μg each of A/Chile/83(H1N1), A/Philippines/82(H3N2), and B/USSR/83 hemagglutinin antigens in each 0 5 ml Manufacturers include Parke-Davis (Fluogen® split), Squibb-Connaught (Fluzone® whole or split), Wyeth Laboratories (Influenza Virus Vaccine, Trivalent split)
[b]From Centers for Disease Control (1985a)
[c]Because of the lower potential for causing febrile reactions, only split (subvirion) vaccine should be used in children. Immunogenicity and reactogenicity of split and whole-virus vaccines are similar in adults when used according to the recommended dosage
[d]Pneumococcal vaccine and influenza vaccine can be given at the same time at different sites without an increase in side effects, but it should be emphasized that, whereas influenza vaccine is given annually, pneumococcal vaccine should be given only once to adults Detailed immunization records should be provided to each patient to help ensure that additional doses of pneumococcal vaccine are not given
[e]The recommended site of vaccination is the deltoid muscle for adults and older children The preferred site for infants and young children is the anterolateral aspect of the thigh musculature
[f]Four weeks or more between doses, both doses recommended for maximum protection However, if the individual received at least one dose of any influenza vaccine recommended from 1978–1979 to 1984–1985, one dose is sufficient

Table 11-8. Amantadine Hydrochloride (Symmetrel®) Dosage by Age of Patient and Level of Renal Function[a]

Age group	Dosage[b]
No recognized renal disease	
1–9 years[c]	4.4–8.8 mg/kg per day once daily or divided twice daily, total dosage should not exceed 150 mg/day
10–64 years[d]	200 mg once daily or divided twice daily
≥64 years	100 mg once daily[e]
Recognized renal disease	
Creatinine clearance ($ml/min{\cdot}1.73m^2$)	
≥80	100 mg twice daily
60–79	200 mg/100 mg on alternate days
40–59	100 mg once daily
30–39	200 mg twice weekly
20–29	100 mg thrice weekly
10–19	200 mg/100 mg alternating every 7 days

[a]From Centers for Disease Control (1985a)
[b]For prophylaxis, amantadine must be taken each day for the duration of influenza A activity in the community (generally 6–12 weeks) For therapy, amantadine should be started as soon as possible after onset of symptoms and should be continued for 24–48 hr after the disappearance of symptoms (generally 5–7 days)
[c]Use in children under 1 year has not been evaluated adequately in one study, a dose of 6 6 mg/kg per day was reportedly well tolerated by children over 2 years of age
[d]Reduction of dosage to 100 mg/day is also recommended for persons with an active seizure disorder, because such persons may be at risk of experiencing an increase in the frequency of their seizures when given amantadine at 200 mg/day
[e]The reduced dosage of 100 mg/day for person 65 years of age or older without recognized renal disease is recommended to minimize the risk of toxicity, because renal function normally declines with age, and because side effects have been reported more frequently in the elderly

do originate from animal sources (Chapter 10), it is not unreasonable to think of control of animal influenza as a strategem for human protection. It is most likely that putative animal virus–human virus reassortants would arise during human contact with domestic species so that only those species need be considered in any plan of barrier immunization. However, at present, systematic vaccination programs for the containment of animal influenza do not exist, although equine immunization is carried out on a sporadic basis but is not always effective (Mumford et al., 1983).

In 1976 serious consideration was given to the idea of vaccinating pigs as a reservoir of 1918-like influenza virus (Boffey, 1976). The use of live-virus vaccines for the control of animal influenza gives pause, because vaccine viruses, like natural strains, could be donors of new antigens to animal virus–human virus hybrids (Lang, 1981).

At our present stage of knowledge, it seems unlikely that any measures short of eradication of swine, horses, and domestic fowl would eliminate the source of pandemic antigens.

SOCIOLOGICAL AND ECONOMIC PROBLEMS IN THE CONTROL OF INFLUENZA

Implementation of Present Vaccination Policy

Present American vaccination policy is reflected in the recommendations of the CDC Advisory Committee presented above (Centers for Disease Control, 1985a). The major emphasis of these recommendations is on the reduction of mortality in the chronically ill and the elderly in the absence of a vaccine that will confer lasting immunity in all members of the population. It should not be forgotten, however, that the major impact of influenza is its pervasive and acutely disruptive morbidity and that this burden can be reduced in any individual or population by the administration of present vaccines. The reason for the lack of a clearly articulated policy on the use of vaccines for the prevention of illness will become apparent as we discuss the problems in the implementation of the present vaccine policy. First, it is important to review briefly the evidence for vaccine efficacy in the high-risk target population.

Vaccine Efficacy in Different Populations

The inactivated whole-virus and subunit vaccines presently licensed in the United States are variably effective in the prevention of influenza for reasons already discussed. In planning vaccination policy, however, it must be appreciated that these vaccines are generally 70–90% effective when used against antigenically similar virus. In primed populations, the single-dose administration of vaccine containing 14 μg of HA stimulates adequate antibody response (Gremillion et al., 1983), but in unprimed subjects a two-dose regimen is necessary (La Montagne et al., 1983).

In view of earlier reservations about the efficacy of influenza vaccine in the elderly (in whom influenza may be life threatening), it is important to note that most recent studies have demonstrated the value of vaccination in this group (Serie et al., 1977; Barker and Mullooly, 1980). Vaccinated elderly patients have reduced risk of illness, hospitalization, and pneumonia (Patriarca et al., 1985) and mortality (Barker and Mullooly, 1980). In a geriatic hospital, herd immunity was demonstrated at a vaccination rate of 79% (Serie et al., 1977). To be most effective, vaccination should precede exposure by no more than 2 to 4 months (MacKenzie, 1977). Only minimal protection was observed in a nursing home 1 year after vaccination (Budnick et al., 1984).

Immunologic response to influenza vaccine is curtailed in some cancer patients, especially those on immunosuppressive therapy (Stiver and Weinerman, 1978; Gross et al., 1985). However, fourfold increases in antibody titers were found in 50% of patients with cancer even though they responded less well than control subjects (Ganz et al., 1978), and patients off chemotherapy for 30 days or more responded like normal controls in one study (Steinherz et al., 1980). Hematology patients and those undergoing organ transplantation responded poorly to vaccination. Cardiology and hemodialysis patients responded normally (Hodges et al., 1979), but in another report the response of dialysis patients was impaired (Versluis et al., 1985).

Antibody response in pregnant women (Murray et al., 1979) and in patients with lupus (Anonymous, 1978) or sickle-cell disease (Steinberg et al., 1978) is unimpaired.

Strategies for Vaccine Administration

Although influenza occurs each year in the north temperate zone, the magnitude of its impact is so variable that its potential is easily forgotten by physician and patient alike. This fact coupled with the ephemeral nature of vaccine-induced immunity has made it difficult to establish regular immunization programs such as those for the common infections of childhood. It has been estimated that only about 20% of the 40 million in whom influenza is a life-threatening disease receive vaccine (Kavet, 1976). Fedson and Kessler (1983) have emphasized that the low immunization rate in high-risk patients is not simply explained by lack of knowledge or by erroneous health beliefs, nor does it usually reflect physician ignorance. In a hospital setting substantial numbers of general medical patients were immunized because there was an organized approach to offering vaccine. A history of recent hospital care may identify many high-risk persons who are candidates for immunization (Fedson and Kessler, 1983; Barker and Mullooly, 1982), and more than 50% may be immunized (Ratner and Fedson, 1983).

Although a logical place for annual vaccinations against influenza is in speciality clinics dealing with cardiac and pulmonary disease, such clinics may be delinquent in their responsibility (Fedson and Kessler, 1983).

In industry or government, when an organized approach to immunization has been employed, there has been significant reduction in morbidity in vaccinated personnel (Smith and Pollard, 1979; Wesselius-deCasparis et al., 1973).

In Japan, school children are immunized under a national program undertaken to benefit the community by reduction in disease transmission. Whether or not this goal is accomplished is not known, as the program does not contain control groups (Dowdle et al., 1980).

What can be accomplished in implementing vaccination is illustrated by the "swine flu" national immunization program of 1976, in which 43 million people were vaccinated in a 2-month period under threat of a pandemic. Analyzed by state, participation in the program ranged from less than 10% to more than 70% (Fig. 11-6). Schoenbaum (1979) has pointed out that these differences are too large to be explained as differences in individual attitudes; rather, organizational factors within states must have played an important role. Compliance was variable within states (Monto and Ross, 1979) and in Vermont was positively correlated with amount of education, income level, past history of influenza vaccination, and recent consultation with a physician (MacPherson et al., 1980).

Public acceptance of vaccine is probably not a major factor in achieving immunization. Even in the wake of the swine influenza immunization program with its negative publicity (see following section), 59% of adults in a Pennsylvania community indicated a willingness to participate in future immunization programs, and 25% said they would not (Pearman, 1978). Education about side effects is apparently most important in changing attitudes and behavior with respect to acceptance of a "flu shot" (Buchner et al., 1985).

An aggressive approach to influenza vaccination was recommended by experts convened at the Symposium on Options for the Control of Influenza, held at Keystone, Colorado in April, 1985. Among the recommendations made were:

1. The use of an immunization record card for high-risk noninstitutionalized adults.
2. An approved immunization program for residents and staff as a requirement for institutional licensure.
3. An approved immunization program as a requirement for institutional eligibility for Medicare reimbursement.
4. Establishment of an influenza immunization policy by the American Medical Association (Centers for Disease Control, 1985b).

The Perennial Problem in Vaccine Supply

Demand for influenza vaccine is inconstant, driven by variable perceptions of epidemic threat, by magnitude of antigenic change in the virus, by aggressiveness of vaccine promotion, and by vaccine availability.

It is understandable that private industry is reluctant to increase production of a chronically underutilized vaccine. In turn, local health officials may not promote use of vaccine that may be in short supply. Except for 1957 and 1976, industry in the United States has produced annually no more than 20 million doses of vaccine, one-half of the amount needed to protect the high-risk group alone. The problem is not unique to the United States; a recent Canadian study showed that influenza control in high risk populations is constrained by influenza vaccine supply (Sheps et al., 1985).

It has been proposed that some form of government subsidization of production be provided (Kilbourne et al., 1974; Riddiough et al., 1983) to insure adequate amounts of vaccine in even the worst epidemic years. Government participation may prove necessary in any event because of increasing problems with liability and personal compensation of victims of vaccine-associated injury (*Journal of the American Medical Association*, 1984).

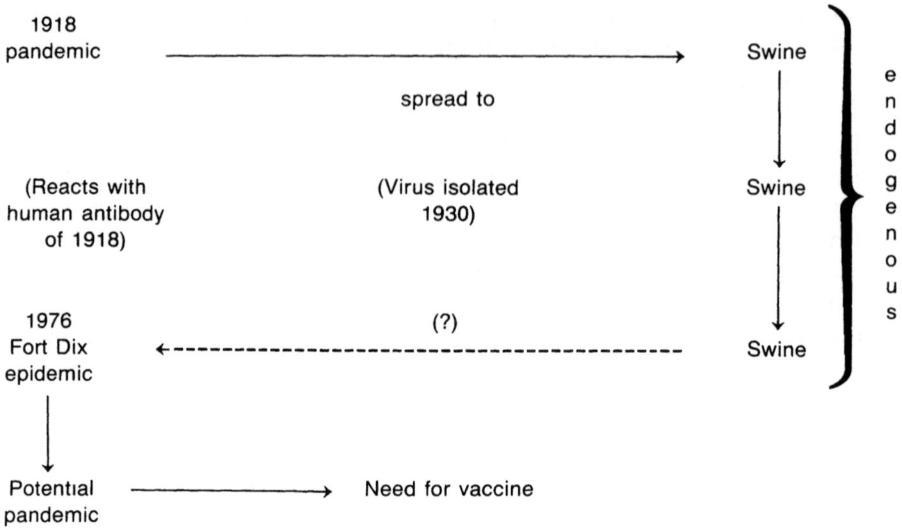

Figure 11-6. Swine harbored the antigenic prototype of the 1918 virus for 58 years. In 1976, swine virus of demonstrable human virulence was transmitted to, and subsequently among, recruits at Fort Dix, N.J.

Mass Vaccination against Pandemic Disease: The Swine Influenza Vaccination Program as Paradigm

As it is now conceived and implemented, United States vaccination policy addresses only the risk of interpandemic influenza associated with minor or intrasubtypic influenza virus variation. In this, the usual situation, most of the population has some degree of partial immunity and is primed to respond to vaccination with the development of immunity even to new viruses containing drifted heterovariant antigens.

Introduction of influenza A virus of new HA subtype led in 1957 and 1968 to major pandemic disease affecting by definition people of all ages because all but older members of the population were devoid of prior immunizing experience with the virus. In this situation, not only is the threat of mortal effect of the virus amplified, but morbidity and consequent economic and social effects can become staggering. Consideration must then be given to programs of extended or mass vaccination as a reasonable exercise in public health practice.

Even the most ambitious of such programs cannot reasonably hope to achieve significant herd immunity as a goal but can be justified only for the protection of the vaccinated individual. Protection of unvaccinated persons through vaccination of their peers has been achieved at 84% levels of influenza vaccination (Warburton et al., 1972), but 90% immunization may be ineffective in achieving herd immunity in the intensive exposure of a boarding school population (Watson et al., 1972). With other respiratory viral diseases (measles and rubella), equally high vaccination frequency (84–96%) may be required for the elimination of infection (Anderson and May, 1985).

Given the limitations of present knowledge, the ecology and genetics of the influenza viruses, and the imperfection of present vaccines, eradication of influenza is not a reasonable public health objective. However, the reduction of widescale morbidity and mortality is. In 1976, Schoenbaum et al. (1976) estimated that the economic costs to the United States of a prospective pandemic in the absence of immunization could exceed $6 billion for the whole population and $3 billion for those in the high-risk group. It was calculated that if vaccine administration costs did not exceed 50¢ a person and vaccine acceptance was at least 59%, then a policy of vaccinating the general adult population could be justified. Immunization of the young, who have a higher attack rate, would prevent more cases, but protection of the middle-aged and elderly would yield more economic benefit by preventing more hospitalization and mortality.

In 1957, without coordinated government effort, United States private industry produced 48 million doses of vaccine (albeit of relatively low potency) between June and November after an April alert (Lyons, 1976). However, the Asian virus spread like wildfire through the summer, peaking in early November before general vaccine availability (Murray, 1961). The vaccine available was often inappropriately distributed; it must be concluded that this pandemic, like the Hong Kong pandemic of 1968 that was to follow, was essentially unmodified by human intervention.

When swine influenza virus, never before proved to be epidemic in man, emerged as a cause of morbidity and mortality in Fort Dix, New Jersey in mid-January 1976, the opportunity appeared to be at hand for the blunting, perhaps even the conquest, of an incipient pandemic. The efforts immediately undertaken to attain this goal became enmeshed in national politics and have been recorded by some as a "fiasco" and "debacle" because no further epidemics appeared. However, valuable lessons were learned that can guide future attempts at mass vaccination against influenza in particular and other infections in general. Thus, the National Immunization Program for "swine flu" serves as a paradigm of the scientific, technical, human, and political problems to be expected with any extraordinary effort to contain an acute disease with the potential for rapid and widespread dissemination.

The "swine flu affair" of 1976 has been the subject of several detailed, if not definitive, analyses (Osborn, 1977; Neustadt and Fineberg, 1978; Silverstein, 1981), with only the latter addressing the scientific as well as the political problems of the program. The present brief review emphasizes the scientific basis for the decision for mass immunization and some of the technical problems that have relevance to future attempts to deal with pandemic disease. In this review I strive for objectivity because of my personal involvement in several aspects of the vaccination program: (1) I had warned of an imminent pandemic (Kilbourne, 1976), (2) my laboratory produced the reassortant viruses used in the vaccines, and (3) I was advisor to the CDC in decisions on the National Immunization Program. There are, however, unchallengeable facts concerning the abortive epidemic of swine influenza at Fort Dix, although the implications of these facts and the responses to them can be, and have been, widely debated (Kilbourne, 1977, 1979).

The Influenza Epidemic at Fort Dix, New Jersey in 1976

An explosive outbreak of acute febrile respiratory disease began at Fort Dix in early January after return of personnel and importation of new recruits following the Christmas holidays.* After initial indications that the epidemic might be caused by adenovirus type 21, influenza virus isolates were obtained from throat washings taken January 29 on hospitalized patients. Of these virus isolates, seven were identified as A/Victoria/75 (H3N2), and the remaining four were later found to be antigenically similar to swine influenza virus. An additional swine influenza virus isolate was recovered from tracheal swabs from a fatal case of influenza in a 19-year-old recruit at postmortem examination on February 5 (Goldfield et al., 1977). During the period February 14–24, 55 isolates, all of H3N2 (Victoria) subtype, were obtained. No additional isolates of swine influenza virus were made, but significant increases in antibody to swine influenza virus were demonstrated in eight additional patients between January 18 and February 14 (Gaydos et al., 1977a). Additional serological evidence, based on the prevalence of swine influenza antibody, indicated that at least 230 infections with the virus had occurred (Hodder et al., 1977). It was concluded that swine influenza virus had been introduced at Fort Dix in early January and was not transmitted after mid-February. Notably, epidemiologic evidence indicated that person-to-person transmission of the virus had occurred in distinct military units (Gaydos et al., 1977b).

It is clear that the Fort Dix epidemic was of mixed etiology, with the A/Victoria/75 strain dominating and persisting longer. Elsewhere in New Jersey, only A/Victoria infections were found. Indeed, except for sporadic rural swine-related cases, no further demonstrations of swine influenza virus in man have since been found anywhere in the world.

Swine influenza virus infection comprised an epidemic within an epidemic and was focused in six of 14 basic combat training companies that entered the reception center in the month following the Christmas holidays. These units were maintained in relative isolation for training purposes during the very time that the swine influenza virus was present in their personnel. The origin of the virus could not be traced, although it appeared most probable that it was introduced into basic combat training companies by members who were infected within the reception center (Top and Russell, 1977). Paradoxically, infection rates within some companies were high (40–45%), yet general transmission of the virus throughout the post did not occur.

In summary, early in 1976 at least 230 recruits at Fort Dix were infected with virus indistinguishable antigenically or otherwise from contemporary strains of swine influenza virus. In 13 documented clinically studied cases, pneumonia occurred in five patients, with fatal outcome in one. The epidemic was brief and less pervasive than a concomitant epidemic of H3N2 virus infection. In this epi-

*Details of the influenza epidemic at Fort Dix, New Jersey in 1976 have been published in a group of companion articles in a supplement to the *J Inf Dis* **136**:S341–385, 1977 demic, person-to-person transmission of virus undoubtedly occurred through an estimated four to seven passages of virus (Top and Russell, 1977).

Swine Influenza Virus Infection at Fort Dix as a Pandemic Threat

Concern about the initially unidentified Fort Dix [A/New Jersey/76 (H1N1)] virus began with its isolation, when samples of the isolate were flown to CDC for typing. Heretofore, the appearance of new or major HA antigenic mutants had always presaged the disappearance of the prevalent subtype of virus and rapid spread of the new strain coincident with pandemic disease. When the new virus was identified as swine influenza virus, the putative agent of 1918, concern evoked caution. Initial studies in my laboratory were restricted to a single isolation room and conducted only by my technician and myself. Early supplies of experimental vaccine were targeted for laboratory or clinical workers in contact with the virus (Infectious Disease Branch, NIAID, 1976).

Decisions on what response to take to the reappearance of a virus at least antigenically related to swine influenza virus had to be made with dispatch on the basis of the following limited information and precedent.

New Information. A transmissible influenza virus antigenically identical with contemporary swine (H1N1) influenza viruses had caused epidemic disease, pneumonia, and death in recruits in a military installation.

Concurrent circulation of two different influenza A subtype viruses (H1N1 and H3N2) had been demonstrated for the first time in man.

Unknown at That Time. The source of the outbreak (still unknown) and the prevalence of the swine influenza virus in the human population were unknown.

The "swine" virus genotype was unknown. Was it a reassortant derived from a combination of human and swine influenza viruses, or did it differ from contemporary pig viruses in any distinguishable way?

Known at That Time. Assuming earlier prevalence of swine influenza virus during 1918–1929, all of the human population less than 45 years of age would lack specific immunity to the virus, and older people would not have experienced homologous antigenic stimulation for 45 years or more. Those less than 19 years of age had had no contact with any H1N1 virus and therefore lacked even heterovariant immunity.

Influenza vaccines were 70–90% effective and free of major risk. Industry had been able to produce 48 million doses in 1957 without the use of genetically engineered high-yield reassortant viruses. There was precedent for the use of swine influenza virus in man. Polyvalent vaccines containing the Shope strain of swine influenza virus had been used from 1955 to 1969 in immunization of the military (Takafuji et al., 1976).

Epidemiologic Precedent. A recent pattern of 11-year prevalence of influenza A subtype viruses had been noted: 1976 was only 8 years after the 1968 pandemic, but the previous shift had been thought of as approximately decennial (Kilbourne, 1977).

Previous appearances of new subtype viruses in 1957 and 1968 (antigenic shift) had heralded the disappearance of the earlier subtype and its rapid replacement by the new virus in susceptible populations.

The National Immunization Program

The Decision to Undertake Mass Immunization. There is no doubt that the decision to undertake mass immunization of the American public was influenced not only by the emergence of an antigenically novel and hence potentially pandemic virus but by the chance that the virus might possess the virulence attributed to the virus of 1918. It is also likely that although scientific advisors and public health personnel were motivated principally to prevent the effects of another 1957 or 1968, the ultimate persuasion of the Congress and the President rested on the implicit threat of 1918. Those responsible for the program have been accused of brandishing that threat to spur progress of the program.

As a scientific advisor to, and advocate of the program, I found it difficult to convey accurately and understandably a scientifically informed perception of the relation of the new virus to the putative 1918 agent and of the hazard that it presented. On initial evidence, the virus could be characterized as an H1N1 virus identical in surface antigens to a 1975 swine influenza virus (Kendal et al., 1977a). Later analysis by RNA gel genotyping disclosed that the Fort Dix virus closely resembled other recent isolates of virus from swine and showed no evidence of gene transfer from human virus (Palese and Schulman, 1976).

Although these data pointed to swine as the definitive source of the virus (Fig. 11-6), they provided no basis for reassurance about the virus' pandemic and virulence potential. It has since been shown that a single point mutation in the HA gene of the Fort Dix virus can significantly influence its infectivity and transmission in its natural host (Kilbourne, 1978, 1983; Kilbourne et al., 1981; Both et al., 1983). There are no genetic markers for influenza virus virulence for man, and, most critically, the virus of 1918 was not at hand for comparative analysis. Therefore, one could only say that the Fort Dix virus might be more virulent, as virulent, or less virulent than the 1918 (and subsequent) strains. The limited clinical information from the authenticated Fort Dix cases was inadequate to judge the virus' potential, but its association with pneumonia and death in young recruits certainly provided no comfort. Furthermore, viruses can increase in virulence with extended human transmission. The message that finally could be stressed was that this transmissible new virus was present in a nonimmune population, and, because its antigenically similar predecessor might have been uniquely virulent, it should not be taken lightly.

In this context, the recommendation of the Advisory Committee for Immunization Practice of the CDC for a mass immunization program was carried to the Secretary of HEW, to the President, and to Congress through a memorandum from the Director of CDC (see Viseltear in Osborne, 1977; Neustadt and Fineberg, 1978; Silverstein, 1981).

Implementation of the Program. Supported by a \$35 million appropriation from the Congress, manufacture of vaccine and intensive surveillance for new swine influenza cases began. The initial goal of immunizing the entire U.S. population called for production of 200 million doses of vaccine before the fall. Retrenchment from this goal followed the most extensive trials ever accorded an influenza vaccine. They were conducted on 7500 volunteers studied between April and October, 1976. These trials confirmed earlier indications with other influenza vaccines of unacceptable toxicity in young children given large single doses and therefore removed all but high-risk children with chronic disease from the target population (Gross, 1977).

In 10-year retrospect, it is remarkable that in the absence of an epidemic and with the disappearance of swine influenza virus from man, the subsequent vaccination program fared as well as it did. There was reasonably prompt presidential and congressional action in response to the pandemic threat. Federal interagency cooperation was superb, and the cooperation from the private sector, the pharmaceutical industry and voluntary health organizations, the American Medical Association, and individual physicians was generally good. Press coverage, in general, was well informed and responsible. Vaccine manufacture proceeded rapidly, aided by an even higher-yielding reassortant virus (Kilbourne, 1978) than had been available for the vaccine trials.

Some problems arose from misinformation, lack of information, or just plain politics. President Ford was seen by some as using the program as a political ploy in an election year, although others saw it as a no-win situation in which you were "damned if you do and damned if you don't" (Sabin, 1977). Those who viewed the program as a drug house bonanza or "rip-off" (Silverstein, 1981, p. 70) did not appreciate the increasing reluctance of pharmaceutical manufacturers to be involved in vaccine production of any sort because of problems of liability.

An unforseen technical problem related to the nature of the Fort Dix virus was the apparent inactivation of the neuraminidase antigen during vaccine manufacture. Opponents of the program seized on this problem, although, in fact, the hemagglutinin is the critical antigen in the vaccine, and neuraminidase potency had not been a requirement in previous influenza vaccines. Public confidence in the program was not reinforced by the additional news that one of the manufacturers had produced trial lots of vaccine employing the wrong (i.e., early Shope strain) virus. Ironically, the neuraminidase of that virus proved stable in production.

As the immunization program finally got under way in October, a new set of problems emerged as misadventures temporally associated with vaccine administration to large numbers of people, including the preferentially targeted chronically ill. These problems had been foreseen by the vaccine manufacturers and their insurers, who had finally forced the government to assume primary liability by the enactment of Public Law 94-380 (94th Congress S.3735).

But, notwithstanding the occurrence of highly publicized and circumstantially related deaths of three cardiac patients in Pittsburgh that led to temporary interruption in vaccination in nine states, the National Immunization Program proceeded and in 2 months resulted in the immunization of 43 million people. The highly variable rates of immunization in different localities (Fig. 11-7) deserve careful study as a guide to future programs.

An event associated with vaccination that could not be ignored was an apparently increased incidence of Guillain-Barré syndrome (see earlier discussion

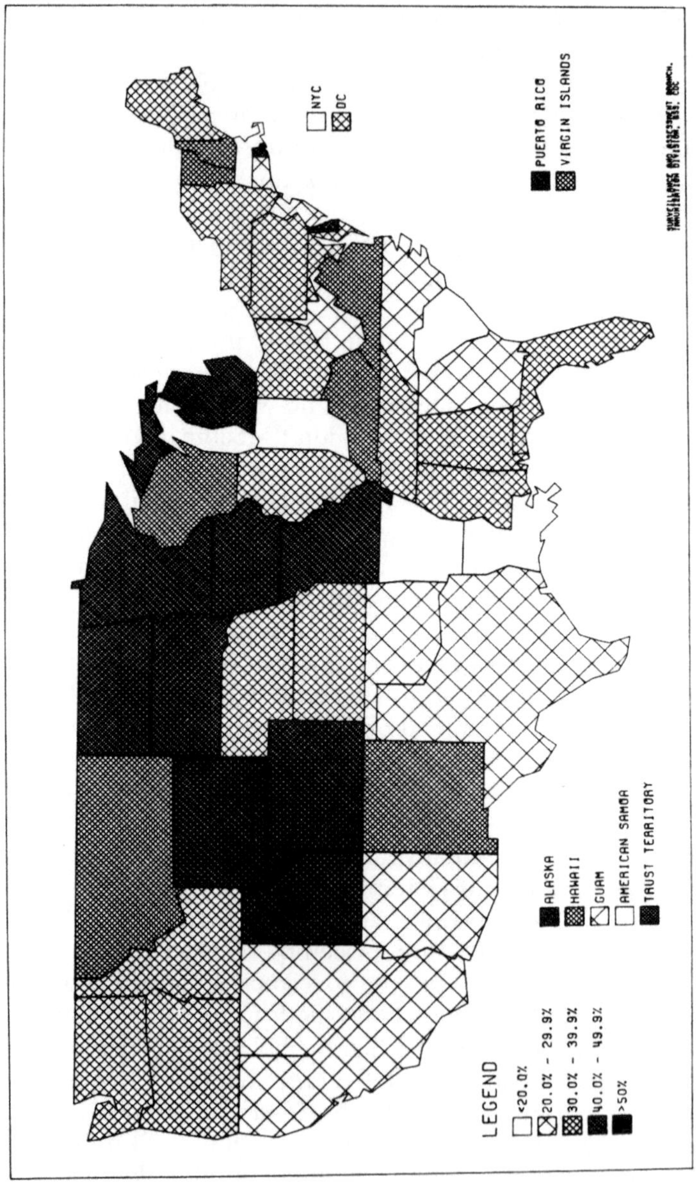

Figure 11-7. Influenza vaccination coverage in the National Immunization Program by state as of June 23, 1977. (Courtesy of Centers for Disease Control.)

in this chapter) in vaccinated persons. This event, coupled with the continuing absence of swine influenza virus in man, made it necessary to abandon the program in mid-December.

The Lessons of 1976

About Influenza. In 1976 we learned new and highly significant things about the epidemiology of influenza. (1) A major antigenic variant of influenza A virus can appear in epidemic form and yet not replace the prevalent subtype or lead to pandemic disease. (2) Cocirculation in man of different influenza A subtype viruses can occur. Both of these findings have been corroborated since with the reappearance of Russian H1N1 virus in 1977 and the persistence of H3N2. (3) Swine influenza virus, long suspected to be the living fossil of the 1918 pandemic virus, can produce transmissible infection in man with apparently minimal changes in its genome and without the hypothetical requirement for hybridization with a human influenza virus. (Reassortment may have indeed occurred in this double epidemic, but if so, it was undetected in the small number of isolates examined.)

About Mass Immunization: A Retrospective Assessment of the National Immunization Program. The National Immunization Program failed in not meeting even its modified goal of immunizing the entire adult population. A price was paid in some loss of public confidence in government-sponsored programs in preventive medicine and by the unfortunate victims of Guillain–Barré syndrome. A monetary cost of settlement of claims against the government has amounted to $73 million to date (1985).

The National Immunization Program succeeded in immunizing 43 million, including nearly 50% of the high-risk group, the greatest protection ever afforded that group and twice the number of persons in that group usually immunized. This population received a bivalent vaccine containing both swine and A/Victoria viruses. (Dowdle, 1977).

Forty-three million people lined up for vaccine despite numerous vicissitudes in the implementation of the program, increasingly hostile press and media coverage, and in the absence of epidemic disease.

Response to the pandemic threat was reasonable and prudent, given contemporary knowledge: better a vaccine without an epidemic than an epidemic without a vaccine.

Planning for Future Mass Immunization Programs in Response to Threats of Pandemic Disease

From new knowledge of influenza gained during and since the 1976 outbreak and during the planning and implementation of the National Immunization Program, the following conclusions and recommendations for future mass immunizations seem justified:

1. In planning for influenza control, it must be assumed that major antigenic variation of influenza virus occurs in no predictable pattern.

2. Although single subtype dominance is the usual epidemic pattern, cocirculation of different subtype viruses can occur either locally or globally.
3. Intensive surveillance for the emergence of new major variants must continue, worldwide.
4. Mechanisms and plans must be established before the fact to deal with the emergence of new major antigenic variants.
5. These plans should include:
   a. Recognition that influenza is a unique public health problem comprising inevitable but variable interpandemic disease and the constant threat of pandemic expression.
   b. Organization for intensive local and national surveillance after recognition of major antigenic variation.
   c. Stand-by facilities, private or governmental, for the production and storage of large supplies of vaccine.
   d. Maintenance of laboratory, clinical, and epidemiologic competence and facilities for the continuing study of influenza as a disease and for conduct of ad hoc vaccine field trials as the occasion may demand.
   e. Resolution in advance of problems in liability associated with vaccine administration.
   f. Organization in advance for the education of the public and prompt frequent communication with medical care personnel and the media at every step of the immunization program.

The cost of pandemic influenza in the United States was 60,000 deaths in 1957–58 and 30,000 deaths in 1968–69. In 1968–69, 51 million Americans were ill, the direct cost of medical care was $750 million, and the total economic burden was $3.9 billion (Kavet, 1972). The anticipated cost of the potential pandemic in 1976 was $6 billion (Schoenbaum et al., 1976). The cost of human misery cannot be estimated.

To face future disasters without preparation seems an untenable position. It is unlikely that we have seen our last pandemic.

REFERENCES

Allan, W. H., Madeley, C. R., and Kendal, A. P., 1971, Studies with avian influenza A viruses: Cross protection experiments in chickens, *J. Gen. Virol.* **12**:79–84.

Anderson, R. M., and May, R. M., 1985, Vaccination and herd immunity to infectious diseases, *Nature* **318**:323–329.

Anonymous, 1978, Influenza immunization in lupus erythematosus: Safe, effective? *Ann. Intern. Med.* **88**:833–834.

Armerding, D., Rossiter, H., Ghazzouli, I., et al., 1982, Evaluation of live and inactivated influenza A virus vaccines in a mouse model, *J. Infect. Dis.* **145**:320–330.

Atassi, M. Z., and Webster, R. G., 1983, Localization, synthesis, and activity of an antigenic site on influenza virus hemagglutinin, *Proc. Natl. Acad. Sci. U.S.A.* **80**:840–844.

Baez, M., Palese, P., and Kilbourne, E. D., 1980, Gene composition of high-yielding influenza vaccine strains obtained by recombination, *J. Infect. Dis.* **141**:362–365.

Banks, J., Bevan, C., Fennerty, A., et al., 1985, Association between rise in antibodies and increase in airway sensitivity after intramuscular injection of killed influenza virus in asthmatic patients, *Eur. J. Respir. Dis.* **66**:268–272.

Barker, W H , and Mullooly, J. P , 1980, Influenza vaccination of elderly persons, *J A.M A.* **244**:2547-2549
Barker, W. H , and Mullooly, J. P., 1982, Pneumonia and influenza deaths during epidemics: Implications for prevention, *Arch. Intern Med.* **142**:85-89.
Barry, D. W , Mayner, R E., Hochstein, H. D., et al., 1976a, Comparative trial of influenza vaccines. II. Adverse reactions in children and adults, *Am. J. Epidemiol.* **104**:47-59.
Barry, D. W., Mayner, R E., Staton, E., et al., 1976b, Comparative trial of influenza vaccines. I. Immunogenicity of whole virus and split product vaccines in man, *Am. J. Epidemiol.* **104**:34-46.
Baskerville, A , Thomas, G., and Peacock, S., 1975, Pathological aspects of immunization of mice against influenza virus infection, *Br. J. Exp Pathol* **56**:17-27.
Beare, A S , 1975, Live viruses for immunization against influenza, *Prog. Med. Virol.* **20**:49-83.
Beare, A S., and Bynoe, M. L , 1969, Attenuation of human influenza A viruses, *Br. Med. J* **4**:198-201
Beare, A. S , Bynoe, M. L , and Tyrrell, D A. J., 1968a, Investigation into the attenuation of influenza viruses by serial passage, *Br Med J* **4**:482-484.
Beare, A. S , Hobson, D , Reed, S. E., et al., 1968b, A comparison of live and killed influenza-virus vaccines, *Lancet* **2**:418-420.
Beare, A S., Schild, G. C , and Craig, J. W , 1975, Trials in man with live recombinants made from A/PR/8/34 (H0N1) and wild H3N2 influenza viruses, *Lancet* **2**:729-738.
Becht, H , Huang, R T. C., Fleisher, B., et al., 1984, Immunogenic properties of the small chain HA2 of the haemagglutinin of influenza viruses, *J. Gen Virol.* **65**:173-183.
Beebe, G. W., Simon, A H , and Vivona, S , 1964, Follow-up study on army personnel who received adjuvant influenza virus vaccine, 1951-1953, *Am. J. Med. Sci.* **247**:385-406.
Bell, T. D., Chai, H., Berlow, B., et al., 1978, Immunization with killed influenza virus in children with chronic asthma, *Chest* **73**:140-145.
Bennink, J R., Yewdell, J. W., Smith, G. L , et al., 1984, Recombinant vaccinia virus primes and stimulates influenza hemagglutinin-specific cytotoxic T cells, *Nature* **311**:578-579.
Bernstein, D. I., and Cherry, J. D., 1983, Clinical reactions and antibody responses to influenza vaccines, *Am J Dis Child* **137**:622-626.
Bernstein, D. I., Zahradnik, J. M., DeAngelis, C. J., et al., 1982, Influenza immunization in children and young adults, *Am. J. Dis. Child.* **36**:513-517.
Betts, R. F , Douglas, R. G., Jr., Roth, F. K., et al., 1977, Efficacy of live attenuated influenza A/Scotland/74 (H3N2) virus vaccine against challenge with influenza A/Victoria/3/75 (H3N2) virus, *J. Infect. Dis.* **136**:746-753.
Beutner, K. R., Chow, T., Rubi, E., et al., 1979, Evaluation of a neuraminidase-specific influenza A virus vaccine in children: Antibody responses and effects on two successive outbreaks of natural infection, *J. Infect. Dis.* **140**:844-850.
Bierman, C. W., Shapiro, G. G., Pierson, W. E., et al., 1977, Safety of influenza vaccination in allergic children, *J. Infect. Dis.* **136**:S652-S655.
Biro, J., and Beregi, E., 1980, The influence of influenza vaccinations on human peripheral lymphocytes relating to aging, *Akt. Gerontol.* **10**:319-322.
Blumberg, S., Bienfang, D., and Kantrowitz, F. G., 1980, A possible association between influenza vaccination and small-vessel vasculitis, *Arch. Intern. Med.* **140**:847-848.
Boffey, P. M., 1976, Swine flu campaign: Should we vaccinate the pigs? *Science* **192**:870.
Both, G. W., Shi, C. H., and Kilbourne, E. D., 1983, Hemagglutinin of swine influenza virus: A single amino acid change pleiotropically affects viral antigenicity and replication, *Proc. Natl Acad Sci. U.S A.* **80**:6996-7000.
Brooks, R., and Reznik, A., 1980, Guillain-Barre syndrome following trivalent influenza vaccine in an elderly patient, *Mt. Sinai J. Med.* **47**:190-191.
Brown, L. A., Dopheide, T. A. A., Ward, C. W., et al., 1980, Antigenic determinants of influenza virus hemagglutinin. V. Antigenicity of the HA2 chain, *J Immunol.* **125**:1583-1588.
Bruyn, H. B., Meiklejohn, G , and Brainerd, H. D., 1949, Influenza vaccination: A comparison of antibody response obtained by various methods of administration, *J. Immunol.* **62**:1-11.
Buchner, D. M., Carter, W. B., and Inui, T. S., 1985, The relationship of attitude changes to compliance with influenza immunization, *Med. Care* **23**:771-779.
Budnick, L. D., Stricof, R. L., and Ellis, F., 1984, An outbreak of influenza A in a nursing home, 1982, *N.Y State J. Med.* **84**:235-238.

Bukowskyj, M., Munt, P. W., Wigle, R., et al., 1984, Theophylline clearance, *Am. Rev. Respir. Dis.* **129**:672–675.

Burlington, D. B., Wright, P. F., van Wyke, K. L., et al., 1985, Development of subtype-specific and heterosubtypic antibodies to the influenza A virus hemagglutinin after primary infection in children, *J. Clin. Microbiol.* **21**:847–849.

Burnet, F. M., 1943, Immunization against epidemic influenza with living attenuated virus, *Med. J. Aust.* **1**:385–389.

Buttar, K., Chernesky, M., and Dolovich, J., 1981, Cutaneous responses to tests with influenza vaccines, *Ann. Allergy* **46**:183–188.

Cannata, J., Cuesta, V., Peral, V., et al., 1981, Reactivation of vasculitis after influenza vaccination, *Br. Med. J.* **83**:526.

Caspary, E. A., Currie, S., Walton, J. N., et al., 1971, Lymphocyte sensitization to nervous tissues and muscle in patients with the Guillain–Barre syndrome, *J. Neurol. Neurosurg. Psychiatry* **34**:179–181.

Centers for Disease Control, 1985a, Immunization Practices Advisory Committee, recommendations, *Morbid. Mortal. Weekly Rep.* **34**:262–267.

Centers for Disease Control, 1985b, Implementation of recommendations for influenza control, *Morbid. Mortal. Weekly Rep.* **34**:639–643.

Chenoweth, A., Waltz, A. D., Stokes, J., Jr., et al., 1936, Active immunization with the viruses of human and swine influenza, *Am. J. Dis. Child.* **52**:757–758.

Clark, A., Potter, C. W., Jennings, R., et al., 1983, A comparison of live and inactivated influenza A (H1N1) virus vaccines, *J. Hyg. (Camb.)* **90**:361–370.

Clements, M. L., and Murphy, B. R., 1986, Development and persistence of local systemic antibody responses in adults given live attenuated or inactivated influenza A virus vaccine, *J. Clin. Microbiol.* **23**:66–72.

Clements, M. L., O'Donnell, S., Levine, M. M., et al., 1983, Dose response of A/Alaska/6/77 (H3N2) cold-adapted reassortant vaccine virus in adult volunteers: Role of local antibody in resistance to infection with vaccine virus, *Infect. Immun.* **40**:1044–1051.

Clements, M. L., Betts, R. F., Maassab, H. F., et al., 1984a, Dose response of influenza A/Washington/897/80 (H3N2) cold-adapted reassortment virus in adult volunteers, *J. Infect. Dis.* **149**:814.

Clements, M. L., Betts, R. F., and Murphy, B. R., 1984b, Advantage of live attenuated cold-adapted influenza A virus over inactivated vaccine for A/Washington/80 (H3N2) wild-type virus infection, *Lancet* **1**:705.

Clements, M. L., Betts, R. F., Tierney, E. L., et al., 1986, Resistance of adults to challenge with influenza A wild-type virus after receiving live or inactivated virus vaccine, *J. Clin. Microbiol.* **23**:73–76.

Cole, P. J., and Molyneux, M. E., 1980, Immunological sensitization of human lymphocytes after influenza vaccination, *J. Infect.* **2**:67–77.

Cook, S D., Dowling, P. C., and Whitaker, J. N., 1970, The Guillain–Barre syndrome, *Arch. Neurol.* **22**:470–474.

Couch, R. B., Douglas, R. G., Fedson, D. S., et al., 1971, Correlated studies of a recombinant influenza-virus vaccine. III. Protection against experimental influenza in man, *J. Infect. Dis* **124**:473–480.

Couch, R. B., Kasel, J. A., Gerin, J. L., et al., 1974, Induction of partial immunity to influenza by a neuraminidase-specific vaccine, *J. Infect. Dis.* **129**:411–420.

Couch, R. B., Webster, R. G., Kasel, J. A., et al., 1979, Efficacy of purified influenza subunit vaccines and relation to the major antigenic determinants on the hemagglutinin molecule, *J. Infect. Dis.* **140**:553–559.

Cox, N. J., and Kendal, A. P., 1984, Genetic stability of A/Ann Arbor/6/60 cold-mutant (temperature-sensitive) live influenza virus genes: Analysis by oligonucleotide mapping of recombinant vaccine strains before and after replication in volunteers, *J. Infect. Dis.* **149**:194–200.

Cox, N. J., Kendal, A. P., Shilov, A. A., et al., 1985, Comparative studies of A/Leningrad/134/57 wild-type and 47-times passaged cold-adapted mutant influenza viruses: Oligonucleotide mapping and RNA–RNA hybridization studies, *J. Gen. Virol.* **66**:1697–1704.

Crawford, C. R., Jennings, R., Bradford, N., et al., 1982, The effect of bacterial peptidoglycans on the immune response of hamsters to influenza virus vaccines, *Clin. Exp. Immunol.* **48**:739–746.

Crawford, C. R., Faiza Mukhlis, F. A., Jennings, R., et al., 1984, Use of zwitterionic detergent for the preparation of an influenza virus vaccine. 1. Preparation and characterization of disrupted virions, *Vaccine* **2**:193–198.

Davenport, F M. 1961, Applied immunology of mineral oil adjuvants, *J. Allergy* **32**:177–189.

Davis, A. R , Bos, T , Ueda, M., et al., 1983, Immune response to human influenza virus hemagglutinin expressed in *Escherichia coli, Gene* **21**:273–284.

Dierckx, J. P., Raindorf, M., and Gillard, C., 1976, Surveillance da la fonction respiratoire chez des sujects souffrant de bronchopathie chronique, vaccines par une souche vivante attenuee de virus grippal (RIT 4025), *Acta Tuberc, Pneumon. Belg* **67**:23–31

Dolin, R., Wise, T. G., Mazur, M. H , et al., 1977, Immunogenicity and reactogenicity of influenza A/New Jersey/76 virus vaccines in normal adults, *J. Infect Dis. (Suppl.)* **136**:435–442

Dolin, R , Reichman, R C., Madore, H P , et al., 1982, A controlled trial of amantadine and rimantadine in the prophylaxis of influenza A infection, *N. Engl. J. Med* **307**:580–584.

Dowdle, W. R , 1977, The swine flu vaccine program, *Am. Soc. Microbiol. News* **43**:243–246.

Dowdle, W. R , Millar, J. D., Schonberger, L B., et al , 1980, Influenza immunization policies and practices in Japan, *J. Infect Dis.* **141**:258–264

Downie, J C., 1984, A genetic and monoclonal analysis of high-yielding reassortants of influenza A virus used for human vaccines, *J Biol Stand.* **12**:101–110.

Dreesman, G. R., and Kennedy, R C., 1985, Anti-idiotypic antibodies. Implications of internal image-based vaccines for infectious diseases, *J. Infect. Dis.* **151**:761–765.

Dull, W. L , Seria, J., Bloomfield, J., et al., 1980, Effects of influenza vaccination on pulmonary function, *Bull Eur Physiopathol. Respir* **16**:655–661.

Dunn, G F N , and Wormald, P J , 1973, Influenza-A Hong-Kong vaccine and the new variant A/ENG/42/72, *Lancet* **1**:95.

Ennis, R. A., Dowdle, W. R., Barry, D. W , et al , 1977a, Endotoxin content and clinical reactivity to influenza vaccines, *J Biol Stand.* **5**:165–172.

Ennis, F. A., Mayner, R. E., Barry, D N , et al., 1977b, Correlation of laboratory studies with clinical responses to A/New Jersey influenza vaccines, *J. Infect. Dis.* **136**:S397–406.

Ennis, F. A , Yi-Hua, Q., and Schild, G. C , 1982, Antibody and cytotoxic T lymphocyte responses of humans to live and inactivated influenza vaccines, *J. Gen. Virol.* **58**:273–281.

Erickson, A. H., 1977, Different trypsin-sensitivity of the hemagglutinin from antigenically similar recombinant influenza vaccines, Doctoral dissertation, The City University of New York.

Evans, A. S , and Kraus, G., 1978, Immunization against swine influenza in the Yale University community, *Yale J. Biol. Med.* **51**:165–176.

Fedson, D. S., and Kessler, H. A., 1983, A hospital-based influenza immunization program, 1977–78, *Am J Public Health* **73**:442–445.

Felix, J. K., Schwartz, R. H., and Myers, G. J., 1976, Isolated hypoglossal nerve paralysis following influenza vaccination, *Am. J. Dis. Child.* **130**:82–83.

Fell, P. J , O'Donnell, H. F., Watson, N. P., et al., 1977, Longer term effects of live influenza vaccine in patients with chronic pulmonary disease, *Lancet* **1**:1282–1284.

Foy, H. M., Cooney, M., McMahan, R., et al., 1969, A field trial of Hong Kong-strain influenza vaccine in Seattle schoolchildren, *Bull. WHO* **41**:564–566.

Francis, T., Pearson, H. E , Salk, J. E., et al., 1944, Immunity in human subjects artificially infected with influenza virus, type B, *Am. J. Public Health* **34**:317–334.

Frank, A. L , Webster, R. G., Glezen, W. P., et al., 1981, Immunogenicity of influenza A/USSR (H1N1) subunit vaccine in unprimed young adults, *J. Med. Virol* **7**:135–142.

Ganz, P. A., Shanley, J. D., and Cherry, J. D., 1978, Responses of patients with neoplastic diseases to influenza virus vaccine, *Cancer* **42**:2244–2247.

Gaydos, J C., Hodder, R A., Top, F. H , et al., 1977a, Swine influenza at Fort Dix, New Jersey (January–February 1976). I. Case finding and clinical study of cases, *J. Infect. Dis.* **136**:S356–S362.

Gaydos, J C., Hodder, R A., Top, F. H., et al., 1977b, Swine influenza at Fort Dix, New Jersey (January–February 1976). II. Transmission and morbidity in units with cases, *J. Infect. Dis.* **136**:S363–S375.

Ghendon, Y. Z , Polezhaev, F. I., Lisovskaya, K. V., et al., 1984, Recombinant cold-adapted attenuated influenza A vaccines for use in children. Molecular genetic analysis of the cold-adapted donor and recombinants, *Infect. Immun.* **44**:730–733.

Goldfield, M., Bartley, J. D., Pizzuti, W., et al., 1977, Influenza in New Jersey in 1976: Isolations of influenza A/New Jersey/76 virus at Fort Dix, *J. Infect. Dis.* **136**:S347–S355.

Gomolin, I. H., Chapron, D. J., and Luhan, P. A., 1985, Lack of effect of influenza vaccine on theophylline levels and warfarin anticoagulation in the elderly, *J. Am. Geriatr. Soc.* **33**:269–272.

Goodeve, A., Potter, C. W., Clark, A., et al., 1983, A graded-dose study of inactivated, surface antigen influenza B vaccine in volunteers: Reactogenicity, antibody response and protection to challenge virus infection, *J. Hyg. (Camb.)* **90**:107–115.

Goodeve, A. C., Jennings, R., and Potter, C. W., 1985, Reassortants of influenza B viruses for use in vaccines: An evaluation, *Arch. Virol.* **83**:169–179.

Graves, P. N., Schulman, J. L., Young, J. F., et al., 1983, Preparation of influenza virus subviral particles lacking the HA1 subunit of hemagglutinin: Unmasking of cross-reactive HA2 determinants, *Virology* **126**:106–116.

Green, N., Alexander, H., Olson, A., et al., 1982, Immunogenic structure of the influenza virus hemagglutinin, *Cell* **28**:477–487.

Greenstreet, R. L., 1984, Estimation of the probability that Guillain–Barre syndrome was caused by the swine flu vaccine: U.S. experience (1976–1977), *Med. Sci. Law* **24**:61–67.

Gremillion, D. H., Meiklejohn, G., Graves, P., et al., 1983, Efficacy of single-dose influenza immunization in Air Force recruits, *J. Infect. Dis.* **147**:1099.

Gross, P. A., 1977, Reactogenicity and immunogenicity of bivalent influenza vaccine in one- and two-dose trials in children: A summary, *J. Infect. Dis.* **136**:S616–S625.

Gross, P. A., Ennis, F. A., Gaerlan, P. F., et al., 1981, Comparison of new triton X-100 and tween-ether-treated split-product vaccines in children, *J. Clin. Microbiol.* **14**:534–538.

Gross, P. A., Gould, A. L., and Brown, A. E., 1985, Effect of cancer chemotherapy on the immune response to influenza virus vaccine: Review of published studies, *Rev. Infect. Dis.* **7**:613–618.

Grossebauer, K., Langmaack, H., Schmidt, B., et al., 1969, Enhancement and neutralization of pyrogenicity of influenza viruses by biologically active substances, *Arch. Ges. Virusforsch.* **28**:151–164.

Gwaltney, J. M., DeSanctis, A. N., Metzgar, D. P., et al., 1977, Systematic reactions to influenza B vaccine. *Am. J. Epidemiol.* **105**:252–260.

Hackett, C. J., Taylor, P. M., and Askonas, B. A., 1983, Stimulation of cytotoxic T cells by liposomes containing influenza virus or its components, *Immunology* **49**:255–263.

Hackett, C. J., Hurwitz, J. L., Dietzschold, B., et al., 1985, A synthetic decapeptide of influenza virus hemagglutinin elicits helper T cells with the same fine recognition specificities as occur in response to whole virus, *J. Immunol.* **135**:1391–1394.

Hall, W. J., Douglas, R. G., Jr., Zaky, D. A., et al., 1976, Attenuated influenza virus in normal adults: Role of pulmonary function studies in vaccine trials, *J. Infect. Dis.* **133**:145–152.

Halperin, W., Weiss, W. I., Altman, R., et al., 1979, A comparison of intradermal and subcutaneous routes of influenza vaccination with A/New Jersey/76 (swine flu) and A/Victoria/75: Report of a study and review of the literature, *Am. J. Public Health* **69**:1247–1251.

Hamzawi, J., Jennings, R., and Potter, C. W., 1981, Antigenicity in hamsters of inactivated vaccines prepared from recombinant influenza viruses, *J. Hyg. (Camb.)* **87**:453–464.

Hasselbacher, P., 1977, Neuropathy after influenza vaccination, *Lancet* **1**:551–552.

Helenius, A., and Simons, K., 1975, Solubilization of membranes by detergents, *Biochim. Biophys. Acta* **415**:29.

Henle, W., and Henle, G., 1945, Effects of adjuvants on vaccination of human beings against influenza, *Proc. Soc. Exp. Biol. Med.* **59**:179–181.

Herbert, A. A., Larke, R. P. B., and Markstad, E. L., 1979, Comparison of responses to influenza A/New Jersey/76-A/Victoria/75 virus vaccine administered intradermally or subcutaneously to adults with chronic respiratory disease, *J. Infect. Dis.* **140**:234–238.

Higashihara, M. 1975, Studies on genetic recombinants of influenza virus, *Kitasato Arch. Exp. Med.* **48**:183–188.

Hilleman, M. R., 1966, Critical appraisal of emulsified oil adjuvants applied to viral vaccines in: *Progress in Medical Virology,* Volume 8 (J. L. Melnick, ed.), S. Karger, Basel, pp. 131–182.

Hilleman, M. R., 1977, Serologic responses to split and whole swine influenza virus vaccines in light of the next influenza pandemic, *J. Infect. Dis.* **136**:S683–S685.

Hirst, G K., Rickard, E. R , Whitman, L., et al , 1942, Antibody response of human beings following vaccination with influenza viruses, *J Exp. Med.* **75**:495–511.
Hodder, R. A., Gaydos, J C., Allen, R. G., et al., 1977, Swine influenza A at Fort Dix, New Jersey (January–February 1976). III. Extent of spread and duration of the outbreak, *J. Infect. Dis* **136**:S369–S375.
Hodges, G R., Davis, J. W , Lewis, D , et al , 1979, Response to influenza A vaccine among high-risk patients, *South. Med. J.* **72**:29–32.
Hoskins, T. W., 1977, Systemic reactions to influenza vaccine containing killed influenza A/New Jersey/8/76, *Practitioner* **219**:400–401.
Hoskins, T W , Davies, J. R., Allchin, A., et al , 1973, Controlled trial of inactivated influenza vaccine containing the A/Hong Kong strain during an outbreak of influenza due to the A/England/42/72 strain, *Lancet* **2**:116–120
Hoskins, T W., Davies, J. R., Smith, A J., et al., 1979, Assessment of inactivated influenza-A vaccine after three outbreaks of influenza A at Christ's Hospital, *Lancet* **1**:33–35
Houston, T P., 1983, Small-vessel vasculitis following simultaneous influenza and pneumococcal vaccination, *N.Y State J Med* **83**:1182–1183
Hruskova, J , Syrucek, L., Zaradova, H , et al., 1978, Secretory antibody response to haemagglutinin of influenza virus A/Swine and A/Victoria in humans after administration of split and whole-virion vaccine, *J Biol Stand.* **6**:181–185.
Hurwitz, E S , Schonberger, L. B., Nelson, D B , et al., 1981, Guillain-Barre syndrome and the 1978–1979 influenza vaccine, *N. Engl. J. Med.* **304**:1557–1561.
Infectious Disease Branch, NIAID, 1976, Memo: Immunizations to protect individuals working with influenza virus, April 28, NIH, Bethesda.
Iwasaki, T., and Nozima, T., 1977, Defense mechanisms against primary influenza virus infection in mice. I. The roles of interferon and neutralizing antibodies and thymus dependence of interferon and antibody production, *J Immunol* **118**:256–263.
Jennings, R , and Potter, C. W., 1973, Enhanced response to influenza A vaccines in hamsters primed by prior heterotype influenza infection, *Arch Ges. Virusforsch.* **42**:197–206.
Jennings, R , Smith, T. L., Spencer, R. C., et al., 1984, Inactivated influenza virus vaccines in man: A comparative study of subunit and split vaccines using two methods for assessment of antibody responses, *Vaccine* **2**:75–80.
Jennings, R , Smith, T. L., Mellersh, A. R., et al., 1985, Antibody response and persistence in volunteers following immunization with varying dosages of a trivalent surface antigen influenza virus vaccine, *J. Hyg. (Camb.)* **94**:87–95.
Johannsen, R., Moser, H., Hinz, J., et al., 1983, The quantification of the haemagglutinin content of influenza whole virus and Tween-ether split vaccines, *J Biol. Stand.* **11**:341–352.
Johnson, P R., Feldman, S., Thompson, J. M., et al., 1985, Comparison of long-term systemic and secretory antibody responses in children given live, attenuated, or inactivated influenza A vaccine, *J Med. Virol.* **17**:325–335.
Journal of the American Medical Association, 1984, AMA offers recommendations for vaccine injury compensation, *J.A.M.A.* **252**:2937–2946.
Kaplan, J. E., Katona, P., Hurwitz, E. S., et al., 1982, Guillain–Barre syndrome in the United States, 1979–1980 and 1980–1981, *J.A.M.A.* **248**:698–700.
Kasel, J. A., Six, H. R., Couch, R. B., et al., 1979, Variant-specific antihemagglutinin serum response to type A influenza natural infection and inactivated vaccines in adults, *Proc. Soc. Exp. Biol. Med.* **161**:519–521.
Kava, T., and Laitinen, L. A., 1985, Effects of killed and live attenuated influenza vaccine on symptoms and specific airway conductance in asthmatics and healthy subjects, *Allergy* **40**:42–47
Kavet, J., 1972, *Influenza and Public Health,* Masters thesis, Harvard School of Public Health, Boston
Kavet, J., 1976, Vaccine utilization: Trends in the implementation of public policy in the U.S.A., in: *Influenza: Virus Vaccines Strategy* (P Selby, ed.), Academic Press, New York, pp. 297–308.
Kendal, A P., Goldfield, M., Noble, G. R., 1977a, Identification and preliminary antigenic analysis of swine influenza-like viruses isolated during an influenza outbreak at Fort Dix, New Jersey, *J Infect Dis* **136**:S381–S385
Kendal, A. P., Noble, G R., and Dowdle, W., 1977b, Neuraminidase content of influenza vaccines

and neuraminidase antibody responses after vaccination of immunological primed and unprimed populations, *J. Infect. Dis.* **136:**S415–424.

Khan, M. W., Gallagher, M., Bucher, D., et al., 1982, Detection of influenza virus neuraminidase-specific antibodies by an enzyme-linked immunosorbent assay, *J. Clin. Microsc.* **16:**115–122.

Kilbourne, E. D., 1969, Future influenza vaccines and the use of genetic recombinants, *Bull. WHO* **41:**643–645.

Kilbourne, E. D., 1976a, Op-ed page, *New York Times,* February 13, p. 33.

Kilbourne, E. D., 1976b, Comparative efficacy of neuraminidase-specific and conventional influenza virus vaccines in the induction of anti-neuraminidase antibody in man, *J. Infect. Dis.* **134:**384–394.

Kilbourne, E. D., 1977, Influenza pandemics in perspective, *J.A.M.A.* **237:**1225–1228.

Kilbourne, E. D., 1978, Opinion: Influenza as a problem in immunology, *J. Immunol.* **120:**1447–1452.

Kilbourne, E. D., 1979, Swine flu: The virus that vanished, *Human Nature* **2:**68–73.

Kilbourne, E. D., 1984, Immunization strategy: Infection-permissive vaccines for the modulation of infection, in: *Modern Approaches to Vaccines* (R. M. Chanock and R. A. Lerner, eds.), Cold Spring Harbor Laboratory, New York, pp. 269–274.

Kilbourne, E. D., 1985, Problems inherent in artificial immunization, *Microbiology*–1985, 243–246.

Kilbourne, E. D., and Murphy, J. S., 1960, Genetic studies of influenza viruses. I. Viral morphology and growth capacity as exchangeable genetic traits. Rapid *in ovo* adaptation of early passage Asian strain isolates by combination with PR8. *J. Exp. Med.* **111:**387–406.

Kilbourne, E. D., Laver, W. G., Schulman, J. L., et al., 1968, Antiviral activity of antiserum specific for an influenza virus neuraminidase. I. *In vitro* effects, *J. Virol.* **2:**281–288.

Kilbourne, E. D., Schulman, J. L., Schild, G. C., et al., 1971, Correlated studies of a recombinant influenza-virus vaccine. I. Derivation and characterization of virus and vaccine. *J. Infect. Dis.* **124:**449–462.

Kilbourne, E. D., Chanock, R. M., Choppin, P. W., et al., 1974, Influenza vaccine: summary of influenza workshop V, *J. Infect. Dis.* **129:**770–771.

Kilbourne, E. D., McGregor, S., and Easterday, B. C., 1981, Transmission of swine of hemagglutinin mutants of swine influenza virus, in: *The Replication of Negative Strand Viruses* (D. H. Bishop and R. W. Compans, eds.), Elsevier/North Holland, Amsterdam, pp. 449–453.

Kilbourne, E. D., Gerhard, W., and Whitaker, C. W., 1983, Monoclonal antibodies to the hemagglutinin Sa antigenic site of A/PR/8/23 influenza virus distinguish biologic mutants of swine influenza virus, *Proc. Natl. Acad. Sci. U.S.A.* **80:**6996–7000.

Klimov, A., Ghendon, Y., Zavadova, H., et al., 1983, High reproduction capacity of recombinants between H3N2 human influenza and fowl plague viruses is due to the gene coding for M proteins, *Acta Virol.* **27:**434–438.

Kramer, P., and McClain, C. J., 1981, Depression of aminopyrine metabolism by influenza vaccination, *N. Engl. J. Med.* **305:**1262–1264.

Kramer, P., Tsuru, M., Cook, C. E., et al., 1984, Effect of influenza vaccine on warfarin anticoagulation, *Clin. Pharmacol. Ther.* **35:**416–418.

Kreuter, J., Mauler, R., Gruschkau, H., et al., 1976, The use of new polymethylmethacrylate adjuvants for split influenza vaccines, *Exp. Cell Biol.* **44:**12–19.

Kuwert, E. K., Werner, J., Scheiermann, N., et al., 1979, Local and systemic antibody response after vaccination with 3 different types of vaccines against influenza. I. Hemagglutination inhibiting antibodies. *Zentralbl. Bakteriol. Hyg.* **244:**155–174.

Lamb, J. R., Eckels, D. D., Lake, P., et al., 1982, Human T-cell clones recognize chemically synthesized peptides of influenza hemagglutinin, *Nature* **300:**66–69.

La Montagne, J. R., Noble, G. R., Quinnan, G. V., et al., 1983, Summary of clinical trials of inactivated influenza vaccine—1978, *Rev. Infect. Dis.* **5:**723–736.

Lang, G., 1981, Turkey influenza vaccination. *Vet. Rec.* **108:**425.

Langmuir, A. D., 1979, Guillain–Barre syndrome: The swine influenza virus vaccine incident in the United States of America, 1976–77: Preliminary communication, *J. R. Soc. Med.* **72:**660–669.

Langmuir, A. D., Bregman, D. J., Kurland, L. T., et al., 1984, An epidemiologic and clinical evaluation of Guillain–Barre syndrome reported in association with the administration of swine influenza vaccines, *J. Epidemiol.* **119:**841–879.

Lazzell, V., Waldman, R. H., Rose, C., 1984, Immunization against influenza in humans using an oral enteric-coated killed virus vaccine, *J. Biol Stand.* **12**:315–321.
Leibovitz, A., Coultrip, R. L., Kilbourne, E. D., et al, 1971, Correlated studies of a recombinant influenza-virus vaccine. IV. Protection against naturally occurring influenza in military trainees, *J. Infect. Dis.* **124**:481–487.
Lerman, S. J., Wright, P F., and Patil, K. D, 1980, Antibody decline in children following A/New Jersey/76 influenza virus immunization, *J. Pediatr.* **96**:271–274
Levine, M., Jones, M W, and Gribble, M, 1984, Increased serum phenytoin concentration following influenza vaccination, *Clin. Pharm* **3**:505–509.
Lyons, J. E., 1976, The manufacturers' role in the preparation of swine flu vaccine, in: *Meeting of the New Jersey Public Health Association, June 23, Fort Dix, N J.*
Maassab, H. F., 1967, Adaptation and growth characteristics of influenza virus at 25°C, *Nature* **213**:612–614.
Maassab, H. F, Monto, A S, and DeBorde, D. C., et al., 1981, Development of cold recombinants of influenza virus as live virus vaccines, in: *Genetic Variation among Influenza Viruses* (D P. Nayak, ed), *ICN–UCLA Symposia on Molecular and Cellular Biology,* Vol XXI, Academic Press, New York, pp 617–637.
Maassab, H R., Kendal, A. P., Abrams, G D, et al., 1982, Evaluation of a cold-recombinant influenza virus vaccine in ferrets, *J. Infect. Dis.* **146**:780–790.
MacKenzie, J. S, 1977, Influenza subunit vaccine· Antibody responses to one and two doses of vaccine and length of response, with particular reference to the elderly, *Br. Med. J.* **1**:200–202.
Macoul, K L., 1982, Bilateral optic nerve atrophy and blindness following swine influenza vaccination, *Ann Ophthalmol.* **14**:398–399.
MacPherson, B. V., Ashikaga, T., Dennis, S. M., et al., 1980, A cohort study of preventive health behavior: Swine flu vaccination acceptance in Vermont, *Del. Med. J.* **52**:427–432.
Marine, W. M., and Stuart-Harris, C., 1976, Reactions and serologic responses in young children and infants after administration of inactivated monovalent influenza A vaccine, *J Pediatr* **88**:26–30.
Marine, W. M., and Thomas, J. E., 1979, Antigenic memory to influenza A viruses in man determined by monovalent vaccines, *Postgrad. Med. J.* **55**:98–108.
Massicot, J. G., Murphy, B. R., Thierry, F., et al., 1980a, Temperature-sensitive mutants of influenza virus, *Virology* **101**:242–249.
Massicot, J. G., Murphy, B. R., Van Wyke, K., et al., 1980b, P1 and P3 genes are responsible for satisfactory level of attenuation of ts-142 recombinants bearing H1N1 or H3N2 surface antigens of influenza A virus, *Virology* **106**:187–190.
Masurel, N., Ophof, P., and deJong, P, 1981, Antibody response to immunization with influenza A/USSR/77 (H1N1) virus in young individuals primed or unprimed for A/New Jersey/76 (H1N1) virus, *J. Hyg. (Camb.)* **87**:201–209.
Matzkin, H., and Nili, E, 1984, Accidental tenfold overdose of influenza A vaccine: A clinical and serological study, *Isr J. Med Sci.* **20**:411–415.
McCahon, D., and Schild, G. C., 1972, Segregation of antigenic and biological characteristics during influenza virus recombination, *J. Gen Virol.* **15**:73–77.
McCarroll, J. R., and Kilbourne, E. D, 1958, The equivalence of the subcutaneous and intradermal routes in immunization with Asiatic strain influenza vaccine, *N. Engl. J. Med* **259**:618–621.
McDonald, J. C., Zuckerman, A J., Beare, A. S., et al., 1962, Trials of live influenza vaccine in the Royal Air Force, *Br. Med. J.* **1**:1036–1042.
McLaren, C, Potter, C. W, and Jennings, R., 1974, Immunity to influenza in ferrets, *Med. Microbiol. Immunol.* **160**:33–45.
McLean, I. W, Jr., Beard, D., Taylor, A. R., et al., 1945, The relation of antibody response in swine to dose of the swine influenza virus inactivated with formalin and with ultraviolet light, *J. Immunol* **51**:65–99.
McMichael, A. J., Gotch, F., Cullen, P., et al., 1981, The human cytotoxic T cell response to influenza A vaccination, *Clin Exp Immunol.* **43**:276–284.
McMichael, A. J., Askonas, B. A, Webster, R G, et al., 1982, B-cell or T-cell immunity? *Immunol Today* **3**:255–260.

Meiklejohn, G., 1958, Effectiveness of monovalent influenza A-prime vaccine during 1957 influenza A-prime epidemic, *Am. J. Hyg.* **67:**237–249.

Meiklejohn, G., 1961, Asian influenza vaccination: Dosage, routes, schedules of inoculation and reactions, *Am. Rev. Respir. Dis.* **83:**175–177.

Meiklejohn, G., Eickhoff, T. C., Graves, P., et al., 1978, Antigenic drift and efficacy of influenza virus vaccines, 1976–1977, *J. Infect. Dis.* **138:**618–624.

Melchizedek, A. N., Wabuke-Bunoti, Taku, A., et al., 1984, Cytolytic T lymphocyte and antibody responses to synthetic peptides of influenza virus hemagglutinin, *J. Immunol.* **133:**2194–2201.

Meredith, C. G., Christian, C., Johnson, R. F., et al., 1985, Effects of influenza virus vaccine on hepatic drug metabolism, *Clin. Pharmacol. Ther.* **37:**396–401.

Mihail, A., Popa, L. M., Petrescu, A., et al., 1981, Study of recombinants derived from autochthonous influenza virus type A strains, *Rev. Roum, Med. Virol.* **32:**225–229.

Miles, R., Potter, C. W., Clarke, A., et al., 1982, A comparative study of the reactogenicity and immunogenicity of two inactivated influenza vaccines in children, *J. Biol. Stand.* **10:**59–68.

Mills, J., and Chanock, R. M., 1971, Temperature-sensitive mutants of influenza virus. I. Behavior in tissue culture and in experimental animals, *J. Infect. Dis.* **123:**145–157.

Mitchell, D. M., Fitzharris, P., Knight, R. A., et al., 1982, Kinetics of specific *in vitro* antibody production following influenza immunization, *Clin. Exp. Immunol.* **48:**491–498.

Monto, A. S., and Ross, H. W., 1979, Swine influenza vaccine program in the community: Acceptability, reactions and responses, *Am. J. Public Health* **69:**233–237.

Monto, A. S., Miller, F. D., and Maassab, H. F., 1982, Evaluation of an attenuated, cold-recombinant influenza B virus vaccine, *J. Infect. Dis.* **145:**57–64.

Morein, B., and Simons, K., 1985, Subunit vaccines against enveloped viruses: Virosomes, micelles and other protein complexes, *Vaccine* **3:**83–93.

Mostow, S. R., Schoenbaum, S. C., Dowdle, W. R., et al., 1970, Studies on inactivated influenza vaccines. II. Effect of increasing dosage on antibody response and adverse reactions in man, *Am. J. Epidemiol.* **92:**248–256.

Mostow, S. R., Dowdle, W. R., Schoenbaum, S. C., et al., 1973, Inactivated vaccines. 1. Volunteer studies with very high doses of influenza vaccine purified by zonal ultracentrifugation, *Postgrad. Med. J.* **49:**152–158.

MRC Advisory Group on Pulmonary Function Tests in Relation to Live Influenza Vaccines, 1984, Trials of live attenuated influenza virus vaccine in patients with chronic obstructive airways disease, *Br. J. Dis. Chest* **78:**236–247.

Muldoon, R. L., Stanley, E. D., and Jackson, G. G., 1976, Use and withdrawal of amantadine chemoprophylaxis during epidemic influenza A, *Am. Rev. Respir. Dis.* **113:**487–491.

Mumford, J., Wood, J. M., Scott, A. M., et al., 1983, Studies with inactivated equine influenza vaccine, *J. Hyg. (Camb.)* **90:**385–395.

Murphy, B. R., and Chanock, R. M., 1985, Immunization against viruses, in: *Virology* (B. N. Fields, ed.), Raven Press, New York, pp. 349–370.

Murphy, B. R., Chalhub, E. G., Nusinoff, S. R., et al., 1972, Temperature-sensitive mutants of influenza virus. II. Attenuation of ts recombinants for man, *J. Infect. Dis.* **126:**170–178.

Murphy, B. R., Chalhub, E. G., Nusinoff, S. R., et al., 1973, Temperature sensitive mutants of influenza virus. III. Further characterization of the ts-1[E] influenza A recombinant, *J. Infect. Dis.* **128:**479–487.

Murphy, B. R., Richman, D. D., Spring, S. B., et al., 1976, Use of temperature-sensitive mutants of influenza A virus as live virus vaccine strains: Evaluation in laboratory animals, adults and children, *Postgrad. Med.* **52:**381–388.

Murphy, B. R., Markoff, L. J., Chanock, R. M., et al., 1980, Genetic approaches to attenuation of influenza A viruses for man, *Phil. Trans. R. Soc. Lond.* [*Biol.*] **288:**401–415.

Murphy, B. R., Buckler-White, A. J., London, W. T., et al., 1984, Avian–human reassortant influenza A viruses derived from mating avian and human influenza A viruses, *J. Infect. Dis.* **150:**841–850

Murphy, K. R., and Strunk, R. C., 1985, Safe administration of influenza vaccine in asthmatic children sensitive to egg proteins, *J. Pediatr.* **106:**931–933.

Murray, D. L., Imagawa, D. T., Kada, D. M., et al., 1979, Antibody response to nonvalent A/New

Jersey/8/76 influenza vaccine in pregnant women, *J Clin Microbiol.* **10**:184–187.
Murray, R., 1961, Some problems in the standardization and control of influenza vaccine in 1957, *Am. Rev Respir. Dis.* **83**:160–167.
Nature, 1985, Editorial: Statutory compensation urged, **316**:476–477.
Nestorowicz, A., Tregear, G. W., Southwell, C. N., et al., 1985, Antibodies elicited by influenza virus hemagglutinin fail to bind to synthetic peptides representing putative antigenic sites, *Mol Immunol.* **22**:145–154
Neustadt, R. E , and Fineberg, H. V., 1978, *The Swine Flu Affair, Decision Making on a Slippery Disease.* U. S. Department of HEW, Washington.
Newton Dunn, G. F , and Wormald, P. J., 1973, Influenza A Hong Kong vaccine and the new variant A/Eng/42/72, *Lancet* **1**:95.
Noble, G R , Kaye, H S., Kendal, A. P., et al., 1977, Age-related heterologous antibody responses to influenza virus vaccination, *J. Infect Dis.* **136**:S686–692.
Odagiri, T , DeBorde, D C., and Maassab, H. F , 1982, Cold-adapted recombinants of influenza A virus in MDCK cells. I Development and characterization of A/Ann Arbor/6/60 × A/Alaska/6/77 recombinant viruses, *Virology* **119**:82–95
Odagiri, T., Smitka, C W , and Maassab, H G., 1983, Cold-adapted reassortants of influenza A virus in MDCK cells. II Role of the temperature-sensitive property of cold-adapted reassortants in mice, *Microbiol Immunol* **27**:203–206.
O'Donoghue, J M., Ray, C. G , Terry, D. W., Jr., et al , 1973, Prevention of nosocomial influenza infection with amantadine, *Am J. Epidemiol.* **97**:276–282.
Ogra, P L., Chow, T., Beutner, K R , et al., 1977, Clinical and immunologic evaluation of neuraminidase-specific influenza A virus vaccine in humans, *J Infect Dis.* **135**:499–506
Okuno, Y., and Nakamura, K , 1966, Prophylactic effectiveness of live influenza vaccine in 1965, *Biken J* **9**:89–95
Osborn, J. (ed.), 1977, *Influenza in America, 1918–1976,* Prodist, New York.
Palese, P , and Schulman, J L., 1976, RNA pattern of "swine" influenza virus isolated from man is similar to those of other swine influenza viruses, *Nature* **263**:528–530
Panicali, D., Davis, S W., Weinberg, R. L., et al., 1983, Construction of live vaccines by using genetically engineered poxviruses: Biological activity of recombinant vaccinia virus expressing influenza virus haemagglutinin, *Proc. Natl. Acad. Sci. U.S A* **80**:5364.
Paoletti, E., Weinberg, R. L , Davis, S. W., et al., 1984, Genetically engineered poxviruses: A novel approach to the construction of live vaccines, *Vaccine* **2**:204–205.
Parkman, P. D., Hopps, H E., Rastogi, S. C., et al., 1977, Summary of clinical trials of influenza virus vaccines in adults, *J Infect Dis* **136**:S722–730.
Patriarca, P A , Kendal, A. P., Stricof, R. L., et al., 1983, Influenza vaccination and warfarin or theophylline toxicity in nursing-home residents, *N. Engl. J. Med.* **308**:1601–1602.
Patriarca, P. A., Kater, N A., Kendal, A. P., et al., 1984, Safety of prolonged administration of rimantadine hydrochloride in the prophylaxis of influenza A virus infections in nursing homes, *Antimicrob. Agents Chemother.* **26**:101–103.
Patriarca, P. A , Weber, J A., Parker, R. A., et al., 1985, Efficacy of influenza vaccine in nursing homes, *J A.M A* **253**:1136–1139.
Payler, D K., and Purdham, P A., 1984, Influenza A prophylaxis with amantadine in a boarding school, *Lancet* **1**:502–504.
Pearman, W A., 1978, Participation in flu immunization projects· What can we expect in the future? *Am. J. Public Health* **68**:674–675.
Pemberton, R M , Jenning, R., and Smith, T. L , 1985, Morphology and antigenicity studies on reassortant influenza (H3N2) viruses for use in inactivated vaccines, *J. Hyg. (Camb)* **94**:229–239.
Perkins, F T., 1972, Experience in the United Kingdom with oil adjuvant influenza vaccines, *Ann Allergy* **30**:288–291
Perry, H D., Mallen, F J , Grodin, R. W., et al., 1979, Reversible blindness in optic neuritis associated with influenza vaccination, *Ann Ophthalmol* **11**:545–550
Petrescu, A., Cajal, N., Bronitki, A., et al., 1977, The experience of the Stefan S Nicolau Institute of Virology in the preparation and administration of inactivated influenza vaccines applicable by

nasal or oral route, *Rev. Roum. Med. Virol.* **28**:213–217.
Petrov, R. V., Zhdanov, V. M., Kabanov, V. A., et al., 1985, [Enhancement of the immune response to surface antigens of influenza virus as a result of their covalent binding with a synthetic carrier polymer], *Biull. Eksp. Biol. Med.* **99**:184–186.
Phillips, C. A., Forsyth, B. R., Christmas, W. A., et al., 1970, Purified influenza vaccine: Clinical and serologic responses to varying doses and different routes of immunization, *J. Infect. Dis.* **122**:26–32.
Poser, C. M., Roman, G., and Emery, E. S. III, 1978, Recurrent disseminated vasculomyelinopathy, *Arch. Neurol.* **35**:166–170.
Potter, C. W., Oxford, J. S., Shore, S. L., et al., 1972, Immunity to influenza in ferrets. I. Response to live and killed virus, *Br. J. Exp. Pathol.* **53**:153–167.
Powers, R. D., Hayden, F. G., Samuelson, J., et al., 1984, Immune response of adults to sequential influenza vaccination, *J. Med. Virol.* **14**:169–175.
Quarles, J. M., Couch, R. B., Cate, T. R., et al., 1981, Comparison of amantadine and rimantadine for prevention of type A (Russian) influenza, *Antivir. Res.* **1**:149–155.
Quinnan, G. V., Schooley, R., Dolin, R., et al., 1983, Serologic responses and systemic reactions in adults after vaccination with monovalent A/USSR/77 and trivalent A/USSR/77, A/Texas/77, B/Hong Kong/72 influenza vaccines, *Rev. Infect. Dis.* **5**:748–757.
Ratner, E. R., and Fedson, D. S., 1983, Influenza and pneumococcal immunization in medical clinics, 1978–1980, *Arch. Intern. Med.* **143**:2066–2069.
Reeve, P., Pibermann, M., Bachmayer, H., et al., 1982, Studies in man with a cold-recombinant live influenza B virus vaccine, *J. Med. Virol.* **9**:1–9.
Reiss, C. S., and Schulman, J. L., 1980, Cellular immune responses of mice to influenza virus vaccines, *J. Immunol.* **125**:2182–2188.
Renton, K. W., Gray, J. D., and Hall, R. I., 1980, Decreased elimination of theophylline after influenza vaccination, *Can. Med. J.* **123**:288–290.
Riddiough, M. A., Sisk, J. E., and Bell, J. C., 1983, Influenza vaccination. Cost-effectiveness and public policy, *J.A.M.A.* **249**:3189–3195.
Rodda, S. J., Gallichio, H. A., and Hampson, A. W., 1981, The single radial immunodiffusion assay highlights small antigenic differences among influenza virus hemagglutinins, *J. Clin. Microbiol.* **14**:479–482.
Rogers, G. N., Daniels, R. S., Skehel, J. J., et al., 1985, Host-mediated selection of influenza virus receptor variants, *J. Biol. Chem.* **260**:7362–7367.
Rose, H. J., 1980, The use of amantadine and influenza vaccine in a type A influenza epidemic in a boarding school, *J. R. Coll. Gen. Pract.* **30**:619–621.
Rosen, C. M., Roman, G., and Emery, E. S. III, 1978, Recurrent disseminated vasculomyelinopathy, *Arch. Neurol.* **35**:166–170.
Rosenzweig, D. Y., Dwyer, D. J., Ferstenfeld, J. E., et al., 1975, Changes in small airway function after live attenuated influenza vaccination, *Am. Rev. Respir. Dis.* **111**:399–403.
Rott, R., Becht, H., and Orlich, M., 1974, The significance of influenza virus neuraminidase in immunity, *J. Gen. Virol.* **22**:35–41.
Ruben, F. L., Potter, C. W., and Stuart-Harris, C. H., 1975, Humoral and secretory antibody responses to immunization with low and high dosage split influenza virus vaccine, *Arch. Virol.* **47**:157–166.
Rytel, M. W., Jackson, L. J., Ferstenfeld, J. E., et al., 1975, New live attenuated influenza A/England/42/72(H3N2) vaccine ("Alice") reactogenicity, immunogenicity, and protection efficacy, *J. Infect. Dis.* **132**:652–659.
Sabin, A. B., 1977, Swine flu: What happened. *Sciences* **17**:14–30.
Salk, J. E., 1948, Reactions to concentrated influenza vaccines, *J. Immunol.* **58**:369–395.
Salk, J. E., Bailey, M. L., and Laurent, A. M., 1952, The use of adjuvants in studies on influenza immunization. II. Increased antibody formation in human subjects inoculated with influenza virus vaccine in a water-in-oil emulsion, *Am. J. Hyg.* **55**:439.
Schiff, G. M., Linnemann, C. C., Shea, L., et al., 1975, Evaluation of a live, attenuated recombinant influenza vaccine in high school children, *Infect. Immun.* **11**:754–757.
Schoenbaum, S. C., 1979, Influenza vaccine—unacceptable or unaccepted, *Am. J. Public Health* **69**:219–221.
Schoenbaum, S. C., McNeil, B. J., and Kavet, J., 1976, The swine-influenza decision, *N. Engl. J. Med.* **295**:759–765.

Scholtissek, C., Vallbracht, A., Flehmig, B., et al., 1979, Correlation of pathogenicity and gene constellation of influenza A viruses. II. Highly neurovirulent recombinants derived from nonneurovirulent or weakly neurvirulent parent virus strains, *Virology* **95**:492–500

Schonberger, L B, Bregman, D J, Sullivan-Bolyai, J. Z., et al., 1979, Guillain-Barre syndrome following vaccination in the National Influenza Immunization Program, United States, 1976-1977, *Am J Epidemiol.* **110**:105–123.

Schulman, J. L., 1967, Experimental transmission of influenza virus infection in mice. III Differing effects of immunity induced by infection and by inactivated influenza virus vaccine on transmission of infection, *J. Exp. Med* **125**:467–478

Schulman, J L., 1975, Immunology of influenza, in *The Influenza Viruses and Influenza* (E D Kilbourne, ed), Academic Press, New York, pp 373–390.

Schulman, J L, and Kilbourne, E D., 1971, Correlated studies of a recombinant influenza-virus vaccine. II Definition of antigenicity in experimental animals, *J Infect Dis* **124**:463–472

Schulman, J L., and Palese, P., 1978, Biological properties of recombinants of A/Hong Kong and A/PR8 viruses Effects of genes for matrix protein and nucleoprotein in virus yield in embryonated eggs, in *Negative Strand Viruses and the Host Cell* (B W J. Mahy and R D. Barry, eds), Academic Press, New York, pp 663–674

Schulman, J L, Khakpour, M, and Kilbourne, E. D, 1968, Protective effects of specific immunity to viral neuraminidase on influenza virus infection of mice, *J Virol* **2**:778–786

Schutze, M -P., Lerclerc, C, Jolivet, M, et al., 1985, Carrier-induced epitopic suppression, a major issue for future synthetic vaccines, *J Immunol* **135**:2319–2322

Serie, C, Barme, M, Hannoun, C., et al., 1977, Effects of vaccination on an influenza epidemic in a geriatic hospital, *Dev Biol. Stand.* **39**:317–321.

Seyal, M., Ziegler, D K., and Couch, J R, 1978, Recurrent Guillain-Barre syndrome following influenza vaccine, *Neurology (Minneap.)* **28**:725–726.

Shapira, M., Jibson, M., Muller, G., et al, 1984, Immunity and protection against influenza virus by synthetic peptide corresponding to antigenic sites of hemagglutinin, *Proc Natl. Acad Sci U S A* **81**:2461–2465

Sheps, S B, Mathias, R G, Wilson, R, et al, 1985, Influenza vaccine utilization in a metropolitan health department supply side constraints, *Can J Public Health* **76**:378–383

Shore, S. L., Potter, C. W., and Stuart-Harris, C. H., 1973, Antibody response to inactivated influenza vaccine given by different routes in patients with chronic bronchopulmonary disease, *Thorax* **28**:721–728

Siegert, R, and Braune, P., 1964, The pyrogens of myxoviruses II. Resistance of influenza A pyrogens to heat, ultraviolet and chemical treatment, *Virology* **24**:218–224.

Silverstein, A. M, 1981, *Pure Politics and Impure Science*, The Johns Hopkins University Press, Baltimore, London

Slepushkin, A. N., Ellengorn, N. S., Russina, A. E., et al., 1968, Assessment of the efficacy of mass influenza vaccinations. [Assessment of the efficacy of mass influenza vaccinations Report II], (English summary), *J Microbiol Epidemiol. Immunobiol.* **10**:26–32.

Smith, G. L., Murphy, B R., and Moss, B, 1983, Construction and characterization of an infectious vaccinia virus recombinant that expresses the influenza hemagglutinin gene and induces resistance to influenza virus infection in hamsters, *Proc Natl. Acad. Sci U S A* **80**:7155–7159.

Smith, J. W. G., and Pollard, R, 1979, Vaccination against influenza: A five-year study in the Post Office, *J. Hyg. (Camb)* **83**:157–170.

Smorodintsev, A. A., Tushinsky, M D, Drobyshevskaya, A I, et al, 1937, Investigation on volunteers infected with influenza virus, *Am J. Med Sci* **194**:159–170

Smorodintsev, A. A, Mikutskaya, B. A, Alexandrova, G. I, 1967, Data on safety and epidemiological efficiency of anti-influenza vaccine studied on pre-school children, *J Hyg Epidemiol Microbiol Immunol* **11**:265–277

Smorodintsev, A A, Karpuhin, G. I, Zlydnikov, D. M., et al., 1970, The prophylactic effectiveness of amantadine hydrochloride in an epidemic of Hong Kong influenza in Leningrad in 1969, *Bull* **42**:865–872.

Soloviev, V D, Orlova, T. G, Porubel, L A, et al, 1961, Study of the genetic characters of vaccine strains of type A-2 influenza virus, *Prob Virol* **6**:743–750.

Spencer, M J, Cherry, J D, and Terasaki, P I, 1976 HL-A antigens and antibody response after

influenza A vaccination, *N. Engl. J. Med.* **294**:13–16.
Spring, S. B., Maassab, H. F., Kendal, A. P., et al., 1977, *Arch. Virol.* **55**:233–246.
Steinberg, E., Overturf, G. D., Portnoy, B., et al., 1978, Serologic and clinical response of children with sickle cell disease to bivalent influenza A split virus vaccine, *J. Pediatr.* **92**:823–825.
Steinherz, P. G., Brown, A. E., Gross, P. A., et al., 1980, Influenza immunization of children with neoplastic diseases, *Cancer* **45**:750–756.
Stewart-Tull, D. E. S., 1985, Immunopotentiating conjugates, *Vaccine* **3**:40–44.
Stiver, H. G., and Weinerman, B. H., 1978, Impaired serum antibody response to inactivated influenza A and B vaccine in cancer patients, *Can. Med. Assoc. J.* **119**:733–735.
Stiver, H. G., Graves, P., Eickhoff, T. C., et al., 1973, Efficacy of "Hong Kong" vaccine in preventing "England" variant influenza A in 1972, *N. Engl. J. Med.* **289**:1267–1271.
Storms, W. W., Dick, E. C., and Busse, W. W., 1976, Intranasal immunization with attenuated live influenza vaccine in asthma, *J. Allergy Clin. Immunol.* **58**:284–290.
Takafuji, E. T., Johnson, D. E., and Segal, H. E., 1976, The swine antigen in previous influenza vaccines, *N. Engl. J. Med.* **295**:1018.
Tian, S., Zhang, L., and Murphy, B. R., 1984, Characterization of the genotype and level of attenuation of an influenza A reassortant virus produced by mating the Xia-ts donor virus with A/Beijing/79 (H1N1) wild type virus, *Vaccine* **2**:189–192.
Tian, S.-F., Buckler-White, A. J., London, W. T., et al., 1985, Nucleoprotein and membrane protein genes are associated with restriction of replication of influenza A/Mallard/NY/78 virus and its reassortants in squirrel monkey respiratory tract, *J. Virol.* **53**:771–775.
Tolpin, M. D., Massicot, J. G., Mullinix, M. G., et al., 1981, Genetic factors associated with loss of the temperature-sensitive phenotype of the influenza A/Alaska/77-ts-1A2 recombinant during growth in vivo, *Virology* **112**:505–517.
Tolpin, M. D., Clements, M. L., Levine, M. M., et al., 1982, Evaluation of a phenotypic revertant of the A/Alaska/77-ts-1A2 reassortant virus in hamsters and in seronegative adult volunteers: Further evidence that the temperature-sensitive phenotype is responsible for attenuation of ts-1A2 reassortant viruses, *Infect. Immun.* **36**:645–650.
Top. F. H., Jr., and Russell, P. K., 1977, Swine influenza A at Fort Dix, New Jersey (January–February 1976), IV. Summary and speculation, *J. Infect. Dis.* **136**:S376–S379.
van de Donk, H. J. M., de Jong, J. C., van Olderen, M. F., et al., 1984, Monoclonal antibodies for the control of influenza virus vaccines, *Dev. Biol. Stand.* **57**:251–255.
Versluis, D. J., Beyer, W. E. P., Masurel, N., et al., 1985, Influenza vaccination in dialysis and transplant patients, *Antiviral Res. (Suppl.)* **1**:289–292.
Wabuke-Bunoti, M. A. N., Taku, A., Fan, D. P., et al., 1984, Cytolytic T lymphocyte and antibody responses to synthetic peptides of influenza virus hemagglutinin, *J. Immunol.* **133**:2194–2201.
Wagner, R. R., Bennett, I. L., Jr., and LeQuire, V. S., 1949, The production of fever by influenzal viruses. I. Factors influencing the febrile response to single injections of virus, *J. Exp. Med.* **90**:321–334.
Waldman, R. H., Mann, J. J., and Small, P. A., Jr., 1969a, Immunization against influenza, *J.A.M.A.* **207**:520–524.
Waldman, R. H., Bond, J. O., Levitt, L. P., et al., 1969b, An evaluation of influenza immunization, *Bull. WHO* **41**:543–548.
Waldman, R. H., Wood, S. H., Torres, E. J., et al., 1970, Influenza antibody response following aerosol administration of inactivated virus, *Am. J. Epidemiol.* **91**:575–584.
Waksman, B. H., and Adams, R. D., 1955, Allergic neuritis: An experimental disease of rabbits induced by the injection of peripheral nervous tissue and adjuvants, *J. Exp. Med.* **102**:213–235.
Warburton, M. F., Jacobs, D. S., Langsford, W. A., et al., 1972, Herd immunity following subunit influenza vaccine administration, *Med. J. Aust.* **2**:67–70.
Watson, W. H., Parks, M., Porter, D. A., et al., 1972, Influenza vaccination in schoolboys, *Lancet* **1**:1336.
Webster, R. G., 1966, Original antigenic sin in ferrets: The response to sequential infections with influenza viruses, *J. Immunol.* **97**:177–183.
Webster, R. G., and Askonas, B. A., 1980, Cross-protection and cross-reactive cytotoxic T cells induced by influenza virus vaccines in mice, *Eur. J. Immunol.* **10**:396–401.

Webster, R. G , Kasel, J. A., Couch, R. B., et al , 1976, Influenza virus subunit vaccines II Immunogenicity and original sin in humans, *J Infect Dis.* **134:**48–58.

Webster, R G , Kawaoka, Y , Bean, W. J , et al , 1985, Chemotherapy and vaccination: A possible strategy for the control of highly virulent influenza virus, *J. Virol.* **55:**173–176

Weibel, R E , McLean, A , Woodhour, A. F , et al., 1973, Ten-year follow-up study for safety of adjuvant 65 influenza vaccine in man, *Proc. Soc Exp Biol Med* **143:**1053–1056.

Weintraub, M I , and Chia, D T S., 1977, Paralytic brachial neuritis after swine flu vaccination, *Arch Neurol* **34:**518

Wenzel, R. P , Hendley, O., Sande, M. A et al., 1973, Revised (1972–1973) bivalent influenza vaccine, *J.A.M.A.* **226:**435–438

Wesselius-de Casparis, A., Hogerzeil, H. H W., and van Beek, A., 1973, Eleven years of experiences in industry with vaccination against influenza. Reduction in absenteeism, in *International Symposium on Influenza Vaccines for Men and Horses, London, 1972 (Symposium Series Immunobiology Standard,* Vol. 20), S. Karger, Basel, pp. 268–274

Wiederholt, W. C , Mulder, D. W , and Lambert, E. H., 1964, The Landry Guillain–Barre–Strohl syndrome or polyradiculoneuropathy: Historical review, report on 97 patients, and present concepts, *Mayo Clin Proc* **39:**427–451

Wiley, D C , Wilson, I A , and Skehel, J J , 1981, Structural identification of the antibody-binding sites of Hong Kong influenza haemagglutinin and their involvement in antigenic variation, *Nature* **289:**373–378.

Wilson, I. A , Niman, H L , Houghten, R A , et al., 1984, The structure of an antigenic determinant in a protein, *Cell* **37:**767–778

Wood, J M , Schild, G C , Newman, R W , et al , 1977, An improved single-radial-immunodiffusion technique for the assay of influenza haemagglutinin antigen Application for potency determinations of inactivated whole virus and subunit vaccines, *J. Biol. Stand.* **5:**237–247

Wood, J. M , Seagroatt, V , Schild, G C., et al., 1981, International collaborative study of single-radial-diffusion and immunoelectrophoresis techniques for the assay of haemagglutinin antigen of influenza virus, *J. Biol. Stand.* **9:**317–330

Woodhour, A F , Friedman, A , Tytell, A A., et al , 1969, Hyperpotentiation by synthetic double-stranded RNA of antibody responses to influenza virus vaccine in adjuvant 65, *Proc Soc Exp Biol. Med* **131:**809–817

World Health Organization, 1985, Recombinant vaccinia viruses as live virus vectors for vaccine antigens Memorandum from a WHO/USPHS/NIBSC Meeting, *Bull WHO* **63:**471–477

Wright, P. F , Sell, S H W , Thompson, J., et al , 1976, Clinical reactions and serologic response following inactivated monovalent influenza type B vaccine in young children and infants, *J Pediatr* **88:**31–35

Wright, P F , Bhargava, M , Johnson, P R., et al., 1985, Simultaneous administration of live, attenuated influenza A vaccines representing different serotypes, *Vaccine* **3:**305–308

Yamane, N , Nakamura, Y , Yuki, M., et al., 1984, Serological evaluation of an influenza A virus cold-adapted reassortant live vaccine, CR-37 (H1N1) in Japanese adult volunteers, *J. Hyg (Camb.)* **92:**231–242

Zahradnik, J M , Kasel, J A., Martin, R R , et al , 1983, Immune responses in serum and respiratory secretions following vaccination with a live cold-recombinant (CR35) and inactivated A/USSR/77 (H1N1) influenza virus vaccine, *J. Med Virol* **11:**277–285

Zeck, R , Solliday, N., Kehoe, T , et al., 1976, Respiratory effects of live influenza virus vaccine: Healthy older subjects and patients with chronic respiratory disease, *Am Rev Respir Dis* **114:**1061–1067

Index

GPSR Compliance

The European Union's (EU) General Product Safety Regulation (GPSR) is a set of rules that requires consumer products to be safe and our obligations to ensure this.

If you have any concerns about our products, you can contact us on ProductSafety@springernature.com

In case Publisher is established outside the EU, the EU authorized representative is:

Springer Nature Customer Service Center GmbH
Europaplatz 3
69115 Heidelberg, Germany

Batch number: 10620881

Printed by Printforce, the Netherlands